Lyapunov-Based
Control of
Robotic Systems

AUTOMATION AND CONTROL ENGINEERING

A Series of Reference Books and Textbooks

Series Editors

FRANK L. LEWIS, Ph.D.,
FELLOW IEEE, FELLOW IFAC

Professor
Automation and Robotics Research Institute
The University of Texas at Arlington

SHUZHI SAM GE, Ph.D.,
FELLOW IEEE

Professor
Interactive Digital Media Institute
The National University of Singapore

Lyapunov-Based Control of Robotic Systems

Aman Behal
University of Central Florida
Orlando, Florida, U.S.A.

Warren Dixon
University of Florida
Gainesville, Florida, U.S.A.

Darren M. Dawson
Clemson University
Clemson, South Carolina, U.S.A.

Bin Xian
Tianjin University
Tianjin, China

CRC Press
Taylor & Francis Group
Boca Raton London New York

CRC Press is an imprint of the
Taylor & Francis Group, an **informa** business

Published 2010 by CRC Press
Taylor & Francis Group
6000 Broken Sound Parkway NW, Suite 300
Boca Raton, FL 33487-2742

© 2010 by Taylor & Francis Group, LLC
CRC Press is an imprint of Taylor & Francis Group, an Informa business

First issued in paperback 2019

No claim to original U.S. Government works

ISBN-13: 978-0-367-45242-1 (pbk)
ISBN-13: 978-0-8493-7025-0 (hbk)

Library of Congress Cataloging-in-Publication Data

Lyapunov-based control of robotic systems / Aman Behal ... [et al.].
    p. cm. -- (Automation and control engineering)
    Includes bibliographical references and index.
    ISBN 978-0-8493-7025-0 (hardcover : alk. paper)
    1. Robots--Control systems. 2. Nonlinear control theory. 3. Lyapunov functions. I. Behal, Aman.

TJ211.35.L83 1009
629.8'92--dc22   2009040275

To my loving wife,

Hina Behal

A.B.

To my parents,

Dwight and Belinda Dixon

W.E.D

To my children,

Jacklyn and David

D.M.D.

To my parents,

Kaiyong Xian and Yanfang Liu

B.X.

Contents

Preface

In one form or another, robots have been around for many hundreds of years and have caught the fancy of nearly all segments of society including but not limited to lay people, novelists, engineers, and computer scientists. While sophisticated algorithms for intelligent integration of robotic systems into society have abounded for many decades, it is only recently that the explosion of computational speed and its cheap availability have made many of these ideas implementable in real-time so that the benefit of robots can now extend beyond the confines of automated assembly lines and impact people directly. Our increasing understanding of human sensory systems and how they play into our superior cognition have led to the development of specialized hardware setups that locally process data at the site of its acquisition and pass on low bandwidth information (in the form of features or keywords) that can be processed at one or more central locations for higher level decision making.

This book aims to describe possible frameworks for setting up nonlinear control design problems that need to be solved in the context of robots trying to understand, interact with, and manipulate their environments. Of course, the environment considered could be static or dynamic, structured or unstructured. Similarly, the robot may suffer from kinematic or dynamic uncertainties. Finally, the measurements available for controller use could be limited (e.g., lack of velocity feedback on robots) or excessive (e.g., video frames captured at a 640×480 resolution by a 30 Hz video camera) — interestingly, both these scenarios require extra preprocessing and computation to extract the required information. All the aforementioned factors can conspire to make for highly challenging control design scenarios. As the title suggests, our preferred framework for addressing these issues will be Lyapunov-based nonlinear control design. Robots are highly nonlinear systems and even though they can be linearized under some restrictive assumptions, most practical scenarios require the design of nonlinear controllers to work around uncertainty and measurement-related issues. It has been our experience over the years that Lyapunov's direct method is an extremely effective tool to both design and analyze controllers for robotic systems.

Chapter 1 begins by providing a brief history of robotics. It is followed by an introduction to the Lyapunov-based design philosophy — pros and cons are discussed. The chapter ends on a practical note by describing the evolution of real-time control design systems and the associated operating environments and hardware platforms that they are based upon. In Chapter 2, we provide the reader with a quick introduction to a host of standard control design tools available for robotic systems. In order to prepare for the chapters ahead, all these techniques are analyzed in a common Lyapunov-based framework. The chapter begins by discussing computed torque methods, where the model nonlinearities are canceled through exact model knowledge, and the system robustness to unmodeled disturbances is discussed under PD control, continuous robust control, and sliding mode control. Next, adaptive control techniques are discussed when the model is uncertain. A variety of adaptive update laws are discussed including the design of a NN-based strategy when the model cannot be expressed as linear in the unknown/uncertain parameters. The chapter closes by discussing the challenges of designing control laws for redundant link robot manipulators when the control objectives are stated in the task-space of the robot.

When robots need to navigate through and/or interact with unstructured or dynamically changing environments, camera-based vision systems are utilized to provide adequate sensing of the environment. Chapter 3 discusses some problems in visual servoing control. The first problem addressed is that of a robot end-effector tracking a prerecorded time-varying reference trajectory under visual feedback from a monocular camera — the use of a single camera results in uncertainty in the depth parameter which needs to be adaptively corrected. The next problem that we address in this chapter is that of estimating the shape of a continuum robot. Traditional position sensing devices such as encoders cannot be used in this situation since it is not easy to define links and joints for such robots —

instead, a vision based solution is proposed and then validated through the design of kinematic controllers that regulate the pose of the robot end-effector via feedback from a sequence of images from a fixed monocular camera. The third problem dealt with in this chapter is that of designing homography-based visual servo control methods for solving tracking and regulation problems in the context of wheeled mobile robots. The final problem addressed in Chapter 3 is the classic Structure from Motion (SFM) problem, specifically the development of an adaptive nonlinear estimator in order to identify the Euclidean coordinates of feature points on an object based upon relative movement between that object and a single fixed camera.

Chapter 4 deals with the problems of path planning and control for manipulator arms and wheeled mobile robots, both when the obstacle locations are known a priori and when they need to be determined in real time using fixed or in-hand vision as an active feedback element. The first problem addressed is path following using velocity field control (VFC) — this technique can be applied when it is more critical to follow a contour exactly than it is to track a desired time-varying trajectory (which incidentally is the standard problem solved in most robotics literature). Another application of path following is when a navigation function (NF) approach is utilized to create a path around obstacles to a desired goal location. As an extension, we also show how VFC- and NF-based techniques can be utilized to solve the obstacle avoidance problem for mobile robots. We then shift gears and address the problem of hybrid servoing control under visual feedback, which may be required to manipulate a robot in unstructured environments — the challenge here is the design of a desired trajectory in the image space based on an image space NF that ensures that the features on the object stay in the camera's field-of-view through the course of the robot's motion. The final portion of this chapter deals with the design of an image space extremum seeking path planner such that a singularity free PBVS controller can be designed that works on visual feedback and is able to reject lens distortion and uncertainties in the camera calibration.

In Chapter 5, we deal with the emerging research area of human-machine interaction. While the primary control objective during human-machine interaction is application specific, the secondary objective invariably is to ensure the safety of the user — to this end, we illustrate the design of control schemes based on passivity such that the machine is a net energy sink. The chapter begins by exploring smart exercise machines that provide optimal physical training for the user by altering the machine's resistance based on user performance. Steer-by-wire control of vehicles is discussed next, with the focus being on locking the steering response of the vehicle to the user input as well as ensuring that the road feel experienced by the user can be appropriately adjusted. The third problem addressed in this chapter is that of teleoperator systems where the focus is on both facilitating the application of desired motion and desired force in the remote environment by the user, as well as ensuring that the system is safely able to reflect desired forces back to the user. The final topic addressed in Chapter 5 is that of a rehabilitation robot which is safely able to direct user limb motions along selectable trajectories in space that optimize their rehabilitation after disease or injury.

The material in this book (unless noted otherwise) has been derived from the authors' research work during the past several years in the area of controls and robotics. This book is aimed at graduate students and researchers who would like to understand the application of Lyapunov-based control design techniques to emerging problems in robotics. This book assumes a background in undergraduate level linear controls theory. Some knowledge of nonlinear systems and Lyapunov-based design techniques for such systems may be desirable – however, the book contains adequate background material in Chapter 2 and Appendix A as well as references to textbooks that deal with these subjects should they be of interest to the reader.

The authors would like to thank our colleagues and collaborators for their valuable assistance and counsel without which this work would not have been possible. We are especially grateful to Dr. Jian Chen, whose work early on as a graduate student at Clemson University and later as a collaborator of the authors, has influenced many of the issues examined in this book. We would also like to acknowledge the unselfish support of the following past and present members of the Controls and Robotics group in the Department of Electrical Engineering at Clemson University: Dr. Vilas Chitrakaran, Dr. Michael McIntyre, Dr. Pradeep Setlur, Dr. Ximing Zhang, Dr. David Braganza, Dr. Enver Tatlicioglu, and Mr. Abhijit Baviskar. Finally, we would like to thank Mr. Jonathan Plant, Senior Editor at CRC Press for his patience and counsel as well as Marsha Pronin, Project Coordinator, and Karen Simon, Project Editor, for their help with reviewing samples and assistance with preparation of the camera-ready manuscript.

1

Introduction

1.1 History of Robotics

From time immemorial, human beings have been fascinated with building and employing machines with human-like capabilities. As far back as 300 B.C., Greek philosophers, mathematicians, and inventors were postulating or designing mechanical contraptions that had steam or water-based actuation as well as some level of autonomy. Around the middle of the last millennium, Leonardo da Vinci and others built mechanical devices primarily for amusement purposes. Cut to the industrial age, the great scientist Nikola Tesla built and demonstrated a remote controlled submersible robot boat at Madison Square Garden in the year 1898. The word "robot" (which actually means "forced labor" in Czech) was introduced into our vocabulary by playwright Karel Capek in a satirical 1920 play entitled *Rossum's Universal Robots*. Science fiction fans are, of course, very familiar with the work of Isaac Asimov who first popularized the term "Robotics" and was responsible for proposing the three fundamental laws of robotics.

Robotics has always attracted the fancy of moviemakers, and robots have been an integral part of popular culture in the United States. The first robot to appear on film was "Maria" in Fritz Lang's 1926 movie named *Metropolis*. In 1951, an alien and his robot appeared on the silver screen in the movie *The Day the Earth Stood Still*. Arthur Clarke's novel *2001: A Space Odyssey* was made into a movie in 1968 which featured a high

functioning robot named HAL that turns rogue and is eventually discon-
nected. A movie that received a great deal of acclaim was Ridley Scott's
Blade Runner that was released in 1982 and which featured Harrison Ford
as a hunter of illegal mutinous androids known as Replicants. Other movies
like *The Terminator* and the *Matrix* series have dealt with sophisticated,
high-functioning humanoids. Most recently, Pixar produced the smash hit
animated robotics film *WALL-E* which features a sentimental robot of the
same name that is designed to clean up the pollution created by mankind.

Space exploration has advantageously employed manipulator arms and
mobile robots over the years. Lunokhod 1 and 2 were the first robotic exploration vehicles (rovers) to be launched to an extraterrestrial body, the
moon, by the Soviets in 1970. After a long gap, the rover *Sojourner* landed
on Mars in 1997 as part of the Pathfinder Mission; it had vision assisted
autonomous navigation and was successful at obtaining and analyzing rock
and soil samples. This was followed in 2004 by the rovers *Spirit* and *Op-
portunity* that are still active and continue to analyze the Martian geology,
as well as its environment, to assess the possibility that life may have been
supported on Mars in the past. Most recently, the *Phoenix* lander executed
the first successful polar landing on the Martian surface and is currently
exploring the possibility of water existing or having existed on the red
planet.

The first commercial robotics company, named Unimation, was started
in 1956 by George Devol and J. Engelberger. As a result of this venture, the
first industrial robot was manufactured and marketed in the United States.
Unimate began work in a General Motors automobile plant in New Jersey
in 1961. This manipulator arm performed spot welding operations as well
as unloading of die casts. This was followed in 1978 by the Programmable
Universal Machine for Assembly, a.k.a. PUMA. Since that time, quite a few
other robot manufacturers have come and gone with only a few achieving
commercial success or longevity in the market. In recent years, personal and
professional service robots have picked up steam. For example, Lego has
achieved success with its Mindstorms Robotics Invention System as has
Sony with its AIBO robot pets. Most recently, Honda's humanoid robot
ASIMO has hogged media limelight with its ability to perform a wide
variety of service and human interaction tasks.

Robotics in research settings has steadily continued to experience an up-
ward spiral since its inception. Robotics research got its academic start in
1959 with the inauguration of the Artificial Intelligence Laboratory at the
Massachusetts Institute of Technology by John McCarthy and Marvin Minsky. Other inaugurations of note were the establishment of the Artificial

Intelligence Laboratory at Stanford University in 1963, and the Robotics Institute at Carnegie Mellon University in 1979. The first computer-controlled mechanical hand was developed at MIT in 1961 followed by the creation of the Stanford Arm in the Stanford Artificial Intelligence Laboratory by Victor Scheinman in 1969. Early flexible robots of note were Minsky's octopus-like Tentacle Arm (MIT, 1968) and Shigeo Hirose's Soft Gripper (Tokyo Institute of Technology, 1976). An eight-legged walking robot named Dante was built at Carnegie Mellon University which was followed by a more robust Dante II that descended into the crater of the volcano Mt. Spurr in Alaska in 1994. Demonstrations of planning algorithms for robots began in the late 1960s when Stanford Research Institute's Shakey was able to navigate structured indoor environments. A decade later, the Stanford cart attempted navigation of natural outdoor scenes as well as cluttered indoor environments. Modern robotics research is focused on higher dimensional robots, modular robots, and the planning issues associated with these types of devices. Simultaneously, robotics is making great strides in medicine and surgery as well as assistance for individuals with disabilities.

1.2   Lyapunov-Based Control Philosophy

The requirements for increasing levels of autonomy and precision in robots have necessitated the development of sophisticated control strategies. Multiple link robots have presented complex, coupled nonlinear dynamics that have inspired the design of numerous output and state feedback control designs, especially the global output feedback problem for robots has been very challenging. Wheeled mobile robots have inspired the design of set-point and tracking controllers for nonholonomic systems. Other applications that have challenged control designers have been rigid link flexible joints in the late nineties as well as higher dimensional and continuum robots of late.

Linear control design is often inadequate outside narrow operating regimes where linearized system models are valid. Nonlinear control strategies can take advantage of full or partial knowledge of the structure and/or parameters of the system in order to craft techniques that are robust to exogenous disturbances, measurement noise, and unmodeled dynamics. Research investigators have utilized a variety of tools for analyzing nonlinear systems arising from nonlinear controllers, nonlinear models, or a combination thereof — singular perturbation, describing functions, and phase plane analysis are some of the popular tools. However, Lyapunov-based techniques (in particular, the so-called direct method of Lyapunov) offer the distinct advantage that they allow both design and analysis under a common framework with one stage motivating the other in an iterative fashion. Lyapunov theory and its derivatives are named after the Russian mathematician and engineer Aleksander Mikhailovich Lyapunov (1857–1918).

Lyapunov stability theory has two main directions — the linearization method and the aforementioned direct method of Lyapunov. The method of linearization provides the fundamental basis for the use of linear control methods [1]. It states that a nonlinear system is locally stable if all the eigenvalues of its linear approximation (via a Taylor series expansion about a nominal operating point) are in the open left half plane and is unstable if at least one is in the open right half plane. Furthermore, the stability cannot be determined without further analysis if the linearization is marginally stable. The direct method of Lyapunov relies on the physical property that a system whose total energy is continuously being dissipated must eventually end up at an equilibrium point [1, 2]. Given a scalar, non-negative energy (or energy-like) function $V(t)$ for a system, it can be shown that if its time derivative $\dot{V}(t) \leq 0$, the system is stable in the sense of Lyapunov in that the system states (energy) can be constrained for all future time to lie inside a ball that is directly related to the size of the initial states of the system.

While a lot of results have been derived in the last fifty years in order to deduce stability properties based on the structure of the Lyapunov function $V(t)$ and its time derivative, we are no closer to understanding how one may choose an appropriate $V(t)$, i.e., it is not clear how closely the scalar function $V(t)$ should mimic the physical (kinetic and potential) energy of the system. What is clear is that the objectives of the control design and the constraints on the measurements lead to the definition of system states that often guide the development of the Lyapunov function. Furthermore, the control design itself is impacted by the need to constrain the time derivative of the Lyapunov function to be negative definite or semi-definite along the closed loop system trajectories. Thus, the control design and the development of the Lyapunov function are intertwined, even though the presentation may tend to indicate a monotonic trajectory from control design toward stability analysis. In the ensuing chapters, one will be able to gain an insight into the variety of choices for Lyapunov functions as well as the appearance of non-intuitive terms in the control input signals that will likely indicate an influence of the Lyapunov-based analysis method on the control design.

While Lyapunov's direct method is good at characterizing the stability of equilibrium points for autonomous and nonautonomous systems alike, it works equally well in showing the boundedness and ultimate boundedness of solutions when no equilibrium points exist [3]. Furthermore, the analysis not only provides a guarantee of stability and the type of stability result (uniform asymptotic, exponential, semi-global ultimately bounded, etc.), it is also able to point out bounds on the regions where the results are guaranteed to be valid. This is in sharp contrast to linearization based approaches where regions of convergence are not easily obtained. Finally, Lyapunov-based design leads to faster identifiers and stronger controllers that are able to prevent catastrophic instabilities associated with traditional estimation based methods such as certainty equivalence [4]. While traditional methods work well with linear systems, they can lead to troubling results such as finite escape times in the case of nonlinear systems. A shortcoming of the Lyapunov-based analysis techniques is that the chosen parameters (while guaranteed to produced closed-loop stability) may be too conservative, thereby compromising the transient response of the system. Moreover, Lyapunov stability theorems only provide sufficient conditions for stability, i.e., without further work, it is not possible to say which of those conditions are also necessary [3].

1.3   The Real-Time Computer Revolution

As we will see in the ensuing chapters, the nonlinear control, estimation, and observation schemes emanating from an application of Lyapunov's direct method tend to have a complex structure and are generally computationally intense compared to their linear counterparts [2]. Thus, there is a requirement for the use of microprocessors, microcontrollers, and/or computers to crunch the numbers. Furthermore, there is a requirement for fast interface hardware for allowing the physical (generally analog) world to interact bidirectionally with the digital domain without creating instabilities and uncertainties due to factors such as phase lags from slow computation, quantization noise due to finite precision, aliasing due to slow sampling, uncertain order of execution of various computation modules, etc. In the last decade, a multitude of control environments have been created in the academic and industrial research and development communities to serve this emerging need for reliable real-time computation.

A real-time implementation is different from a traditional implementation in that the worst-case performance of the hardware and the software is the most important consideration rather than the average performance. In real-time operation, the processing of external data arriving in the computer must be completed within a predetermined time window, failing which the results obtained are not useful even if they are functionally accurate [5]. In real-time applications, two types of predictability have been specified, namely, microscopic and macroscopic predictability [6]. Microscopic predictability is the idea that each layer of the application from data input to control output should operate deterministically and predictably with failure at one layer dooming the entire application. However, a more robust idea is macroscopic or overall or top-layer predictability which is more suited for complex applications – here, failure to meet a deadline for an internal layer is taken care of by specialized handling. As an example, visual processing schemes generally aggregate data (inliers) to reach a threshold of statistical significance, thus, the computation required may vary substantially between different control cycles. Under such variation, a robot running under an external visual-servoing control and an internal encoder based control is normally programmed to extrapolate a setpoint if a visual processing deadline is missed rather than shut down its entire operation — a correction to the prediction (if necessary) is made in the ensuing cycles when the visual information becomes available.

Traditionally, real-time prototyping has been performed on a heterogeneous system comprising a PC host and a DSP single-board computer (SBC) system, where the control executes on the DSP SBC while the host PC is used to provide plotting, data logging, and control parameter adjustments (online or offline) [7]. As explained in [8], the DSP board is designed to very rapidly execute small programs that contain many floating point operations. Moreover, since the DSP board is dedicated to executing the control program, the host computer is not required to perform fast and/or real-time processing — thus, in a heterogeneous architecture, the host computer can run a non real-time operating system such as MS-DOS, Windows, Linux, MacOS, etc. An example of such a system is the popular dSPACE™ Controller Board based on PowerPC technology that sits in the PCI bus of a general purpose computer (GPC). Other examples of Host/DSP systems include the MS-Windows based Winmotor and the QNX based QMotor 1.0 that were developed by the Clemson University Controls and Robotics group in the 1990s. While still enormously popular, there are some disadvantages to the host/DSP architecture. As explained in [7] and [8], these include hardware cost, limited system flexibility, and complexity of the software.

Over the years, developments in hardware and software have taken the edge off the two main reasons for the existence of the host/DSP combination architecture. Increases in computational power in general over the last decade as well as innovations such as pipelining and multiple core CPUs have made it possible for GPCs to be able to execute complex control strategies at rates in excess of tens of KHz. Today's desktop computers can not only run complex control algorithms in the background, they can simultaneously render high-bandwidth data in one or more GUI windows. Furthermore, the other advantage offered by multiprocessor architectures, namely, deterministic response, has been eliminated by the emergence of hard real-time operating systems such as RTLinux and QNX [7]. Examples of systems that harness the PC's computational ability include Opal-RT Technologies' RT-LAB as well as QMotor 2.0 and its successor QMotor 3.0 both of which were developed by the Controls and Robotics group at Clemson University. All of these systems are based on QNX which is a real-time microkernel operating system [9]. In particular, QMotor 3.0 [8] allows easy incorporation of new hardware by employing a client-server architecture where data from hardware can be accessed by the control program (client) by talking to the server for that hardware — this communication is done via message passing or by using shared memory.

Nearly all of the robot path planning and control algorithms derived in the ensuing chapters have been validated through simulation or experiments. While MATLAB^{TM} and SIMULINK^{TM} have been the primary environments for running computer simulations, the experimental work shown here has been performed via control code written in C and compiled and executed in the QMotor environment running on QNX-based desktop computers.

References

[1] J. J. E. Slotine and W. Li, *Applied Nonlinear Control*, Englewood Cliff, NJ: Prentice Hall, Inc., 1991.

[2] M. S. de Queiroz, D.M. Dawson, S. Nagarkatti, and F. Zhang, *Lyapunov-Based Control of Mechanical Systems*, Cambridge, MA: Birkhauser, 2000.

[3] H. K. Khalil, *Nonlinear Systems*, 3rd edition, Prentice Hall, 2002.

[4] M. Krstic, I. Kanellakopoulos, and P. Kokotovic, *Nonlinear and Adaptive Control Design*, New York: John Wiley & Sons, 1995.

[5] M. Colnaric, "State of the Art Review Paper: Advances in Embedded Hard Real-Time Systems Design," *Proceedings of the IEEE International Symposium on Industrial Electronics*, Vol. 1, pp. 37–42, 1999.

[6] J. A. Stankovic and K. Ramamritham, "Editorial: What is predictability for real-time systems?," *Real-Time Systems*, Vol. 2, No. 4, pp. 246–254, November 1990.

[7] M. Loffler, N. Costescu, and D. Dawson, "QMotor 3.0 and the QMotor robotic toolkit: a PC-based control platform," *IEEE Control Systems Magazine*, Vol. 22, No. 3, pp. 12–26, June 2002.

[8] N. Costescu, D. Dawson, and M. Loffler, "QMotor 2.0 – A real-time PC-based control environment," *IEEE Control Systems Magazine*, Vol. 19, pp. 68–76, June 1999.

[9] http://www.qnx.com.

2
Robot Control

2.1   Introduction

This chapter provides background on some established control methods that are available for holonomic robotic systems. The goal is to describe a number of common control strategies that are applied to robotic systems under a common Lyapunov-based analysis framework. The chapter is divided into four technical sections. The first section presents the dynamics for a robot manipulator and several assumptions/properties associated with those dynamics. The second section focuses on feedback mechanisms that yield ultimate boundedness or asymptotic stability of the tracking error. In general, the controllers presented in this section fall under a broad class of methods known as computed torque controllers — these methods are a form of feedback linearization [3, 6, 8] in the sense that an inner-loop controller is used to cancel out (exactly or approximately) the nonlinear dynamics of the robot, resulting in residual dynamics for which a variety of classical (e.g., root-locus based lead/lag compensation, frequency response methods, etc.) and modern control methods may be applied (e.g., loop shaping, linear quadratic regulator, pole placement, etc.). A proportional derivative (PD) controller is first presented and analyzed with respect to robustness to unknown disturbances — it is shown that high gain feedback suffices to reduce the tracking error to a small ball around the origin (i.e., uniform ultimate boundedness (UUB)). A continuous robust controller is then designed and analyzed to show that high frequency feedback can be utilized in lieu of high gain feedback to damp out the disturbance uncertainty. Finally, a discontinuous controller is also presented to work around the issue of steady-state tracking error. Discontinuous controllers (e.g., sliding mode control) provide a method to obtain an asymptotic/exponential stability result in the presence of uncertainties in the dynamics, provided the actuator is able to provide infinite bandwidth.

The problem with high gain and high frequency approaches is that they can lead to reduced stability margins and are susceptible to noise. Adaptive controllers are utilized in conjunction with or as an alternative to robust control methods. Adaptive controllers are feedforward controllers with a self-adjusting mechanism to compensate for uncertainties in the system parameters – this online adjustment of the system weights allows for asymptotic tracking without needing high gain or high frequency actuation. However, the price to be paid for this improved performance is an increase in the order of the overall control design. The Adaptive Control Design Section describes several types of adaptive control methods including: a direct adaptive controller, a desired compensation extension that allows for off-line computation of the regression matrix, and a neural network based controller when a linear in the parameters model is not available for the uncertainty.

Finally, there are unique challenges that emerge when describing the control objectives in the task-space of the robot. Given the fact that the usefulness of robots is generally derived from establishing a desirable pose (i.e., position and orientation) of the robot end-effector with respect to the environment, it is often best to formulate control objectives for the robotic system in the task-space. A potential exists for an over- or underdetermined problem based on the task-space objective and the number of actuated joints in the robot. A transformation is also required because the control objective is formulated in the task-space while the control is implemented in the joint-space. The last section in this chapter highlights these issues and provides some typical solutions.

2.2 Modeling and Control Objective

2.2.1 Robot Manipulator Model and Properties

The system model for an n-link, revolute, direct-drive robot can be written as

$$M(q)\ddot{q} + V_m(q, \dot{q})\dot{q} + G(q) + F_d\dot{q} + \tau_d = \tau \tag{2.1}$$

where $q(t)$, $\dot{q}(t)$, $\ddot{q}(t) \in \mathbb{R}^n$ denote the link position, velocity, and acceleration vectors, respectively, $M(q) \in \mathbb{R}^{n \times n}$ represents the inertia matrix, $V_m(q, \dot{q}) \in \mathbb{R}^{n \times n}$ represents the centripetal-Coriolis matrix, $G(q) \in \mathbb{R}^n$ represents the gravity vector, $F_d \in \mathbb{R}^{n \times n}$ is the constant, diagonal, positive-definite, dynamic friction coefficient matrix, $\tau_d \in \mathbb{R}^n$ is a bounded disturbance vector that represents other unmodeled dynamics (e.g., static friction), and $\tau(t) \in \mathbb{R}^n$ represents the torque input vector.

The dynamic equation of (2.1) is assumed to exhibit the following properties which are employed during the control development and stability analysis in the subsequent sections.

Property 2.1: The inertia matrix $M(q)$ is a symmetric, positive-definite matrix that satisfies the following inequality

$$m_1 \|\xi\|^2 \le \xi^T M(q) \xi \le m_2 \|\xi\|^2 \qquad \forall \xi \in \mathbb{R}^n \qquad (2.2)$$

where m_1, m_2 are known positive constants, and $\|\cdot\|$ denotes the standard Euclidean norm. The induced infinity norm, denoted by $\|\cdot\|_{i\infty}$, of the inverse of the inertia matrix is assumed to be bounded by a known positive constant ζ_M as

$$\left\| M^{-1} \right\|_{i\infty} \le \zeta_M$$

Property 2.2: The inertia and centripetal-Coriolis matrices satisfy the following skew symmetric relationship

$$\xi^T \left(\frac{1}{2} \dot{M}(q) - V_m(q, \dot{q}) \right) \xi = 0 \qquad \forall \xi \in \mathbb{R}^n \qquad (2.3)$$

where $\dot{M}(q)$ denotes the time derivative of the inertia matrix.

Property 2.3: The dynamic equation of (2.1) can be linear parameterized as

$$Y_d(q_d, \dot{q}_d, \ddot{q}_d)\theta = M(q_d)\ddot{q}_d + V_m(q_d, \dot{q}_d)\dot{q}_d + G(q_d) + F_d\dot{q}_d \qquad (2.4)$$

where $\theta \in \mathbb{R}^p$ contains the unknown constant system parameters, and the *desired regression matrix* $Y_d(q_d, \dot{q}_d, \ddot{q}_d) \in \mathbb{R}^{n \times p}$ contains known bounded functions of the desired link position, velocity, and acceleration trajectory signals denoted by $q_d(t)$, $\dot{q}_d(t)$, $\ddot{q}_d(t) \in \mathbb{R}^n$, respectively. It is assumed that $q_d(t)$, $\dot{q}_d(t)$, $\ddot{q}_d(t)$, $\dddot{q}_d(t)$, $Y_d(\cdot)$, and $\dot{Y}_d(\cdot)$ are all bounded functions of time.

Property 2.4: The centripetal-Coriolis, gravity, friction, and disturbance terms of (2.1) can be upper bounded as

$$\|V_m(q,\dot{q})\|_{i\infty} \le \zeta_{c1} \|\dot{q}\|, \quad \|G(q)\| \le \zeta_g,$$
$$\|F_d\|_{i\infty} \le \zeta_{fd}, \quad \|\tau_d\| \le \zeta_{td} \tag{2.5}$$

where ζ_{c1}, ζ_g, ζ_{fd}, ζ_{td} denote known positive bounding constants, and $\|\cdot\|_{i\infty}$ denotes the infinity-norm of a matrix.

Property 2.5: The centripetal-Coriolis matrix satisfies the switching relationship

$$V_m(q,\xi)\nu = V_m(q,\nu)\xi \quad \forall \xi, \nu \in \mathbb{R}^n. \tag{2.6}$$

2.2.2   Control Objective

The objective in this chapter is to develop link position tracking controllers for the robot manipulator model given by (2.1). To quantify the performance of the control objective, the link position tracking error $e(t) \in \mathbb{R}^n$ is defined as

$$e = q_d - q \tag{2.7}$$

where $q_d(t) \in \mathbb{R}^n$ denotes the desired link position trajectory. To facilitate the subsequent control development and stability analysis, the order of the dynamic expression given in (2.1) can be reduced by defining a filtered tracking error-like variable $r(t) \in \mathbb{R}^n$ as

$$r = \dot{e} + \alpha e \tag{2.8}$$

where $\alpha \in \mathbb{R}$ is a positive constant control gain. Specifically, by defining the filtered tracking error, the control objective can be formulated in terms of $r(t)$, because linear analysis tools (see Lemmas A.18 and A.16 in Appendix A) can be used to conclude that if $r(t)$ is bounded then $e(t)$ and $\dot{e}(t)$ are bounded, and if $r(t) \to 0$ then $e(t), \dot{e}(t) \to 0$. The mismatch $\tilde{\theta}(t) \in \mathbb{R}^p$ between the actual parameter vector θ and the estimate vector $\hat{\theta}(t) \in \mathbb{R}^p$ is defined as

$$\tilde{\theta} = \theta - \hat{\theta}. \tag{2.9}$$

2.3   Computed Torque Control Approaches

2.3.1   PD Control

Control Development

After taking the second time derivative of (2.7), premultiplying the resulting expression by $M(q)$, utilizing (2.1), and then performing some algebraic manipulation, the following expression can be obtained

$$M\ddot{e} = M\ddot{q}_d + N + \tau_d - \tau \tag{2.10}$$

where $N(q, \dot{q}) \in \mathbb{R}^n$ is an auxiliary function defined as

$$N = V_m(q, \dot{q})\dot{q} + G(q) + F_d\dot{q}.$$

Based on the open-loop error system in (2.10), a computed torque PD controller can be developed as

$$\tau = M\left(\ddot{q}_d + k_d\dot{e} + k_p e\right) + N \tag{2.11}$$

where $k_p, k_d \in \mathbb{R}$ denote gains for the proportional and derivative errors, respectively, where

$$k_p = \alpha\left(k_d - \alpha\right) + k_{p2} \tag{2.12}$$

where $k_{p2} \in \mathbb{R}$ is an auxiliary proportional gain, and $k_d > \alpha$. The resulting closed-loop error system can be determined by substituting (2.11) into (2.10) as

$$\ddot{e} + k_d\dot{e} + k_p e = M^{-1}\tau_d. \tag{2.13}$$

Based on the closed-loop system of (2.13) and the bounds prescribed in (2.5), it is easy to see the UUB property of the tracking error.

Alternative Control Development and Analysis

The PD controller developed in (2.11) is written in a traditional manner with explicit gains for the proportional and derivative feedback. An alternative approach is provided in this section that makes greater use of the filtered tracking error formulation and illustrates the advantages of using Property 2.2. To facilitate the alternative design, the open-loop error system for $r(t)$ is formulated by taking the time derivative of (2.8), premultiplying the resulting expression by $M(q)$, and substituting (2.1) to obtain

$$M\dot{r} = M\ddot{q}_d + V_m(q, \dot{q})\dot{q} + G(q) + F_d\dot{q} + \tau_d - \tau + \alpha M\dot{e}. \tag{2.14}$$

Motivated by the desire to use Property 2.2, the expression in (2.14) is rewritten as

$$M\dot{r} = -V_m(q, \dot{q})r - F_d r + \tau_d - \tau + N \tag{2.15}$$

where $N(q, \dot{q}, t) \in \mathbb{R}^n$ is now defined as

$$N = M\ddot{q}_d + V_m(q, \dot{q})\left(\dot{q}_d + \alpha e\right) + G(q) + F_d\left(\dot{q}_d + \alpha e\right) + \alpha M\dot{e}.$$

Specifically, the computed torque PD controller developed in (2.11) is re-designed as

$$\tau = (k_1 + k_2)\, r + N. \qquad (2.16)$$

where k_1 and k_2 are control gains. To illustrate why the controller in (2.16) can still be considered as a computed torque PD controller, consider the alternate form

$$\tau = k\dot{e} + \alpha k e + N$$

where $k \triangleq k_1 + k_2$ denotes the derivative gain, and αk denotes the proportional gain. After substituting (2.16) into (2.15) the closed-loop error system can be determined as

$$M\dot{r} = -V_m(q,\dot{q})r - F_d r + \tau_d - kr. \qquad (2.17)$$

Theorem 2.1 *Given the open-loop error system in (2.15), the computed torque PD controller given in (2.16) ensures that the tracking error is globally uniformly ultimately bounded in the sense that*

$$\|e(t)\| \le \sqrt{\zeta_0 \exp(-\zeta_1 t) + \zeta_2} \qquad (2.18)$$

for some positive constants $\zeta_0, \zeta_1, \zeta_2$.

Proof: Let $V(t) \in \mathbb{R}$ denote the non-negative function

$$V = \frac{1}{2} r^T M r. \qquad (2.19)$$

By using (2.17) and Properties 2.1 and 2.2, the time derivative of (2.19) can be expressed as

$$\begin{aligned}
\dot{V} &= \frac{1}{2} r^T \dot{M} r + r^T(-V_m r - F_d r + \tau_d - kr)\\
&= r^T(-F_d r + \tau_d - kr)\\
&\le -k\|r\|^2 + \zeta_{td}\|r\|.
\end{aligned}$$

Completing the squares yields the following inequalities

$$\begin{aligned}
\dot{V} &\le -k_1\|r\|^2 - k_2\left[\|r\|^2 - \frac{\zeta_{td}}{k_2}\|r\| + \left(\frac{\zeta_{td}}{2k_2}\right)^2\right] + \frac{\zeta_{td}^2}{4k_2}\\
&\le -\frac{k_1 m_1}{2} V + \frac{\zeta_{td}^2}{4k_2} \qquad (2.20)
\end{aligned}$$

where m_1 is defined in Property 2.1. The inequality in (2.20) can be solved using standard linear analysis tools (see Lemma A.20 in Appendix A) as

$$V(t) \le V(0)\exp(-\frac{k_1 m_1}{2}t) + \frac{\zeta_{td}^2}{2m_1 k_1 k_2}(1 - \exp(-\frac{k_1 m_1}{2}t)). \qquad (2.21)$$

Based on (2.19), the inequality in (2.21) can be written as

$$
\begin{aligned}
\|r(t)\|^2 &\leq \frac{m_2}{m_1}\|r(0)\|^2 \exp(-\frac{k_1 m_1}{2}t) + \frac{\zeta_{td}^2}{m_1^2 k_1 k_2}(1 - \exp(-\frac{k_1 m_1}{2}t)) \\
&\leq \left(\frac{m_2}{m_1}\|r(0)\|^2 - \frac{\zeta_{td}^2}{m_1^2 k_1 k_2}\right)\exp(-\frac{k_1 m_1}{2}t) + \frac{\zeta_{td}^2}{m_1^2 k_1 k_2}. \quad (2.22)
\end{aligned}
$$

The inequality in (2.18) can be obtained from (2.22) by invoking Lemma A.21 in Appendix A. ∎

2.3.2   Robust Control

The previous section illustrated how a PD controller can be used to yield a UUB stability result in the presence of a bounded disturbance torque. From the expression in (2.22), it is clear that by increasing the proportional and derivative gains arbitrarily large (i.e., high gain feedback), the residual steady state error ζ_2 can be decreased arbitrarily small. Yet, high gain feedback can be problematic if noise is present in the system (e.g., feedback from a vision system, use of backwards differencing to obtain velocities from position measurements, etc.). Motivated by the desire to reduce the steady-state error without increasing the gains arbitrarily large, an alternative robust controller can be developed that relies on high bandwidth from the actuator.

Control Development

Consider the following computed torque controller

$$\tau = kr + u_R + N. \tag{2.23}$$

that is designed the same as the PD controller in (2.16), with the addition of a robustifying feedback component $u_R \in \mathbb{R}^n$. The high frequency (variable structure) robustifying feedback term in (2.23) is defined as

$$u_R = \frac{r\zeta_{td}^2}{\|r\|\zeta_{td} + \epsilon} \tag{2.24}$$

where $\epsilon \in \mathbb{R}$ is an arbitrarily small positive design constant. After substituting (2.23) and (2.24) into (2.15), the following closed-loop error system is obtained

$$M\dot{r} = -V_m(q,\dot{q})r - F_d r + \tau_d - \frac{r\zeta_{td}^2}{\|r\|\zeta_{td} + \epsilon} - kr. \tag{2.25}$$

Stability Analysis

Theorem 2.2 *Given the open-loop error system in (2.15), the robust controller given in (2.23) and (2.24) ensures that the tracking error is uniformly ultimately bounded in the sense that*

$$\|e(t)\| \le \sqrt{\zeta_0 \exp(-\zeta_1 t) + \zeta_2} \tag{2.26}$$

for some positive constants $\zeta_0, \zeta_1, \zeta_2$.

Proof: By using (2.25) and Properties 2.1 and 2.2, the time derivative of (2.19) can be expressed as

$$
\begin{aligned}
\dot{V} &= \frac{1}{2} r^T \dot{M} r + r^T \left(-V_m r - F_d r + \tau_d - \frac{r \zeta_{td}^2}{\|r\| \zeta_{td} + \epsilon} - kr \right) \\
&= r^T \left(-F_d r + \tau_d - \frac{r \zeta_{td}^2}{\|r\| \zeta_{td} + \epsilon} - kr \right) \\
&\le -k \|r\|^2 + \epsilon \left(\frac{\|r\| \zeta_{td}}{\|r\| \zeta_{td} + \epsilon} \right) \\
&\le -k \|r\|^2 + \epsilon.
\end{aligned}
\tag{2.27}
$$

The inequality in (2.27) can be used along with the same stability analysis developed for the PD controllers to conclude that

$$
\begin{aligned}
\|r(t)\|^2 &\le \frac{m_2}{m_1} \|r(0)\|^2 \exp(-\frac{km_1}{2} t) + \frac{2\epsilon}{km_1}(1 - \exp(-\frac{km_1}{2} t)) \\
&\le \left(\frac{m_2}{m_1} \|r(0)\|^2 - \frac{2\epsilon}{km_1} \right) \exp(-\frac{km_1}{2} t) + \frac{2\epsilon}{km_1}. \quad \blacksquare
\end{aligned}
\tag{2.28}
$$

The inequality in (2.26) can be obtained from (2.28) by invoking Lemma A.21 in Appendix A. Comparing the results in (2.22) and (2.28) highlights the benefits of the robust control design. Specifically, the steady state error ζ_2 in (2.28) does not depend on the upper bound on the disturbance term ζ_{td} and the residual error can be made arbitrarily small by either increasing the control gains (as in the PD controller) or by decreasing the design parameter ϵ. Decreasing the design parameter ϵ increases the bandwidth requirements of the actuator.

2.3.3 Sliding Mode Control

Control Development

The last section illustrates how the increasing the frequency of the controller can be used to improve the steady state error of the system. Taken to the extreme, if the design parameter ϵ is set to zero, then the controller requires infinite actuator bandwidth (which is not practical in typical engineering systems), and the state error vanishes (theoretically). This discontinuous controller is called a sliding mode controller. To illustrate the sliding mode control, consider the control design in (2.23) and (2.24) with $\epsilon = 0$ as

$$\tau = kr + sgn(r)\zeta_{td} + N. \tag{2.29}$$

The corresponding closed-loop error system is given by

$$M\dot{r} = -V_m(q, \dot{q})r - F_d r + \tau_d - sgn(r)\zeta_{td} - kr. \tag{2.30}$$

Stability Analysis

Theorem 2.3 *Given the open-loop error system in (2.15), the sliding mode controller given in (2.29) and (2.30) ensures that the tracking error is globally exponentially stable.*

Proof: By using (2.30) and Properties 2.1 and 2.2, the time derivative of (2.19) can be expressed as

$$
\begin{aligned}
\dot{V} &= \frac{1}{2}r^T \dot{M}r + r^T\left(-V_m r - F_d r + \tau_d - sgn(r)\zeta_{td} - kr\right) \quad (2.31) \\
&= r^T(-F_d r + \tau_d - sgn(r)\zeta_{td} - kr) \\
&\leq -k\|r\|^2
\end{aligned}
$$

The inequality in (2.31) can be used to conclude that

$$\|e(t)\| \leq \|r(t)\| \leq \sqrt{\frac{m_2}{m_1}}\|r(0)\|\exp(-\frac{km_1}{4}t). \qquad \blacksquare \tag{2.32}$$

2.4   Adaptive Control Design

Control designs in the previous section use high gain feedback or high (or infinite) frequency to compensate for the bounded disturbances in the robot dynamics. The previous approaches also exploit exact knowledge of the dynamic model in computed torque design. Various estimation methods can be used for estimation of the feedforward terms, thereby reducing the requirement for high gains and high frequency to compensate for the uncertainties/disturbances in the dynamic model. This section focuses on two popular feedforward methods: adaptive control for systems with linear in the parameters uncertainty, and neural network (and fuzzy logic)-based controllers to compensate for uncertainty that does not satisfy the linear in the parameters assumption (Property 2.3).

The controllers in the previous sections were based on the assumption of exact model knowledge (i.e., $N(\cdot)$ was used in the control designs to (partially) feedback linearize the system) with the exception of the added disturbance. In this section, the assumption of exact model knowledge is relaxed, but the added disturbance is neglected for simplicity and without loss of generality in the sense that the previous robust and sliding mode feedback methods can be used in conjunction with the methods in this section to also compensate for added disturbances.

2.4.1 Direct Adaptive Control

Control Development

The open-loop error system in (2.14) (with $\tau_d = 0$) can be written as

$$M\dot{r} = -V_m(q,\dot{q})r - F_d r + Y\theta - \tau \qquad (2.33)$$

where $Y(q,\dot{q},t) \in \mathbb{R}^{n\times p}$ is a nonlinear regression matrix, and $\theta \in \mathbb{R}^p$ is a vector of uncertain constant parameters (i.e., linear in the parameters assumption) defined as

$$Y\theta = M\ddot{q}_d + V_m(q,\dot{q})(\dot{q}_d + \alpha e) + G(q) + F_d(\dot{q}_d + \alpha e) + \alpha M\dot{e}.$$

Based on (2.33), an adaptive feedforward controller can be designed as

$$\tau = Y\hat{\theta} + kr \qquad (2.34)$$

where $\hat{\theta}(t) \in \mathbb{R}^p$ denotes a time varying estimate of θ. There is significant variation in how the adaptive update law is developed to generate $\hat{\theta}(t)$. Typically, a tracking error based gradient update law is designed motivated by the desire to cancel common terms in the Lyapunov analysis. Based on the subsequent stability analysis, a gradient update law for the open-loop error system in (2.33) is

$$\dot{\hat{\theta}} = \Gamma Y^T r \qquad (2.35)$$

where $\Gamma \in \mathbb{R}^{p\times p}$ is a diagonal matrix of adaptation gains. In practice, the initial condition $\hat{\theta}(0)$ are best-guess estimates of the parameters (e.g., information obtained from a manufacturer's specification sheet, or results from some off-line parameter estimation method, etc.). Substituting (2.34) into (2.33) yields the closed-loop error system

$$M\dot{r} = -V_m(q,\dot{q})r - F_d r + Y\tilde{\theta} - kr \qquad (2.36)$$

where $\tilde{\theta}(t) \in \mathbb{R}^p$ denotes the mismatch between the unknown parameters and the estimate vector as

$$\tilde{\theta} = \theta - \hat{\theta}. \tag{2.37}$$

Stability Analysis

Theorem 2.4 *Given the open-loop error system in (2.33), the adaptive controller given in (2.34) and (2.35) ensures global asymptotic tracking in the sense that*

$$e(t) \to 0 \qquad as \qquad t \to \infty. \tag{2.38}$$

Proof: Let $V(t) \in \mathbb{R}$ denote the non-negative function

$$V = \frac{1}{2} r^T M r + \frac{1}{2} \tilde{\theta}^T \Gamma^{-1} \tilde{\theta}. \tag{2.39}$$

By using (2.36), the time derivative of (2.37), and Properties 2.1 and 2.2, the time derivative of (2.39) can be expressed as

$$\dot{V} = r^T \left(Y\tilde{\theta} - F_d r - kr \right) - \tilde{\theta}^T \Gamma^{-1} \dot{\hat{\theta}}. \tag{2.40}$$

Substituting (2.35) into (2.40) and canceling common terms yields

$$\dot{V} = -r^T \left(F_d r + kr \right). \tag{2.41}$$

Since the expression in (2.41) is always negative semi-definite, (2.39) can be used to conclude that $V(t), r(t), \tilde{\theta}(t) \in \mathcal{L}_\infty$. Since $r(t) \in \mathcal{L}_\infty$, linear analysis methods [9] can be applied to (2.8) to prove that $e(t), \dot{e}(t) \in \mathcal{L}_\infty$, and since the desired trajectory $q_d(t)$ and its time derivatives are assumed to be bounded, $q(t), \dot{q}(t) \in \mathcal{L}_\infty$. Given that $q(t), \dot{q}(t) \in \mathcal{L}_\infty$, then Properties 2.2 and 2.4 can be used to conclude that $Y(q, \dot{q}, t) \in \mathcal{L}_\infty$. The fact that $\tilde{\theta}(t) \in \mathcal{L}_\infty$ can be used with (2.37) to conclude that $\hat{\theta}(t) \in \mathcal{L}_\infty$. Since $Y(q, \dot{q}, t), \hat{\theta}(t), r(t) \in \mathcal{L}_\infty$, the control is bounded from (2.34) and the adaptation law $\dot{\hat{\theta}}(t) \in \mathcal{L}_\infty$ from (2.35). The closed-loop error dynamics in (2.36) can be used to conclude that $\dot{r}(t) \in \mathcal{L}_\infty$; hence, $r(t)$ is uniformly continuous from Lemma A.12 in Appendix A. Lemma A.14 in Appendix A can be applied to (2.39) and (2.41) to conclude that $r(t)$ is square integrable (i.e., $r(t) \in \mathcal{L}_2$). Since $r(t), \dot{r}(t) \in \mathcal{L}_\infty$ and $r(t) \in \mathcal{L}_2$, a corollary to Barbalat's Lemma given in Lemma A.1 in Appendix A can be used to conclude that

$$r(t) \to 0 \qquad as \qquad t \to \infty. \tag{2.42}$$

Based on (2.42), Lemma A.18 in Appendix A can be invoked to conclude the result in (2.38). ∎

The motivation for the tracking error-based gradient update given in (2.35) is clear from (2.40). Through some control design and analysis modifications, additional update laws can also be used to obtain the result in (2.38) including least squares update laws and composite adaptive update laws based on both tracking and prediction (of the output) error. An example of a least squares adaptation law based on the tracking error is given by

$$\dot{\hat{\theta}} = PY^T r, \quad \dot{P} = -PY^T Y P \tag{2.43}$$

where $P(t) \in \mathbb{R}^{p \times p}$ is a time-varying symmetric matrix, where $P(0)$ is selected to be a positive definite, symmetric matrix. Composite adaptive update laws are updates based on combining information from both the tracking and the prediction error

$$\dot{\hat{\theta}} = PY_f^T \varepsilon + PY^T r \tag{2.44}$$

where $Y_f(\cdot)$ denotes the regression matrix $Y(\cdot)$ after it has been convolved with a low-pass filter, while $\varepsilon(t)$ denotes the prediction error [12], [28].

DCAL Extension

The adaptive update laws in (2.35), (2.43), and (2.44) depend on $q(t)$ and $\dot{q}(t)$. This dependency means that the regression matrix must be computed on-line and requires velocity feedback. For applications with demanding sampling times or limited computational resources, the need to compute the regression matrix on-line can be problematic. Moreover, velocity measurements may only be available through numerical differentiation (and hence, will likely contain noise), and some control designs require the derivative of the regression matrix, thus requiring acceleration measurements. To address these issues, this section describes the desired compensation adaptation law (DCAL) first developed in [24]. DCAL control designs are based on the idea of formulating a regression matrix that is composed of the desired position and velocity feedback rather than the actual states. Therefore, the regression matrix can be computed off-line (for an *a priori* given trajectory) and does not require velocity measurements. Furthermore, even if an adaptation law uses velocity feedback outside of the regression matrix (e.g., such as $r(t)$ in (2.35)), the DCAL-based desired regression matrix can be integrated by parts so that the actual estimate $\hat{\theta}(t)$ is only a function of position feedback. The following development illustrates how the DCAL strategy can be applied along with a filter mechanism to develop an adaptive output feedback controller (i.e., only $q(t)$ and hence $e(t)$ are measurable).

To facilitate the development of a DCAL based controller, a filter $e_f(t) \in \mathbb{R}^n$ is defined as

$$e_f = -ke + p \tag{2.45}$$

where $k \in \mathbb{R}$ is a positive filter gain, and $p(t) \in \mathbb{R}^n$ is generated from the following differential expression

$$\dot{p} = -(k+1)p + (k^2 + 1)e. \tag{2.46}$$

The filter developed in (2.45) and (2.46) is included in the control design by redefining the filtered tracking error $r(t)$ in (2.8) as

$$r = \dot{e} + e + e_f. \tag{2.47}$$

The filtered tracking error in (2.47) is not measurable due to the dependence on $\dot{e}(t)$, but the definition of $r(t)$ is useful for developing the closed-loop error system and stability analysis. To develop the new open-loop error system, time derivative of (2.47) is premultiplied by the inertia matrix, and (2.1) (with $\tau_d = 0$), (2.7), and (2.45)–(2.47) are used to yield

$$\begin{aligned} M\dot{r} &= M(q)\ddot{q}_d + V_m(q, \dot{q})\dot{q} + G(q) + F_d\dot{q} - \tau \tag{2.48} \\ &+ M(r - e - e_f) + M\left(-k\dot{e} - (k+1)p + (k^2+1)e\right). \end{aligned}$$

After using Property 2.5, (2.7), (2.45), and (2.47), the open-loop dynamics can be expressed as

$$M\dot{r} = -V_m(q, \dot{q})r + Y_d\theta - kMr + \chi - \tau \tag{2.49}$$

where $Y_d(q_d, \dot{q}_d, \ddot{q}_d)\theta$ is defined in Property 2.3, and the auxiliary term $\chi(e, r, q, \dot{q}, \dot{q}_d) \in \mathbb{R}^n$ is defined as

$$\begin{aligned} \chi &= M(q)\ddot{q}_d + V_m(q, \dot{q}_d)(\dot{q}_d - r + e + e_f) + V_m(q, \dot{q})(e + e_f) \tag{2.50} \\ &+ G(q) + F_d\dot{q} + M(r - e - e_f) + M(e - e_f) - Y_d\theta \end{aligned}$$

By using the Mean Value Theorem, the term $\chi(r, e, e_f, q, \dot{q}, q_d, \dot{q}_d, \ddot{q}_d)$ can be upper bounded as

$$\|\chi\| \leq \zeta_1 \|x\| + \zeta_2 \|x\|^2 \tag{2.51}$$

where $\zeta_1, \zeta_2 \in \mathbb{R}$ are some positive bounding constants, and $x(t) \in \mathbb{R}^{3n}$ is defined as

$$x = \begin{bmatrix} r^T & e^T & e_f^T \end{bmatrix}^T. \tag{2.52}$$

Based on (2.49), an adaptive DCAL controller can be designed as

$$\tau = Y_d\hat{\theta} - ke_f + e \tag{2.53}$$

where $\hat{\theta}(t)$ is generated from the following gradient-based DCAL adaptive update law

$$\dot{\hat{\theta}} = \Gamma Y_d^T r. \tag{2.54}$$

By integrating (2.54) by parts, the estimate $\hat{\theta}(t)$ can be expressed as

$$\hat{\theta}(t) = \hat{\theta}(0) + \Gamma Y_d^T e \big|_0^t - \Gamma \int_0^t \dot{Y}_d^T (\sigma) e (\sigma) d (\sigma) \tag{2.55}$$

$$+\Gamma \int_0^t Y_d^T (\sigma) (e (\sigma) + e_f (\sigma)) d (\sigma).$$

From (2.53) and (2.55), the controller does not depend on velocity, and the regression matrix can be computed off-line. Substituting (2.53) into (2.49) yields the closed-loop error system

$$M\dot{r} = -V_m(q, \dot{q})r + Y_d\tilde{\theta} - kMr + ke_f - e + \chi. \tag{2.56}$$

To facilitate the subsequent stability analysis, let k be defined as the constant

$$k = \frac{1}{m_1} \left(\zeta_1^2 k_{n1} + \zeta_2^2 k_{n2} + 1 \right) \tag{2.57}$$

where ζ_1 and ζ_2 are the constants defined in (2.51), and k_{n1} and $k_{n2} \in \mathbb{R}$ are positive constants selected (large enough) to satisfy the following initial condition dependent sufficient condition

$$k_{n2} \left(1 - \frac{1}{4k_{n1}}\right) \geq \frac{1}{4} \frac{\max(1, m_2) \|x(0)\|^2 + \lambda_{\max}(\Gamma^{-1}) \left\|\tilde{\theta}_{\max}(0)\right\|^2}{\min(1, m_1, \lambda_{\min}(\Gamma^{-1}))} \tag{2.58}$$

where $\lambda_{\max}()$ denotes the maximum eigenvalue of the argument, and $\tilde{\theta}_{\max}(0)$ denotes a known upperbound on the parameter estimate mismatch.

Stability Analysis

Theorem 2.5 *Given the open-loop error system in (2.49), the adaptive controller given in (2.53) and (2.54) ensures semi-global asymptotic tracking in the sense that*

$$e(t) \to 0 \quad as \quad t \to \infty \tag{2.59}$$

provided k is selected according to (2.58).

Proof: Let $V(t) \in \mathbb{R}$ denote the non-negative function

$$V = \frac{1}{2}r^T Mr + \frac{1}{2}e^T e + \frac{1}{2}e_f^T e_f + \frac{1}{2}\tilde{\theta}^T \Gamma^{-1}\tilde{\theta} \tag{2.60}$$

that can be upper and lower bounded as

$$\frac{1}{2}\min(1, m_1, \lambda_{\min}(\Gamma^{-1})) \|z\|^2 \leq V \leq \frac{1}{2}\max(1, m_2, \lambda_{\max}(\Gamma^{-1})) \|z\|^2$$
$$(2.61)$$

for some positive constants $\lambda_1, \lambda_2 \in \mathbb{R}$ and $z(t) \in \mathbb{R}^{3n+p}$ is defined as

$$z = \begin{bmatrix} r^T & e^T & e_f^T & \tilde{\theta}^T \end{bmatrix}^T.$$

By using Properties 2.1 and 2.2, (2.45)–(2.47), and (2.56), the time derivative of (2.60), can be expressed as

$$\dot{V} = r^T\left(Y_d\tilde{\theta} - kMr + ke_f - e + \chi\right) + e^T\left(r - e - e_f\right) \tag{2.62}$$

$$+ \frac{1}{2}e_f^T\left(-k\left(r - e - e_f\right) - (k+1)\left(e_f + ke\right) + (k^2 + 1)e\right) - \tilde{\theta}^T\Gamma^{-1}\dot{\hat{\theta}}.$$

Substituting (2.54) into (2.62) and canceling common terms yields

$$\dot{V} = r^T\left(-kMr + \chi\right) - e^Te - e_f^Te_f \tag{2.63}$$

which, by using (2.51) can be upper bounded as

$$\dot{V} \leq -km_1 \|r\|^2 + \zeta_1 \|x\| \|r\| + \zeta_2 \|r\| \|x\|^2 - \|e\|^2 - \|e_f\|^2 \tag{2.64}$$

where m_1 is introduced in (2.2). By using (2.57) and completing the squares on the first three terms in (2.64), the following inequality can be developed

$$\dot{V} \leq -\left(1 - \frac{1}{4k_{n1}} - \frac{\|x\|^2}{4k_{n2}}\right)\|x\|^2. \tag{2.65}$$

By using (2.61), the inequality in (2.65) can be further upper bounded as

$$\dot{V} \leq -\left(1 - \frac{1}{4k_{n1}} - \frac{0.5V}{\min(1, m_1, \lambda_{\min}(\Gamma^{-1}))k_{n2}}\right)\|x\|^2$$
$$\leq -c\|x\|^2 \text{ if } \left(1 - \frac{1}{4k_{n1}} - \frac{0.5V}{\min(1, m_1, \lambda_{\min}(\Gamma^{-1}))k_{n2}}\right) \geq 0$$
$$(2.66)$$

for some positive constant c. The second inequality in (2.66) illustrates a semi-global stability result. That is, if the condition in (2.66) is satisfied, then $V(t)$ will always be smaller or equal to $V(0)$. Therefore, if the condition in (2.66) is satisfied for $V(0)$ then it will be satisfied for all time. As a result, provided the sufficient condition given in (2.58) is satisfied (i.e., k is selected large enough based on the initial conditions in the system), then (2.60) and (2.66) can be used to conclude that $V(t), z(t), x(t), e(t), e_f(t), r(t), \tilde{\theta}(t) \in \mathcal{L}_\infty$. Similar boundedness arguments can now be used to conclude that all closed-loop signals are bounded and that $x(t) \in \mathcal{L}_2$, and Barbalat's Lemma given in Lemma A.1 in Appendix A can be used to conclude that

$$e(t) \to 0 \quad as \quad t \to \infty. \quad \blacksquare$$

2.4.2 Neural Network-Based Control

The adaptive control methods developed in the previous section are based on the linear in the parameters assumption (see Property 2.3). Sometimes, the system uncertainty contains unmodeled effects, or at least uncertainty that does not satisfy Property 2.3. For these systems, function approximation methods (e.g., neural network (NN), fuzzy logic-based approximators, genetic algorithms, etc.) can be used as a feedforward control method. The advantage of function approximation is that the uncertainty does not need to be modeled; however, function approximation methods have inherent reconstruction errors that can degrade the steady-state performance of a system (e.g., resulting in a UUB stability result). In this section, a NN-based feedforward controller is developed (other function approximation methods have a similar control structure).

Function Approximation

As a result of the universal approximation property, multilayer NNs can approximate generic nonlinear continuous functions. Specifically, let $\mathbb{S}$ be a compact simply connected set of $\mathbb{R}^{p_1+1}$. With map $f : \mathbb{S} \to \mathbb{R}^n$, define $\mathbb{C}^n(\mathbb{S})$ as the space where f is continuous. There exist weights and thresholds such that some function $f(\xi) \in \mathbb{C}^n(\mathbb{S})$ can be represented by a three-layer NN as [14], [15]

$$f(\xi) = W^T \sigma\left(V^T \xi\right) + \varepsilon(\xi), \tag{2.67}$$

for some given input $\xi(t) \in \mathbb{R}^{p_1+1}$. In (2.67), $V \in \mathbb{R}^{(p_1+1) \times p_2}$ and $W \in \mathbb{R}^{(p_2+1) \times n}$ are bounded constant ideal weight matrices for the first-to-second and second-to-third layers respectively, where p_1 is the number of neurons in the input layer, p_2 is the number of neurons in the hidden layer, and n is the number of neurons in the output layer. The activation function in (2.67) is denoted by $\sigma(\cdot) \in \mathbb{R}^{p_2+1}$, and $\varepsilon(\xi) \in \mathbb{R}^n$ is the functional reconstruction error. A variety of activation functions can be used for $\sigma(\cdot)$. Some popular activation functions include sigmoid-based radial basis functions and fuzzy logic-based triangle membership functions. Note that augmenting the input vector $\xi(t)$ and activation function $\sigma(\cdot)$ by "1" allows thresholds to be included as the first columns of the weight matrices [14], [15]. Thus, any tuning of W and V then includes tuning of thresholds as well. The computing power of the NN comes from the fact that the activation function $\sigma(\cdot)$ is nonlinear and the weights W and V can be modified or tuned through some learning procedure [15]. Based on (2.67), the typical three-layer NN approximation for $f(\xi)$ is given as [14], [15]

$$\hat{f}(\xi) \triangleq \hat{W}^T \sigma(\hat{V}^T \xi), \tag{2.68}$$

where $\hat{V}(t) \in \mathbb{R}^{(p_1+1) \times p_2}$ and $\hat{W}(t) \in \mathbb{R}^{(p_2+1) \times n}$ are subsequently designed estimates of the ideal weight matrices. The estimation errors for the ideal weight matrices, denoted by $\tilde{V}(t) \in \mathbb{R}^{(p_1+1) \times p_2}$ and $\tilde{W}(t) \in \mathbb{R}^{(p_2+1) \times n}$, are defined as

$$\tilde{V} \triangleq V - \hat{V}, \quad \tilde{W} \triangleq W - \hat{W},$$

and the mismatch for the hidden-layer output error for a given $x(t)$, denoted by $\tilde{\sigma}(x) \in \mathbb{R}^{p_2+1}$, is defined as

$$\tilde{\sigma} \triangleq \sigma - \hat{\sigma} = \sigma(V^T\xi) - \sigma(\hat{V}^T\xi). \tag{2.69}$$

The NN estimate has several properties that facilitate the subsequent development. These properties are described as follows.

Property 2.6 The Taylor series expansion for $\sigma\left(V^T\xi\right)$ for a given ξ may be written as [14], [15]

$$\sigma(V^T\xi) = \sigma(\hat{V}^T\xi) + \sigma'(\hat{V}^T\xi)\tilde{V}^T\xi + O(\tilde{V}^T\xi)^2, \tag{2.70}$$

where $\sigma'(\hat{V}^T\xi) \equiv d\sigma\left(V^T\xi\right)/d\left(V^T\xi\right)|_{V^T\xi=\hat{V}^T\xi}$, and $O(\hat{V}^T\xi)^2$ denotes the higher order terms. After substituting (2.70) into (2.69) the following expression can be obtained:

$$\tilde{\sigma} = \hat{\sigma}'\tilde{V}^T\xi + O(\tilde{V}^T\xi)^2, \tag{2.71}$$

where $\hat{\sigma}' \triangleq \sigma'(\hat{V}^T\xi)$.

Property 2.7 The ideal weights are assumed to exist and be bounded by known positive values so that

$$\|V\|_F^2 \le \bar{V}_B \tag{2.72}$$

$$\|W\|_F^2 \le \bar{W}_B, \tag{2.73}$$

where $\|\cdot\|_F$ is the Frobenius norm of a matrix, and $tr\,(\cdot)$ is the trace of a matrix.

Property 2.8 The estimates for the NN weights, $\hat{W}(t)$ and $\hat{V}(t)$, can be bounded using a smooth projection algorithm (see [21]).

Property 2.9 The typical choice of activation function is the sigmoid function

$$\sigma(\xi) = \frac{1}{1 + e^{\kappa\xi}},$$

where

$$\|\sigma\| < 1 \text{ and } \|\dot{\sigma}\| \le \sigma_n,$$

and $\sigma_n \in \mathbb{R}$ is a known positive constant.

Property 2.10 On a given compact set S, the net reconstruction error $\varepsilon(\xi)$ is bounded as

$$\|\varepsilon(\xi)\| \leq \varepsilon_n,$$

where $\varepsilon_n \in \mathbb{R}$ is a known positive constant.

Closed-Loop Error System

The open-loop error system in (2.14) can be written as

$$M\dot{r} = -V_m(q, \dot{q})r + f + \tau_d - \tau \tag{2.74}$$

where the function $f(t) \in \mathbb{R}^n$ is defined as

$$f \triangleq M(q)\ddot{q}_d + V_m(q, \dot{q})(\dot{q}_d + \alpha e) + G(q) + F_d\dot{q} + \alpha M\dot{e}. \tag{2.75}$$

The auxiliary function in (2.75) can be represented by a three layer NN as described in (2.67) where the NN input $x_1(t) \in \mathbb{R}^{5n+1}$ is defined as $x_1 \triangleq [1, \ddot{q}_d, q, \dot{q}, e, \dot{e}]^T$.

Based on (2.33), a NN-based feedforward controller can be designed as

$$\tau = \hat{f} + kr \tag{2.76}$$

where $\hat{f}(t) \in \mathbb{R}^n$ is the estimate for $f(t)$ and is defined as in (2.68), while the update laws for $\hat{W}(t)$ and $\hat{V}(t)$ are designed (and generated on-line) based on the subsequent stability analysis as

$$\dot{\hat{W}} = \Gamma_1 \hat{\sigma} r^T - \Gamma_1 \hat{\sigma}' \hat{V}^T x_1 r^T \qquad \dot{\hat{V}} = \Gamma_2 x (\hat{\sigma}'^T \hat{W} r)^T \tag{2.77}$$

where $\Gamma_1 \in \mathbb{R}^{(p_2+1)\times(p_2+1)}$ and $\Gamma_2 \in \mathbb{R}^{(p_1+1)\times(5n+1)}$ are constant, positive definite, symmetric gain matrices. The closed-loop error system can be developed by substituting (2.76) into (2.74) and using (2.67) and (2.68) as

$$M\dot{r} = -V_m(q, \dot{q})r + W^T\sigma(V^T x_1) - \hat{W}^T\sigma(\hat{V}^T x_1) + \varepsilon(x_1) + \tau_d - kr \tag{2.78}$$

Simple algebraic manipulations as well as an application of the Taylor series approximation in (2.71) yields

$$M\dot{r} = -V_m(q, \dot{q})r + \tilde{W}^T\hat{\sigma} - \tilde{W}^T\hat{\sigma}'\hat{V}^T x_1 + \hat{W}^T\hat{\sigma}'\tilde{V}^T x_1 + \chi - kr - e, \tag{2.79}$$

where the notations $\hat{\sigma}_1$ and $\tilde{\sigma}_1$ were introduced in (2.69), and $\chi(t) \in \mathbb{R}^n$ is defined as

$$\chi = \tilde{W}^T\hat{\sigma}'V^T x_1 + W^T O(\tilde{V}^T x_1)^2 + \varepsilon(x_1) + \tau_d + e. \tag{2.80}$$

Based on Properties 2.4, 2.6–2.10, $\chi(t)$ can be bounded as

$$\|\chi\| \leq c_1 + c_2 \|e\| + c_3 \|r\|, \tag{2.81}$$

where $c_i \in \mathbb{R}$, $(i = 1, 2, 3)$ are known positive bounding constants. In (2.79), $k \in \mathbb{R}$ is a positive constant control gain defined, based on the subsequent stability analysis, as

$$k \triangleq k_1 + k_{n1}c_1^2 + k_{n2}c_2^2 \tag{2.82}$$

where $k_1, k_{n1}, k_{n2} \in \mathbb{R}$ are positive constant gains.

Stability Analysis

Theorem 2.6 *Given the open-loop error system in (2.74), the NN-based controller given in (2.76) ensures global uniformly ultimately bounded stability in the sense that*

$$\|e(t)\| \leq \varepsilon_0 \exp(-\varepsilon_1 t) + \varepsilon_2 \tag{2.83}$$

where $\varepsilon_0, \varepsilon_1, \varepsilon_2$ are some positive constants, provided k is selected according to the following sufficient conditions

$$\alpha > \frac{1}{4k_{n2}} \qquad k_1 > c_3 \tag{2.84}$$

where c_3 is defined in (2.81).

Proof: Let $V(t) \in \mathbb{R}$ denote a non-negative, radially unbounded function defined as

$$V = \frac{1}{2}r^T M r + \frac{1}{2}e^T e + \frac{1}{2}tr(\tilde{W}^T \Gamma_1^{-1} \tilde{W}) + \frac{1}{2}tr(\tilde{V}^T \Gamma_2^{-1} \tilde{V}). \tag{2.85}$$

It follows directly from the bounds given in Properties 2.1, 2.7, and 2.8, that $V(t)$ can be upper and lower bounded as

$$\lambda_1 \|z\|^2 \leq V(t) \leq \lambda_2 \|z\|^2 + \zeta, \tag{2.86}$$

where $\lambda_1, \lambda_2, \zeta \in \mathbb{R}$ are known positive bounding constants, and $z(t) \in \mathbb{R}^{2n}$ is defined as

$$z = \begin{bmatrix} r^T & e^T \end{bmatrix}^T.$$

The time derivative of $V(t)$ in (2.85) can be determined as

$$\dot{V} = r^T(\frac{1}{2}\dot{M}r - V_m(q,\dot{q})r + \tilde{W}^T\hat{\sigma} - \tilde{W}^T\hat{\sigma}'\hat{V}^T x_1 + \hat{W}^T\hat{\sigma}'\tilde{V}^T x_1 + \chi$$
$$-kr - e) + e^T(r - \alpha e) - tr(\tilde{W}^T\Gamma_1^{-1}\dot{\hat{W}}_1) - tr(\tilde{V}^T\Gamma_2^{-1}\dot{\hat{V}}) \tag{2.87}$$

By using Properties 2.1 and 2.2, (2.77), (2.79), (2.81), and (2.82) the expression in (2.87) can be upper bounded as

$$\dot{V} \le \left[c_1 \|r\| - k_{n1}c_1^2 \|r\|^2 \right] + \left[c_2 \|e\| \|r\| - k_{n2}c_2^2 \|r\|^2 \right] + (c_3 - k_1) \|r\|^2 - \alpha \|e\|^2 \tag{2.88}$$

where the fact that $tr(AB) = tr(BA)$ was used. After applying the nonlinear damping argument provided in Lemma A.19 of Appendix A on the bracketed terms, one obtains

$$\dot{V} \le \frac{1}{4k_{n1}} - \left(\alpha - \frac{1}{4k_{n2}} \right) \|e\|^2 - (k_1 - c_3) \|r\|^2 \, .$$

A further upperbound can be developed after using (2.84) and (2.85) as follows

$$\dot{V}(t) \le -\frac{c}{\lambda_2} V(t) + \varepsilon_x, \tag{2.89}$$

where $c \in \mathbb{R}$ and $\varepsilon_x \in \mathbb{R}$ are positive constants defined as

$$c \triangleq \min \left\{ (k_1 - c_3), \; \left(\alpha - \frac{1}{4k_{n2}} \right) \right\} \quad \text{and} \quad \varepsilon_x \triangleq \frac{1}{4k_{n1}} + \frac{c\zeta}{\lambda_2}. \tag{2.90}$$

The linear differential inequality in (2.89) can be solved as

$$V(t) \le V(0)e^{(-\frac{c}{\lambda_2})t} + \varepsilon_x \frac{\lambda_2}{c} \left[1 - e^{(-\frac{c}{\lambda_2})t} \right]. \tag{2.91}$$

The inequalities in (2.86) can now be used along with (2.90) and (2.91) to conclude that

$$\|e(t)\|^2 \le \left[\frac{\lambda_2 \|z(0)\|^2 + \zeta - \varepsilon_x \frac{\lambda_2}{c}}{\lambda_1} \right] e^{(-\frac{c}{\lambda_2})t} + \varepsilon_x \frac{\lambda_2}{\lambda_1 c}. \tag{2.92}$$

In a similar approach to the one developed in the first section, it can be shown that all signals remain bounded. ∎

2.5   Task-Space Control and Redundancy

The controllers developed in the previous sections are all based on a joint-space-based control objective. Yet, the control objective for many robotic applications is best described in the task-space or Cartesian-space where the relationship between the robot and external objects is relevant. By using inverse kinematics, the desired task-space trajectory can be related to the desired joint-space trajectory for all the controllers derived hitherto. However, the problem becomes challenging when dealing with redundancy.

A kinematically redundant manipulator is a robotic arm that has more degrees of freedom (DOF) than required to perform an operation in the task-space; hence, these extra degrees of freedom allow the robot manipulator to perform more dextrous manipulation and/or provide the robot manipulator system with increased flexibility for the execution of sophisticated tasks. Since the dimension (i.e., n) of the link position variables is greater than the dimension (i.e., m) of the task-space variables, the null space of Jacobian matrix has a minimum dimension of $n - m$. That is, any link velocity in the null space of the manipulator Jacobian will not affect the task-space velocity. This motion of the joints is referred to as self-motion, since it is not observed in the task-space. As stated in [7], [18], [19], and [27], there are generally an infinite number of solutions for the inverse kinematics of a redundant manipulator. Thus, given a desired task-space trajectory, it can be difficult to select a reasonable joint-space trajectory to ensure stability and boundedness of all signals along with satisfying the mechanical constraints such as singularities and obstacle avoidance.

This section considers the nonlinear control of kinematically redundant robot manipulators through the development of a computed torque exponential link position and sub-task tracking controller. An adaptive full-state feedback controller is also developed that achieves asymptotic link position and sub-task tracking despite parametric uncertainty associated with the dynamic model. The developed controllers do not require the computation of the inverse kinematics and do not place any restrictions on the self-motion of the manipulator; hence, the extra degrees of freedom are available for subtasks (i.e., maintaining manipulability, avoidance of joint limits and obstacle avoidance). The reader is referred to [32] and [33] for more details.

2.5.1   Kinematic Model

The end-effector position and orientation in the task-space, denoted by $x(t) \in \mathbb{R}^m$, is defined as follows

$$x = f(q) \tag{2.93}$$

where $f(q) \in \mathbb{R}^m$ denotes the forward kinematics, and $q(t) \in \mathbb{R}^n$ denote the link position. Based on (2.93), the differential relationships between the end-effector position and the link position variables can be calculated as

$$\begin{aligned} \dot{x} &= J(q)\dot{q} \\ \ddot{x} &= \dot{J}(q)\dot{q} + J(q)\ddot{q} \end{aligned} \tag{2.94}$$

where $\dot{q}(t)$, $\ddot{q}(t) \in \mathbb{R}^n$ denote the link velocity and acceleration vectors, respectively, and the manipulator Jacobian, denoted by $J(q) \in \mathbb{R}^{m \times n}$, is defined as

$$J(q) = \frac{\partial f(q)}{\partial q}. \tag{2.95}$$

A pseudo-inverse of $J(q)$, denoted by $J^+(q) \in \mathbb{R}^{n \times m}$, is defined as

$$J^+ = J^T \left(JJ^T\right)^{-1} \tag{2.96}$$

where $J^+(q)$ satisfies the following equality

$$JJ^+ = I_m. \tag{2.97}$$

where $I_m \in \mathbb{R}^{m \times m}$ denotes the $m \times m$ identity matrix. As shown in [18], the pseudo-inverse defined by (2.96) satisfies the Moore-Penrose Conditions

$$\begin{array}{cc} JJ^+J = J & J^+JJ^+ = J^+ \\ \left(J^+J\right)^T = J^+J & \left(JJ^+\right)^T = JJ^+ \end{array} . \tag{2.98}$$

In addition to the above properties, the matrix $(I_n - J^+J)$ satisfies the following useful properties

$$\begin{array}{cc} \left(I_n - J^+J\right)\left(I_n - J^+J\right) = I_n - J^+J & J\left(I_n - J^+J\right) = 0 \\ \left(I_n - J^+J\right)^T = \left(I_n - J^+J\right) & \left(I_n - J^+J\right)J^+ = 0 \end{array} . \tag{2.99}$$

The control development in this section is based on the assumption that the minimum singular value of the manipulator Jacobian, denoted by σ_m is greater than a known small positive constant $\delta > 0$, such that $\max\left\{\|J^+(q)\|\right\}$ is known *a priori* and all kinematic singularities are always avoided. For revolute robot manipulators, the Jacobian and its pseudo-inverse are bounded for all possible $q(t)$ (i.e., these kinematic terms only depend on $q(t)$ as arguments of trigonometric functions).

2.5.2 Control Objective and Error System Formulation

The task-space position error, denoted by $e(t) \in \mathbb{R}^m$, is defined as

$$e = x_d - x \tag{2.100}$$

where $x_d(t) \in \mathbb{R}^m$ denotes the desired task-space trajectory, where the desired trajectory terms $x_d(t)$, $\dot{x}_d(t)$, and $\ddot{x}_d(t)$ are assumed to be bounded functions of time. As in [7], a sub-task tracking error, denoted by $e_N(t) \in \mathbb{R}^n$, can also be defined as

$$e_N = \left(I_n - J^+J\right)(g - \dot{q}) \tag{2.101}$$

where $I_n \in \mathbb{R}^{n \times n}$ denotes the $n \times n$ identity matrix, and $g(t) \in \mathbb{R}^n$ is an auxiliary signal that is constructed according to the sub-task control objective (e.g., joint-limit avoidance, or obstacle avoidance). The subsequent stability analysis mandates that the sub-task control objective be formulated in such a manner that both $g(t)$ and $\dot{g}(t)$ are bounded signals. To provide motivation for the definition of the sub-task control objective given by (2.101), take the time derivative of (2.100) and then substitute (2.94) for $\dot{x}(t)$ to obtain

$$\dot{e} = \dot{x}_d + \alpha e - \alpha e - J\dot{q} \qquad (2.102)$$

where the term αe has been added and subtracted to right-hand side of (2.102) to facilitate the control formulation, and $\alpha \in \mathbb{R}^{m \times m}$ denotes a diagonal, positive definite gain matrix. Using the properties of the pseudo-inverse of the manipulator Jacobian defined in (2.97), the relationship in (2.102) can be rewritten as

$$\dot{e} = -\alpha e + J \left(J^+ \left(\dot{x}_d + \alpha e \right) + \left(I_n - J^+ J \right) g - \dot{q} \right). \qquad (2.103)$$

Based on the structure of (2.103) and the subsequent analysis, the filtered tracking error signal, denoted by $r(t) \in \mathbb{R}^n$, is defined as

$$r = J^+ \left(\dot{x}_d + \alpha e \right) + \left(I_n - J^+ J \right) g - \dot{q}; \qquad (2.104)$$

hence, the closed-loop task-space position tracking error system can now be written as

$$\dot{e} = -\alpha e + Jr. \qquad (2.105)$$

In the following control development, the structure of (2.105) is used to ensure that the task-space error and the filtered tracking error defined by (2.100) and (2.104), respectively, are both regulated. To illustrate how the regulation of the filtered tracking error also ensures regulation of the sub-task tracking error defined by (2.101), the filtered tracking error in (2.104) is pre-multiplied by $(I_n - J^+ J)$ and then the properties given in (2.99) are applied to obtain

$$e_N = (I_n - J^+ J)r \qquad (2.106)$$

where (2.101) was used. From (2.106), if $r(t)$ is regulated then $e_N(t)$ is regulated, and hence, the sub-task control can also be achieved.

The structure of (2.105) also provides motivation to regulate $r(t)$ in order to regulate $e(t)$. Taking the time derivative of (2.104), pre-multiplying by the inertia matrix $M(q)$, and then substituting (2.1) yields the open loop dynamics

$$M\dot{r} = -V_m r + Y\phi - \tau \qquad (2.107)$$

where the regression matrix/parameter vector formulation $Y\phi$ is defined as

$$
\begin{aligned}
Y\phi = \ & M\frac{d}{dt}\{J^+\left(\dot{x}_d + \alpha e\right) + (I_n - J^+J)g\} \\
& +V_m\{J^+\left(\dot{x}_d + \alpha e\right) + (I_n - J^+J)g\} \\
& +G(q) + F(\dot{q})
\end{aligned}
\tag{2.108}
$$

where $Y(\ddot{x}_d, \dot{x}_d, x, q, \dot{q}, \dot{g}, g) \in \mathbb{R}^{n \times r}$ denotes a regression matrix, and $\phi \in \mathbb{R}^r$ denotes the constant system parameters (e.g., mass, inertia, friction coefficients).

2.5.3   Computed Torque Control Development and Stability Analysis

Based on the above error system development and the subsequent stability analysis, the control torque input $\tau(t)$ is designed as

$$
\tau = Y\phi + Kr + J^T e
\tag{2.109}
$$

where $K \in \mathbb{R}^{n \times n}$ is a constant, positive definite, diagonal gain matrix. After substituting (2.109) into (2.107), the closed-loop error system for $r(t)$ can be determined as

$$
M\dot{r} = -V_m r - J^T e - Kr.
\tag{2.110}
$$

Theorem 2.7 *The control law described by (2.109) guarantees global exponential task-space and sub-task tracking in the sense that both signals $e(t)$ and $e_N(t)$ are bounded by an exponential envelope.*

Proof: Let $V(t) \in \mathbb{R}$ denote a non-negative, radially unbounded function defined as

$$
V = \frac{1}{2}e^T e + \frac{1}{2}r^T M r.
\tag{2.111}
$$

After taking the time derivative of (2.111), substituting (2.105) and (2.110), using Properties 2.1 and 2.2, and then canceling common terms, yields

$$
\dot{V} = -e^T \alpha e - r^T Kr \leq -2\frac{\min(\alpha, \lambda_{\min}(K))}{\max(1, \lambda_{\max}(M))}V.
\tag{2.112}
$$

The structure of (2.111) and (2.112) indicate that $e(t)$ and $r(t) \in \mathcal{L}_\infty$. All signals can be shown to remain bounded by employing standard signal chasing arguments, utilizing assumptions that $x_d(t)$, $\dot{x}_d(t)$, $\ddot{x}_d(t)$, $g(t)$, $\dot{g}(t) \in \mathcal{L}_\infty$, and using the fact kinematic and dynamic terms denoted by $M(q)$, $V_m(q, \dot{q})$, $G(q)$, $J(q)$, and $J^+(q)$ are bounded for all possible $q(t)$. Yet, a standard problem associated with redundant manipulators is that the self motion limits the ability to show that $q(t)$ remains bounded; however, all signals in the manipulator kinematics/dynamics and the control remain bounded independent of the boundedness of $q(t)$ because $q(t)$ only appears as the argument of trigonometric functions. The structure of (2.111) and (2.112) indicates that $e(t)$ and $r(t)$ are bounded by an exponential envelope, and hence, due to the boundedness of $J^+(q)$ and $J(q)$, (2.106) can be used to conclude that $e_N(t)$ is also bounded by an exponential envelope.

■

2.5.4 Adaptive Control Extension

The computed torque controller in (2.109) can also be developed as an adaptive controller as

$$\tau = Y\hat{\phi} + Kr + J^T e \tag{2.113}$$

where $\hat{\phi}(t) \in \mathbb{R}^r$ denotes the parameter estimate vector that is generated based on the following update law

$$\dot{\hat{\phi}} = \Gamma_\phi Y^T r \tag{2.114}$$

where $\Gamma_\phi \in \mathbb{R}^{r \times r}$ is a constant, positive definite, diagonal gain matrix. After substituting (2.113) into (2.107), the closed loop dynamics for $r(t)$ can be obtained as

$$M\dot{r} = -V_m r + Y\tilde{\phi} - J^T e - Kr. \tag{2.115}$$

where $\tilde{\phi}(t) = \phi - \hat{\phi}(t) \in \mathbb{R}^r$ denotes the parameter estimation error.

Theorem 2.8 *The control law given by (2.113) and (2.114) guarantees global asymptotic task-space and sub-task tracking in the sense that*

$$\lim_{t \to \infty} e(t),\ e_N(t) = 0. \tag{2.116}$$

Proof: Let $V(t) \in \mathbb{R}$ denote a non-negative, radially unbounded function defined as

$$V_a = \frac{1}{2}e^T e + \frac{1}{2}r^T M r + \frac{1}{2}\tilde{\phi}^T \Gamma_\phi^{-1} \tilde{\phi}. \tag{2.117}$$

After taking the time derivative of (2.117), substituting (2.105), (2.114), and (2.115), using Properties 2.1 and 2.2, and then canceling common terms, yields

$$\dot{V}_a = -e^T \alpha e - r^T K r. \tag{2.118}$$

The structure of (2.117) and (2.118) indicates that $e(t), r(t), \tilde{\phi}(t) \in \mathcal{L}_\infty$. All signals remain bounded by noting that $\dot{\tilde{\phi}}(t) = -\dot{\hat{\phi}}(t)$ (i.e., ϕ is a constant vector) and employing similar arguments to those used in the previous proof. Since all signals are bounded, (2.105) and (2.115) that $\dot{e}(t)$ and $\dot{r}(t)$ are bounded (i.e., $e(t)$ and $r(t)$ are uniformly continuous). From the structure of (2.118), standard arguments can be utilized to show that $e(t)$, $r(t) \in \mathcal{L}_2$. Since $e(t)$, $r(t) \in \mathcal{L}_2$ and uniformly continuous, Lemma A.1 in Appendix A can be invoked to conclude that $\lim_{t \to \infty} \|e(t)\|, \|r(t)\| = 0$; hence, (2.116) follows directly from (2.106). ■

References

[1] R. Colbaugh, and K. Glass, "Robust Adaptive Control of Redundant Manipulators," *Journal of Intelligent and Robotic Systems*, Vol. 14, pp. 68–88, 1995.

[2] J. Craig, P. Hsu, and S. Sastry, "Adaptive Control of Mechanical Manipulators," *Proceedings of the IEEE Conference on Robotics and Automation,* pp. 190–195, San Francisco, CA, Mar. 1986.

[3] W. Dayawansa, W. M. Boothby, and D. L. Elliot, "Global State and Feedback Equivalence of Nonlinear Systems," *Systems and Control Letters*, Vol. 6, pp. 229–234, 1985.

[4] W. E. Dixon, D. M. Dawson, E. Zergeroglu and A. Behal, *Nonlinear Control of Wheeled Mobile Robots*, Springer-Verlag London Limited, 2001.

[5] W. E. Dixon, A. Behal, D. M. Dawson, and S. Nagarkatti, *Nonlinear Control of Engineering Systems: A Lyapunov-Based Approach*, Birkhäuser Boston, 2003.

[6] E. G. Gilbert and I. J. Ha, "An Approach to Nonlinear Feedback Control with Applications to Robotics," *IEEE Transactions on Systems, Man, and Cybernetics*, Vol. SMC-14, No. 6, pp. 879–884, 1984.

[7] P. Hsu, J. Hauser, and S. Sastry, "Dynamic Control of Redundant Manipulators," *Journal of Robotic Systems*, Vol. 6, pp. 133–148, 1989.

[8] L. R. Hunt, R. Su, and G. Meyer, "Global Transformations of Nonlinear Systems," *IEEE Transactions on Automatic Control*, Vol. 28, pp. 24–31, 1983.

[9] T. Kailath, *Linear Systems*, Englewood Cliffs, NJ: Prentice Hall, 1980.

[10] H. K. Khalil, *Nonlinear Systems*, 3rd edition, Prentice Hall, 2002.

[11] O. Khatib, "Dynamic Control of Manipulators in Operational Space," *Proceedings of the 6th IFTOMM Congress on Theory of Machines and Mechanisms*, pp. 1–10, New Delhi, Dec. 1983.

[12] M. Krstić, I. Kanellakopoulos, and P. Kokotović, *Nonlinear and Adaptive Control Design*, New York, NY: John Wiley and Sons, 1995.

[13] F. L. Lewis, C. T. Abdallah, and D. M. Dawson, *Control of Robot Manipulators*, New York: MacMillan Publishing Co., 1993.

[14] F. L. Lewis, "Nonlinear Network Structures for Feedback Control," *Asian Journal of Control*, Vol. 1, No. 4, pp. 205–228, 1999.

[15] F. L. Lewis, J. Campos, and R. Selmic, *Neuro-Fuzzy Control of Industrial Systems with Actuator Nonlincarities*, SIAM, PA, 2002.

[16] S. Y. Lim, D. M. Dawson, and K. Anderson, "Re-examining the Nicosia-Tomei Robot Observer-Controller from a Backstepping Perspective," *IEEE Transactions on Control Systems Technology*, Vol. 4, No. 3, pp. 304–310, May 1996.

[17] R. Middleton and C. Goodwin, "Adaptive Computed Torque Control for Rigid Link Manipulators," *Systems Control Letters*, Vol. 10, pp. 9–16, 1988.

[18] Y. Nakamura, *Advanced Robotics Redundancy and Optimization*, Addison-Wesley Pub. Co., Inc., 1991.

[19] D. N. Nenchev, "Redundancy Resolution through Local Optimization: A Review," *Journal of Robotic Systems*, Vol. 6, pp. 769–798, 1989.

[20] R. Ortega and M. Spong, "Adaptive Motion Control of Rigid Robots: A Tutorial," *Automatica*, Vol. 25, No. 6, pp. 877–888, 1989.

[21] P. M. Patre, W. MacKunis, C. Makkar, W. E. Dixon, "Asymptotic Tracking for Uncertain Dynamic Systems via a Multilayer NN Feedforward and RISE Feedback Control Structure," *IEEE Transactions on Control Systems Technology*, Vol. 16, No. 2, pp. 373–379, 2008.

[22] Z. X. Peng and N. Adachi, "Compliant Motion Control of Kinematically Redundant Manipulators," *IEEE Transactions on Robotics and Automation*, Vol. 9, No. 6, pp. 831–837, Dec. 1993.

[23] C. R. Rao, and S. K. Mitra, *Generalized Inverse of Matrices and Its Applications*, New York: Wiley, 1971.

[24] N. Sadegh and R. Horowitz, "Stability and Robustness Analysis of a Class of Adaptive Controllers for Robotic Manipulators," *International Journal of Robotic Research*, Vol. 9, No. 3, pp. 74–92, June 1990.

[25] S. Sastry, and M. Bodson, *Adaptive Control: Stability, Convergence, and Robustness*, Prentice Hall, Inc: Englewood Cliffs, NJ, 1989.

[26] H. Seraji, "Configuration Control of Redundant Manipulators: Theory and Implementation," *IEEE Transactions on Robotics and Automation*, Vol. 5, No. 4, pp. 472–490, August 1989.

[27] B. Siciliano, "Kinematic Control of Redundant Robot Manipulators: A Tutorial," *Journal of Intelligent and Robotic Systems*, Vol. 3, pp. 201–212, 1990.

[28] J. J. E. Slotine and W. Li, *Applied Nonlinear Control*, Prentice Hall, Inc: Englewood Cliffs, NJ, 1991.

[29] M. W. Spong and M. Vidyasagar, *Robot Dynamics and Control*, John Wiley and Sons, Inc: New York, NY, 1989.

[30] M. W. Spong and M. Vidyasagar, *Robot Dynamics and Control*, New York: John Wiley and Sons, Inc., 1989.

[31] T. Yoshikawa, "Analysis and Control of Robot Manipulators with Redundancy," in *Robotics Research — The First International Symposium*, MIT Press, Cambridge, MA, pp. 735–747, 1984.

[32] E. Zergeroglu, D. M. Dawson, I. Walker, and A. Behal, "Nonlinear Tracking Control of Kinematically Redundant Robot Manipulators," *Proceedings of the American Control Conference*, Chicago, IL, pp. 2513–2517, June 2000.

[33] E. Zergeroglu, D. M. Dawson, I. Walker, and P. Setlur, "Nonlinear Tracking Control of Kinematically Redundant Robot Manipulators," *IEEE/ASME Transactions on Mechatronics*, Vol. 9, No. 1, pp. 129–132, March 2004.

3

Vision-Based Systems

3.1   Introduction

Often robots are required to inspect, navigate through, and/or interact with unstructured or dynamically changing environments. Adequate sensing of the environment is the enabling technology required to achieve these tasks. Given recent advances in technologies such as computational hardware and computer vision algorithms, camera-based vision systems have become a popular and rapidly growing sensor of choice for robots operating in uncertain environments.

Given some images of an environment, one of the first tasks is to detect and identify interesting features in the object. These features can be distinguished based on attributes such as color, texture, motion, and contrast. Textbooks such as [4, 32, 52, 88, 90, 94, 98, 109] provide an excellent introduction and discussion of methods to find and track these attributes from image to image. This chapter assumes that some attributes can be used to identify an object of interest in an image, and points on the object, known as "feature points," can be tracked from one image to another. By observing how these feature points move in time and space, control and estimation algorithms can be developed. The first section of this chapter describes image geometry methods using a single camera. The development of this geometry enables the subsequent sections of the chapter to focus on the control and estimation of the motion of a plane attached to an object.

The terminology *visual servo control* refers to the use of information from a camera directly in the feedback loop of a controller. The typical objective of most visual servo controllers is to force a hand-held camera to a Euclidean position defined by a static reference image. Yet, many practical applications require a robotic system to move along a predefined or dynamically changing trajectory. For example, a human operator may predefine an image trajectory through a high-level interface, and this trajectory may need to be modified on-the-fly to respond to obstacles moving in and out of the environment. It is also well known that a regulating controller may produce erratic behavior and require excessive initial control torques if the initial error is large. The controllers in Section 3.3 focus on the more general tracking problem, where a robot end-effector is required to track a prerecorded time-varying reference trajectory. To develop the controllers, a homography-based visual servoing approach is utilized. The motivation for using this approach is that the visual servo control problem can be incorporated with a Lyapunov-based control design strategy to overcome many practical and theoretical obstacles associated with more traditional, purely image-based approaches. Specifically, one of the challenges of this problem is that the translation error system is corrupted by an unknown depth-related parameter. By formulating a Lyapunov-based argument, an adaptive update law is developed to actively compensate for the unknown depth parameter. In addition, the presented approach facilitates: i) translation/rotational control in the full six degree-of-freedom task-space without the requirement of an object model, ii) partial servoing on pixel data that yields improved robustness and increases the likelihood that the centroid of the object remains in the camera field-of-view [85], and iii) the use of an image Jacobian that is only singular for multiples of 2π, in contrast to the state-dependent singularities present in the image Jacobians associated with many of the purely image-based controllers. The controllers target both the fixed camera and the camera-in-hand configurations. The control development for the fixed camera problem is presented in detail, and the camera-in-hand problem is included as an extension.

Conventional robotic manipulators are designed as a kinematic chain of rigid links that bend at discrete joints to achieve a desired motion at its end-effector. Continuum robots [92] are robotic manipulators that draw inspiration from biological appendages like elephant trunks and squid tentacles, and can bend anywhere along the length of their body. In theory, they have infinite mechanical degrees-of-freedom so that their end-effector can be positioned at a desired location while concurrently satisfying workspace constraints such as tight spaces and the presence of obstacles. However, from an engineering perspective, an important implication of such a design is that although such devices have a high kinematic redundancy, they are infinitely underactuated. A variety of bending motions must be generated with only a finite number of actuators. While there has been considerable progress in the area of actuation strategies for such robots [92], the dual problem of sensing the configuration of such robots has been a challenge. From a controls perspective, a reliable position controller would require an accurate position sensing mechanism. However, internal motion sensing devices such as encoders cannot be used to determine either the shape or the end-effector position of a continuum robot, since there is no intuitive way to define links and joints on such a device.

A literature survey reveals that a few indirect methods have been proposed by researchers to estimate the shape of continuum robots, such as models [31, 56] that relate internal bellow pressures in fluid filled devices, or change in tendon length in tendon driven devices, or position of the end-effector. However, these methods do not have accuracies comparable to position sensing in rigid link robots because of the compliant nature of continuum devices. For example, in a tendon driven continuum robot, due to coupling of actuation between sections, various sections of the robot can potentially change shape without the encoders detecting a change in tendon length or tension. Motivated by a desire to develop an accurate strategy for real-time shape sensing in such robots, Hannan et al. [57] implemented simple image processing techniques to determine the shape of the Elephant Trunk robotic arm, where images from a fixed camera were used to reconstruct the curvatures of various sections of the robot. This technique was only applicable to the case where the motion of the arm was restricted to a plane orthogonal to the optical axis of the camera. However, the result in [57] demonstrated conclusively that there is a large difference in curvature measurements obtained from indirect cable measurements as compared to a vision-based strategy, and hence, the information obtained from *ad hoc* indirect shape measurement techniques is indeed questionable. Section 3.4 addresses this issue through a visual servo control approach. From a decomposition of the homography and from the equations describing the forward kinematics of the robot [68], the curvatures that define the shape of various sections of the robot can be fully determined. Various kinematic control strategies for hyperredundant robots [73, 97, 115] are then used to develop a kinematic controller that accurately positions the robot end-effector to any desired position and orientation by using a sequence of images from a single external video camera.

Homography-based visual servo control methods can also be used to develop relative translation and rotation error systems for the mobile robot regulation and tracking control problems. By using a similar approach as in the previous applications in this chapter, projective geometric relationships are exploited to enable the reconstruction of the Euclidean coordinates of feature points with respect to the mobile robot coordinate frame. By decomposing the homography into separate translation and rotation components, measurable signals for the orientation and the scaled Euclidean position can be obtained. Full Euclidean reconstruction is not possible due to the lack of an object model and the lack of depth information from the on-board camera to the target; hence, the resulting translation error system is unmeasurable. To accommodate for the lack of depth information, the unknown time-varying depth information is related to a constant depth-related parameter. The closed loop error systems are then constructed using Lyapunov-based methods including the development of an adaptive estimate for the constant depth related parameter. Both the setpoint and tracking controllers are implemented on an experimental testbed. Details of the testbed are provided along with experimental results that illustrate the performance of the presented controllers.

In addition to visual servo control, the recovery of Euclidean coordinates of feature points of a moving object from a sequence of images (i.e., image-based motion estimation) is a mainstream research problem with significant potential impact for applications such as autonomous vehicle/robotic guidance, navigation, and path planning. Motion estimation bears a close resemblance to the classical problem in computer vision, known as "Structure from Motion (SFM)," which is the determination of 3D structure of a scene from its 2D projections on a moving camera. In the motion estimation section of this chapter, a unique nonlinear estimation strategy is presented that simultaneously estimates the velocity and structure of a moving object using a single camera. By imposing a persistent excitation condition, the inertial coordinates for all the feature points on an object are determined. A homography-based approach is then utilized to develop the object kinematics in terms of reconstructed Euclidean information and image-space information for the fixed camera system. The development of object kinematics requires *a priori* knowledge of a single geometric length between two feature points on the object. A novel nonlinear integral feedback estimation method is then employed to identify the linear and angular velocity of the moving object. Identifying the velocities of the object facilitates the development of a measurable error system that can be used to formulate a nonlinear least squares adaptive update law. A Lyapunov-based analysis is then presented that indicates if a persistent excitation condition is satisfied then the time-varying Euclidean coordinates of each feature point can be determined. While the problem of estimating the motion and Euclidean position of features on a moving object is addressed in this chapter by using a fixed camera system, the development can also be recast for the camera-in-hand problem where a moving camera observes stationary objects. That is, by recasting the problem for the camera-in-hand, the development in this chapter can also be used to address the Simultaneous Localization and Mapping (SLAM) problem [36], where the information gathered from a moving camera is utilized to estimate both the motion of the camera (and hence, the relative position of the vehicle/robot) as well as position of static features in the environment.

3.2   Monocular Image-Based Geometry

This section focuses on the image geometry obtained between images taken by a single camera at different points in time and space. The geometry obtained from observing four coplanar and non-colinear feature points from a fixed camera is initially described. Euclidean reconstruction of the feature points from image coordinates is then described. The geometry is then extended to the more popular camera-in-hand scenario, where the observed object is stationary and the camera is moving. For both the fixed camera and the camera-in-hand, the relative motion between the viewed object and the camera is encoded through the construction of a homography that relates two spatiotemporal images. Development is provided that summarizes a well known decomposition algorithm that can be used to extract scaled translation and rotation information from the homography construction. Details are also provided that illustrate how a virtual parallax method enables the homography-based techniques to be applied for problems where the observed feature points do not lie in a plane.

3.2.1   Fixed-Camera Geometry

To make the subsequent development more tractable, four feature points located on an object (i.e., the end-effector of a robot manipulator) denoted by O_i $\forall i = 1, 2, 3, 4$ are considered to be coplanar and not colinear (It should be noted that if four coplanar target points are not available, then the subsequent development can exploit the classic eight-points algorithm [84] with no four of the eight target points being coplanar or the subsequently described Virtual Parallax method). Based on this assumption, consider a fixed plane, denoted by π^*, that is defined by a reference image of the object. In addition, consider the actual and desired motion of the plane containing the end-effector target points, denoted by π and π_d, respectively (see Figure 3.1). To develop a relationship between the planes, an inertial coordinate system, denoted by $\mathcal{I}$, is defined where the origin coincides with the center of a fixed camera. The Euclidean coordinates of the target points on π, π_d, and π^* can be expressed in terms of $\mathcal{I}$, respectively, as

$$
\begin{aligned}
\bar{m}_i(t) &\triangleq \begin{bmatrix} x_i(t) & y_i(t) & z_i(t) \end{bmatrix}^T \\
\bar{m}_{di}(t) &\triangleq \begin{bmatrix} x_{di}(t) & y_{di}(t) & z_{di}(t) \end{bmatrix}^T \\
\bar{m}_i^* &\triangleq \begin{bmatrix} x_i^* & y_i^* & z_i^* \end{bmatrix}^T
\end{aligned}
\tag{3.1}
$$

under the standard assumption that the distances from the origin of $\mathcal{I}$ to the target points remains positive (i.e., $z_i(t)$, $z_{di}(t)$, $z_i^* > \varepsilon$ where ε denotes an arbitrarily small positive constant). Orthogonal coordinate systems $\mathcal{F}$, $\mathcal{F}_d$, and $\mathcal{F}^*$ are attached to the planes π, π_d, and π^*, respectively (see Figure 3.1). To relate the coordinate systems, let $R(t)$, $R_d(t)$, $R^* \in SO(3)$ denote the rotation between $\mathcal{F}$ and $\mathcal{I}$, $\mathcal{F}_d$ and $\mathcal{I}$, and $\mathcal{F}^*$ and $\mathcal{I}$, respectively, and let $x_f(t)$, $x_{fd}(t)$, $x_f^* \in \mathbb{R}^3$ denote the respective translation vectors expressed in the coordinates of $\mathcal{I}$. As also illustrated in Figure 3.1, $n^* \in \mathbb{R}^3$ denotes the constant unit normal to the plane π^* expressed in the coordinates of $\mathcal{I}$, and $s_i \in \mathbb{R}^3$ denotes the constant coordinates of the $i-th$ target point. The constant distance from the origin of $\mathcal{I}$ to π^* along the unit normal is denoted by $d^* \in \mathbb{R}$ and is defined as

$$
d^* \triangleq n^{*T}\bar{m}_i^* .
\tag{3.2}
$$

The subsequent development requires that the constant rotation matrix R^* be known. The constant rotation matrix R^* can be obtained *a priori* using various methods (e.g., a second camera, Euclidean measurements). The subsequent development is also based on the assumption that the target points do not become occluded.

From the geometry between the coordinate frames depicted in Figure 3.1, the following relationships can be developed

$$
\begin{aligned}
\bar{m}_i &= x_f + Rs_i \\
\bar{m}_{di} &= x_{fd} + R_d s_i \\
\bar{m}_i^* &= x_f^* + R^* s_i .
\end{aligned}
\tag{3.3}
$$

After solving the third equation in (3.3) for s_i and then substituting the resulting expression into the first and second equations, the following relationships can be obtained

$$
\bar{m}_i = \bar{x}_f + \bar{R}\bar{m}_i^* \qquad \bar{m}_{di} = \bar{x}_{fd} + \bar{R}_d\bar{m}_i^*,
\tag{3.4}
$$

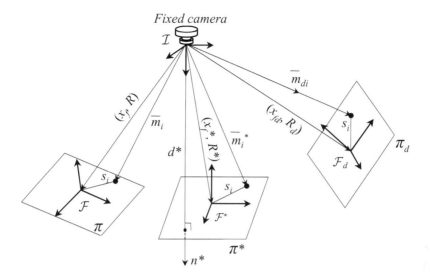

FIGURE 3.1. Coordinate Frame Relationships between a Fixed Camera and the Plane Defined by the Current, Desired, and Reference Feature Points (i.e., π, π_d, and π^*).

where $\bar{R}(t)$, $\bar{R}_d(t) \in SO(3)$ and $\bar{x}_f(t)$, $\bar{x}_{fd}(t) \in \mathbb{R}^3$ are new rotational and translational variables, respectively, defined as

$$\begin{aligned} \bar{R} &= R\left(R^*\right)^T & \bar{R}_d &= R_d\left(R^*\right)^T \\ \bar{x}_f &= x_f - \bar{R}x_f^* & \bar{x}_{fd} &= x_{fd} - \bar{R}_d x_f^* \,. \end{aligned} \tag{3.5}$$

From (3.2), the relationships in (3.4) can be expressed as

$$\bar{m}_i = \left(\bar{R} + \frac{\bar{x}_f}{d^*}n^{*T}\right)\bar{m}_i^* \qquad \bar{m}_{di} = \left(\bar{R}_d + \frac{\bar{x}_{fd}}{d^*}n^{*T}\right)\bar{m}_i^*. \tag{3.6}$$

Geometric insight into the structure of $\bar{R}(t)$ and $\bar{x}_f(t)$ defined in (3.5) can be obtained from Figures 3.1 and 3.2. Consider a fictitious camera that has a frame $\mathcal{I}^*$ attached to its center such that $\mathcal{I}^*$ initially coincides with $\mathcal{I}$. Since $\mathcal{I}$ and $\mathcal{I}^*$ coincide, the relationship between $\mathcal{I}^*$ and $\mathcal{F}^*$ can be denoted by rotational and translational parameters (x_f^*, R^*) as is evident from Figure 3.1. Without relative translational or rotational motion between $\mathcal{I}^*$ and $\mathcal{F}^*$, the two coordinate frames are moved until $\mathcal{F}^*$ aligns with $\mathcal{F}$, resulting in Figure 3.2. It is now evident that the fixed camera problem reduces to a stereo vision problem with the parameters $(x_f - R\left(R^*\right)^T x_f^*,\ R\left(R^*\right)^T)$ denoting the translation and rotation between $\mathcal{I}$ and $\mathcal{I}^*$.

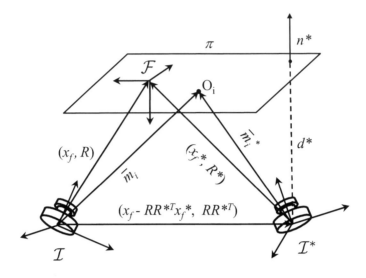

FIGURE 3.2. Geometric relationships for $\bar{R}(t)$ and $\bar{x}_f(t)$.

3.2.2 Euclidean Reconstruction

The relationship given by (3.6) provides a means to quantify a translation and rotation error between $\mathcal{F}$ and $\mathcal{F}^*$ and between $\mathcal{F}_d$ and $\mathcal{F}^*$. Since the Euclidean position of $\mathcal{F}$, $\mathcal{F}_d$, and $\mathcal{F}^*$ cannot be directly measured, a Euclidean reconstruction is developed in this section to obtain the position and rotational error information by comparing multiple images acquired from the fixed, monocular vision system. Specifically, comparisons are made between the current image, the reference image obtained *a priori*, and the *a priori* known sequence of images that define the trajectory of $\mathcal{F}_d$. To facilitate the subsequent development, the normalized Euclidean coordinates of the points on π, π_d, and π^* can be respectively expressed in terms of $\mathcal{I}$ as $m_i(t)$, $m_{di}(t)$, $m_i^* \in \mathbb{R}^3$, as

$$m_i \triangleq \frac{\bar{m}_i}{z_i} = \begin{bmatrix} \dfrac{x_i}{z_i} & \dfrac{y_i}{z_i} & 1 \end{bmatrix}^T \tag{3.7}$$

$$m_{di} \triangleq \frac{\bar{m}_{di}}{z_{di}} = \begin{bmatrix} \dfrac{x_{di}}{z_{di}} & \dfrac{y_{di}}{z_{di}} & 1 \end{bmatrix}^T \tag{3.8}$$

$$m_i^* \triangleq \frac{\bar{m}_i^*}{z_i^*} = \begin{bmatrix} \dfrac{x_i^*}{z_i^*} & \dfrac{y_i^*}{z_i^*} & 1 \end{bmatrix}^T . \tag{3.9}$$

From the expressions given in (3.6)–(3.9), the rotation and translation between the coordinate systems can now be related in terms of the normalized coordinates as

$$m_i = \underbrace{\frac{z_i^*}{z_i}}_{\alpha_i} \underbrace{\left(\bar{R} + \bar{x}_h n^{*T}\right)}_{H} m_i^* \tag{3.10}$$

$$m_{di} = \underbrace{\frac{z_i^*}{z_{di}}}_{\alpha_{di}} \underbrace{\left(\bar{R}_d + \bar{x}_{hd} n^{*T}\right)}_{H_d} m_i^*, \tag{3.11}$$

where $\alpha_i(t)$, $\alpha_{di}(t) \in \mathbb{R}$ denote invertible depth ratios, $H(t)$, $H_d(t) \in \mathbb{R}^{3\times3}$ denote Euclidean homographies [46], and $\bar{x}_h(t)$, $\bar{x}_{hd}(t) \in \mathbb{R}^3$ denote scaled translation vectors that are defined as

$$\bar{x}_h = \frac{\bar{x}_f}{d^*} \qquad \bar{x}_{hd} = \frac{\bar{x}_{fd}}{d^*} . \tag{3.12}$$

Each target point on π, π_d, and π^* will have a projected pixel coordinate expressed in terms of $\mathcal{I}$, denoted by $u_i(t)$, $v_i(t) \in \mathbb{R}$ for π, $u_{di}(t)$, $v_{di}(t) \in \mathbb{R}$ for π_d, and u_i^*, $v_i^* \in \mathbb{R}$ for π^*, that are defined as

$$p_i \triangleq \begin{bmatrix} u_i & v_i & 1 \end{bmatrix}^T \qquad p_{di} \triangleq \begin{bmatrix} u_{di} & v_{di} & 1 \end{bmatrix}^T \qquad p_i^* \triangleq \begin{bmatrix} u_i^* & v_i^* & 1 \end{bmatrix}^T. \tag{3.13}$$

In (3.13), $p_i(t)$, $p_{di}(t)$, $p_i^* \in \mathbb{R}^3$ represent the image-space coordinates of the time-varying target points, the desired time-varying target point trajectory, and the constant reference target points, respectively. To calculate the Euclidean homography given in (3.10) and (3.11) from pixel information, the projected pixel coordinates of the target points are related to $m_i(t)$, $m_{di}(t)$, and m_i^* by the following pinhole camera models [46]

$$p_i = Am_i \qquad p_{di} = Am_{di} \qquad p_i^* = Am_i^*, \tag{3.14}$$

where $A \in \mathbb{R}^{3\times3}$ is a known, constant, and invertible intrinsic camera calibration matrix that is explicitly defined as [84]

$$A = \begin{bmatrix} fk_u & -fk_v\cot(\theta) & u_o \\ 0 & \dfrac{fk_v}{\sin(\theta)} & v_o \\ 0 & 0 & 1 \end{bmatrix}. \tag{3.15}$$

In (3.15), $u_o, v_o \in \mathbb{R}$ denote the pixel coordinates of the principal point (i.e., the image center that is defined as the frame-buffer coordinates of the intersection of the optical axis with the image plane), $k_u, k_v \in \mathbb{R}$ represent camera scaling factors, $\theta \in \mathbb{R}$ is the angle between the axes of the imaging elements (CCD) in the camera, and $f \in \mathbb{R}$ denote the focal length of the camera. After substituting (3.14) into (3.10) and (3.11), the following relationships can be developed

$$p_i = \alpha_i \underbrace{\left(AHA^{-1}\right)}_{G} p_i^* \qquad p_{di} = \alpha_{di} \underbrace{\left(AH_dA^{-1}\right)}_{G_d} p_i^*, \tag{3.16}$$

where $G(t) = [g_{ij}(t)]$, $G_d(t) = [g_{dij}(t)]$ $\forall i, j = 1, 2, 3 \in \mathbb{R}^{3 \times 3}$ denote projective homographies. From the first relationship in (3.16), a set of 12 linearly independent equations given by the 4 target point pairs $(p_i^*, p_i(t))$ with 3 independent equations per target pair can be used to determine the projective homography up to a scalar multiple (i.e., the product $\alpha_i(t)G(t)$ can be determined). From the definition of $G(t)$ given in (3.16), various techniques can then be used (e.g., see [47, 116]) to decompose the Euclidean homography, to obtain $\alpha_i(t)$, $G(t)$, $H(t)$, and the rotation and translation signals $\bar{R}(t)$ and $\bar{x}_h(t)$, and n^*. Likewise, by using the target point pairs $(p_i^*, p_{di}(t))$, the desired Euclidean homography can be decomposed to obtain $\alpha_{di}(t)$, $G_d(t)$, $H_d(t)$, and the desired rotation and translation signals $\bar{R}_d(t)$ and $\bar{x}_{hd}(t)$. The rotation matrices $R(t)$ and $R_d(t)$ can be computed from $\bar{R}(t)$ and $\bar{R}_d(t)$ by using (3.5) and the fact that R^* is assumed to be known. Hence, $R(t)$, $\bar{R}(t)$, $R_d(t)$, $\bar{R}_d(t)$, $\bar{x}_h(t)$, $\bar{x}_{hd}(t)$, and the depth ratios $\alpha_i(t)$ and $\alpha_{di}(t)$ are all known signals that can be used for control synthesis.

3.2.3 Camera-in-Hand Geometry

Based on the development provided for the fixed camera problem in the previous sections, the geometry for the camera-in-hand problem can be developed in a similar manner. Consider the geometric relationships depicted in Figure 3.3, where the camera is held by a robot end-effector (not shown). The coordinate frames $\mathcal{F}$, $\mathcal{F}_d$, and $\mathcal{F}^*$ depicted in Figure 3.3 are attached to the camera and denote the actual, desired, and reference locations for the camera, respectively. From the geometry between the coordinate frames, $\bar{m}_i^*$ can be related to $\bar{m}_i(t)$ and $\bar{m}_{di}(t)$ as

$$\bar{m}_i = x_f + R\bar{m}_i^* \qquad \bar{m}_{di} = x_{fd} + R_d\bar{m}_i^*, \tag{3.17}$$

where $\bar{m}_i(t)$, $\bar{m}_{di}(t)$, and $\bar{m}_i^*$ now denote the Euclidean coordinates of O_i expressed in $\mathcal{F}$, $\mathcal{F}_d$, and $\mathcal{F}^*$, respectively. In (3.17), $R(t)$, $R_d(t) \in SO(3)$ denote the rotation between $\mathcal{F}$ and $\mathcal{F}^*$ and between $\mathcal{F}_d$ and $\mathcal{F}^*$, respectively, and $x_f(t)$, $x_{fd}(t) \in \mathbb{R}^3$ denote translation vectors from $\mathcal{F}$ to $\mathcal{F}^*$ and $\mathcal{F}_d$ to $\mathcal{F}^*$ expressed in the coordinates of $\mathcal{F}$ and $\mathcal{F}_d$, respectively. By utilizing (3.2), (3.7)–(3.9), and a relationship similar to (3.12), the expressions in (3.17) can be written as

$$m_i = \alpha_i \underbrace{\left(R + x_h n^{*T}\right)}_{H} m_i^*$$

(3.18)

$$m_{di} = \alpha_{di} \underbrace{\left(R_d + x_{hd} n^{*T}\right)}_{H_d} m_i^*$$

(3.19)

In (3.18) and (3.19), $x_h(t)$, $x_{hd}(t) \in \mathbb{R}^3$ denote the following scaled translation vectors

$$x_h = \frac{x_f}{d^*} \qquad x_{hd} = \frac{x_{fd}}{d^*},$$

$\alpha_i(t)$ and $\alpha_{di}(t)$ are introduced in (3.10) and (3.11), and $m_i(t)$, $m_{di}(t)$, and m_i^* now denote the normalized Euclidean coordinates of O_i expressed in $\mathcal{F}$, $\mathcal{F}_d$, and $\mathcal{F}^*$, respectively. Based on the development in (3.17)–(3.19), the Euclidean reconstruction and control formulation can be developed in the same manner as for the fixed camera problem. Specifically, the signals $R(t)$, $R_d(t)$, $x_h(t)$, $x_{hd}(t)$, and the depth ratios $\alpha_i(t)$ and $\alpha_{di}(t)$ can be computed. The error systems for the camera-in-hand problem are defined the same as for the fixed camera problem (i.e., see (3.36)–(3.39)); however, $u(t)$, $u_d(t)$, $\theta(t)$, and $\theta_d(t)$ are defined as in (3.45) in terms of $R(t)$ and $R_d(t)$, respectively, for the camera-in-hand problem.

3.2.4 Homography Calculation

The previous development was based on the assumption that four corresponding coplanar but non-collinear feature points could be determined and tracked between images. This section presents methods to estimate the collineation $G(t)$ and the scaled Euclidean homography by solving a set of linear equations (3.16) obtained from the four corresponding coplanar and non-collinear feature points.

Based on the arguments in [60], a transformation is applied to the projective coordinates of the corresponding feature points to improve the accuracy in the estimation of $G(t)$. The transformation matrices, denoted by $P(t)$, $P^* \in \mathbb{R}^{3 \times 3}$, are defined in terms of the projective coordinates of three of the coplanar non-collinear feature points as

$$P \triangleq \begin{bmatrix} p_1 & p_2 & p_3 \end{bmatrix} \qquad P^* \triangleq \begin{bmatrix} p_1^* & p_2^* & p_3^* \end{bmatrix}.$$

(3.20)

From (3.16) and (3.20), it is easy to show that

$$P\tilde{G} = GP^*,$$

(3.21)

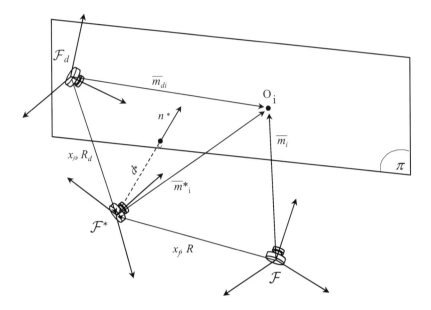

FIGURE 3.3. Coordinate frame relationships between the fixed feature point plane and the camera-in-hand at the current, desired, and reference position and orientation (i.e., $\mathcal{F}$, $\mathcal{F}_d$, and $\mathcal{F}^*$).

where

$$\tilde{G} = P^{-1}GP^* = diag\left(\alpha_1^{-1},\, \alpha_2^{-1},\, \alpha_3^{-1}\right) \triangleq diag\left(\tilde{g}_1,\, \tilde{g}_2,\, \tilde{g}_3\right). \qquad (3.22)$$

In (3.22), $diag(.)$ denotes a diagonal matrix with arguments as the diagonal entries. Utilizing (3.22), the relationship in (3.16) can be expressed in terms of $\tilde{G}(t)$ as

$$q_i = \alpha_i \tilde{G} q_i^*, \qquad (3.23)$$

where

$$q_i = P^{-1}p_i \qquad (3.24)$$

$$q_i^* = P^{*^{-1}}p_i^* \qquad (3.25)$$

define the new transformed projective coordinates. Note that the transformation normalizes the projective coordinates, and it is easy to show that

$$\begin{bmatrix} q_1 & q_2 & q_3 \end{bmatrix} = \begin{bmatrix} q_1^* & q_2^* & q_3^* \end{bmatrix} = I_3 \in \mathbb{R}^{3\times3},$$

where I_3 is the 3×3 identity matrix. The transformed image coordinates of a fourth matching pair of feature points

$$q_4(t) \triangleq \begin{bmatrix} q_{4u}(t) & q_{4v}(t) & q_{4w}(t) \end{bmatrix}^T \in \mathbb{R}^3 \qquad q_4^* \triangleq \begin{bmatrix} q_{4u}^* & q_{4v}^* & q_{4w}^* \end{bmatrix}^T \in \mathbb{R}^3$$

can be expressed as

$$q_4 = \alpha_4 \tilde{G} q_4^*,$$
(3.26)

where it can be shown that

$$q_{4u} = \frac{q_{4w}}{q_{4w}^*} q_{4u}^* \frac{\alpha_3}{\alpha_1} \qquad q_{4v} = \frac{q_{4w}}{q_{4w}^*} q_{4v}^* \frac{\alpha_3}{\alpha_2} \qquad \frac{q_{4w}}{q_{4w}^*} = \frac{\alpha_4}{\alpha_3}.$$

The above set of equations can be solved for

$$\frac{\alpha_4(t)}{\alpha_3(t)} \qquad \text{and} \qquad \alpha_3(t)\tilde{G}(t) = diag(\frac{\alpha_3(t)}{\alpha_1(t)}, \frac{\alpha_3(t)}{\alpha_2(t)}, 1).$$

Since the camera intrinsic calibration matrix is assumed to be known, the scaled Euclidean homography can be calculated as

$$\alpha_3(t)H(t) = \alpha_3(t)A^{-1}G(t)A.$$

As noted before, $H(t)$ can be decomposed into its constituent rotation matrix, unit normal vector, scaled translation vector, and the depth ratio $\alpha_3(t)$. With the knowledge of $\alpha_3(t)$ and $\frac{\alpha_4(t)}{\alpha_3(t)}$, the depth ratios $\alpha_1, \alpha_2, \alpha_3$, and α_4 can be calculated for all of the feature points.

If some noncoplanar points are also tracked between images, the depth ratios for those feature points can also be determined. Consider a feature point O_j on the object that is not on the plane π^*. The expressions in (3.3)–(3.6) can be used to conclude that

$$m_j = \frac{z_j^*}{z_j} \left(\frac{\bar{x}_f}{z_j^*} + \bar{R}m_j^* \right).$$
(3.27)

Multiplying both sides of the equation with the skew-symmetric form of $\bar{x}_h(t)$, denoted by $[\bar{x}_h(t)]_\times \in \mathbb{R}^{3\times3}$, yields [84]

$$[\bar{x}_h]_\times m_j = \alpha_j \left([\bar{x}_h]_\times \frac{\bar{x}_f}{z_j^*} + [\bar{x}_h]_\times \bar{R}m_j^* \right) = \alpha_j [\bar{x}_h]_\times \bar{R}m_j^*.$$
(3.28)

The signal $\bar{x}_h(t)$ is directly obtained from the decomposition of Euclidean homography matrix $H(t)$. Hence, the depth ratios for feature points O_j not lying on the plane π^* can be computed as

$$\alpha_j = \frac{\left\| [\bar{x}_h]_\times m_j \right\|}{\left\| [\bar{x}_h]_\times \bar{R}m_j^* \right\|}.$$
(3.29)

3.2.5 Virtual Parallax Method

In general, all feature points of interest on the moving object may not be coplanar. In such a case, the virtual parallax method may be used to develop a virtual plane from noncoplanar points. Based on the development in [84], any three feature points on an object may be selected to define the plane π^* shown in Figure 3.4. All feature points O_i on a plane satisfy (3.16). Consider a feature point O_j on the object that is not on the plane π^*. Let us define a virtual feature point O_j', on π^*, defined at the point of intersection of the vector from the optical center of the camera to O_j and the plane π^*. Let p_j^* be the projective image coordinates of the point O_j (and O_j') on the image plane when the object is at the reference position denoted by $\mathcal{F}^*$. As shown in Figure 3.4, when the object is viewed from a different pose, resulting from either a motion of the object or a motion of the camera, the actual feature point O_j and the virtual feature point O_j' projects to $p_j(t)$ and $p_j'(t)$, respectively, on the image plane of the camera. For any feature point O_j, both $p_j(t)$ and $p_j'(t)$ lie on the same epipolar line l_j [84] that is given by

$$l_j = p_j \times p_j', \tag{3.30}$$

where $\times$ denotes the cross product of the two vectors. Since the projective image coordinates of corresponding coplanar feature points satisfy (3.16), then

$$l_j = p_j \times Gp_j^*. \tag{3.31}$$

Based on the constraint that all epipolar lines meet at the epipole [84], a set of any three non-coplanar feature points can be selected such that the epipolar lines satisfy the constraint

$$\begin{vmatrix} l_j & l_k & l_l \end{vmatrix} = 0 \tag{3.32}$$

$$\begin{vmatrix} p_j \times Gp_j^* & p_k \times Gp_k^* & p_l \times Gp_l^* \end{vmatrix} = 0. \tag{3.33}$$

The transformation matrices, denoted by $P(t)$, $P^* \in \mathbb{R}^{3\times3}$, and defined in (3.20), are constructed using the image coordinates of the three coplanar feature points selected to define the plane π^*. After coordinate transformations defined in (3.24) and (3.25), the epipolar constraint of (3.33) now becomes

$$\begin{vmatrix} q_i \times \tilde{G}q_i^* & q_j \times \tilde{G}q_j^* & q_k \times \tilde{G}q_k^* \end{vmatrix} = 0, \tag{3.34}$$

where $\tilde{G}(t) \in \mathbb{R}^{3\times3}$ is defined in (3.22). As shown in [84], the set of homogeneous equations in (3.34) can be written in the form

$$C_{jkl}\bar{X} = 0, \tag{3.35}$$

where $\bar{X} = \left[\tilde{g}_1^2\tilde{g}_2, \ \tilde{g}_1\tilde{g}_2^2, \ \tilde{g}_1^2\tilde{g}_3, \ \tilde{g}_2^2\tilde{g}_3, \ \tilde{g}_1\tilde{g}_3^2, \ \tilde{g}_2\tilde{g}_3^2, \ \tilde{g}_1\tilde{g}_2\tilde{g}_3\right]^T \in \mathbb{R}^7$, and the matrix $C_{jkl} \in \mathbb{R}^{m \times 7}$ is of dimension $m \times 7$ where $m = \frac{n!}{6(n-3)!}$ and n is the number of epipolar lines (i.e., one for image coordinates of each feature point). Hence, apart from three coplanar feature points that define the transformation matrices in (3.20), at least five additional feature points (i.e., $n = 5$) are required in order to solve the set of equations given in (3.35). As shown in [84], $\tilde{G}(t)$ can be determined and used to calculate the scale factors, rotation matrix and normal to the plane as previously explained.

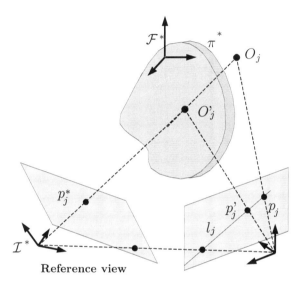

FIGURE 3.4. Virtual parallax.

3.3   Visual Servo Tracking

3.3.1   Control Objective

The objective in this section is to develop a visual servo controller for the fixed camera problem that ensures that the trajectory of $\mathcal{F}$ tracks $\mathcal{F}_d$ (i.e., $\bar{m}_i(t)$ tracks $\bar{m}_{di}(t)$), where the trajectory of $\mathcal{F}_d$ is constructed relative to the reference camera position/orientation given by $\mathcal{F}^*$. To ensure that $\bar{m}_i(t)$ tracks $\bar{m}_{di}(t)$ from the Euclidean reconstruction given in (3.10) and (3.11), the tracking control objective can be stated as: $\bar{R}(t) \rightarrow \bar{R}_d(t)$, $m_1(t) \rightarrow m_{d1}(t)$, and $z_1(t) \rightarrow z_{d1}(t)$ (and hence, $\bar{x}_h(t) \rightarrow \bar{x}_{hd}(t)$). Any point O_i

can be utilized in the subsequent development, so to reduce the notational complexity, the image point O_1 is used without loss of generality; thus, the subscript 1 is utilized in lieu of i in the subsequent development. The 3D control objective is complicated by the fact that only 2D image information is measurable. That is, while the development of the homography provides a means to reconstruct some Euclidean information, the formulation of a controller is challenging due to the fact that the time varying signals $z_1(t)$ and $z_{d1}(t)$ are not measurable. In addition, it is desirable to servo on actual pixel information (in lieu of reconstructed Euclidean information) to improve robustness to intrinsic camera calibration parameters and to increase the likelihood that the object will stay in the camera field-of-view.

To reformulate the control objective in light of these issues, a translation tracking error, denoted by $e_v(t) \in \mathbb{R}^3$, is defined as follows:

$$e_v = p_e - p_{ed} \tag{3.36}$$

where $p_e(t)$, $p_{ed}(t) \in \mathbb{R}^3$ are defined as

$$p_e = \begin{bmatrix} u_1 & v_1 & -\ln(\alpha_1) \end{bmatrix}^T \quad p_{ed} = \begin{bmatrix} u_{d1} & v_{d1} & -\ln(\alpha_{d1}) \end{bmatrix}^T, \tag{3.37}$$

and $\ln(\cdot)$ denotes the natural logarithm. A rotation tracking error, denoted by $e_\omega(t) \in \mathbb{R}^3$, is defined as

$$e_\omega \triangleq \Theta - \Theta_d \tag{3.38}$$

where $\Theta(t)$, $\Theta_d(t) \in \mathbb{R}^3$ denote the axis-angle representation of $\bar{R}(t)$ and $\bar{R}_d(t)$ as [105]

$$\Theta = u(t)\theta(t) \quad \Theta_d = u_d(t)\theta_d(t). \tag{3.39}$$

For the representations in (3.39), $u(t)$, $u_d(t) \in \mathbb{R}^3$ represent unit rotation axes, and $\theta(t), \theta_d(t) \in \mathbb{R}$ denote the respective rotation angles about $u(t)$ and $u_d(t)$ that are assumed to be confined to the following regions

$$-\pi < \theta(t) < \pi \quad -\pi < \theta_d(t) < \pi . \tag{3.40}$$

Based on the error system formulations in (3.36) and (3.38), the control objective can be stated as the desire to regulate the tracking error signals $e_v(t)$ and $e_\omega(t)$ to zero.

If the tracking error signals $e_v(t)$ and $e_\omega(t)$ are regulated to zero then the object can be proven to be tracking the desired trajectory. Specifically, to ensure that $\bar{m}_i(t)$ tracks $\bar{m}_{di}(t)$ from the Euclidean reconstruction given in (3.1), the tracking control objective can be stated as $\bar{R}(t) \to \bar{R}_d(t)$, $m_1(t) \to m_{d1}(t)$, and $z_1(t) \to z_{d1}(t)$. The expressions in (3.13) and (3.37) can be used to conclude that if $\|e_v(t)\| \to 0$ then $p_1(t) \to p_{d1}(t)$ and the ratio $\alpha_1(t)/\alpha_{d1}(t) \to 1$; hence, (3.14) and the definition of the depth ratios in (3.10) and (3.11) can be used to show that $m_1(t) \to m_{d1}(t)$ and $z_1(t) \to z_{d1}(t)$. Given that $m_1(t) \to m_{d1}(t)$ and $z_1(t) \to z_{d1}(t)$, (3.7) and (3.8) can be used to prove that $\bar{m}_1(t) \to \bar{m}_{d1}(t)$. To examine if $\bar{R}(t) \to \bar{R}_d(t)$, the difference between the expressions defined in (3.5) can be determined as (see Lemma B.1 in Appendix B)

$$\bar{R} - \bar{R}_d = \sin\theta \, [u]_\times - \sin\theta_d \, [u_d]_\times + 2\sin^2\frac{\theta}{2}[u]_\times^2 - 2\sin^2\frac{\theta_d}{2}[u_d]_\times^2 \ . \quad (3.41)$$

If $\|e_\omega(t)\| \to 0$, then (3.38) and (3.39) can be used to show that

$$u(t)\theta(t) \to u_d(t)\theta_d(t) \text{ as } t \to \infty, \quad (3.42)$$

which implies that

$$\|u(t)\theta(t)\|^2 \to \|u_d(t)\theta_d(t)\|^2 \text{ as } t \to \infty,$$

and

$$\theta^2(t)\,\|u(t)\|^2 \to \theta_d^2(t)\,\|u_d(t)\|^2 \text{ as } t \to \infty. \quad (3.43)$$

Since $\|u(t)\| = \|u_d(t)\| = 1$, (3.43) can be used to conclude that

$$\theta(t) \to \pm\theta_d(t) \text{ as } t \to \infty.$$

The result in (3.42) indicates that

$$\text{Case 1)} \qquad u(t) \to u_d(t) \text{ when } \theta(t) \to \theta_d(t) \qquad (3.44)$$
$$\text{Case 2)} \qquad u(t) \to -u_d(t) \text{ when } \theta(t) \to -\theta_d(t) \ .$$

After substituting each case given in (3.44) into (3.41) and then passing the limit, it is clear that $\bar{R}(t) \to \bar{R}_d(t)$. Based on the results that $\bar{m}_1(t) \to \bar{m}_{d1}(t)$ and that $\bar{R}(t) \to \bar{R}_d(t)$, it is clear that $\bar{m}_i(t) \to \bar{m}_{di}(t)$. A particular solution for $\theta(t)$ and $u(t)$ can be determined as [105]

$$\theta = \cos^{-1}\left(\frac{1}{2}\left(\text{tr}\left(\bar{R}\right) - 1\right)\right) \qquad [u]_\times = \frac{\bar{R} - \bar{R}^T}{2\sin(\theta)}, \qquad (3.45)$$

where the notation $\text{tr}(\cdot)$ denotes the trace of a matrix, and $[u]_\times$ denotes the 3×3 skew-symmetric expansion of $u(t)$.

To develop a tracking control design, it is typical that the desired trajectory is used as a feedforward component in the control design. Hence, for a kinematic controller the desired trajectory is required to be at least first order differentiable and at least second order differentiable for a dynamic level controller. To this end, a sufficiently smooth function (e.g., a spline function) is used to fit the sequence of target points to generate the desired trajectory $p_{di}(t)$; hence, it is assumed that $p_{ed}(t)$ and $\dot{p}_{ed}(t)$ are bounded functions of time. From the projective homography introduced in (3.16), $p_{di}(t)$ can be expressed in terms of the *a priori* known functions $\alpha_{di}(t)$, $H_d(t)$, $\bar{R}_d(t)$, and $\bar{x}_{hd}(t)$. Since these signals can be obtained from the prerecorded sequence of images, sufficiently smooth functions can also be generated for these signals by fitting a sufficiently smooth spline function to the signals. In practice, the *a priori* developed smooth functions $\alpha_{di}(t)$, $\bar{R}_d(t)$, and $\bar{x}_{hd}(t)$ can be constructed as bounded functions with bounded time derivatives. Based on the assumption that $\bar{R}_d(t)$ is a bounded first order differentiable function with a bounded derivative, (3.45) can be used to conclude that $u_d(t)$ and $\theta_d(t)$ are bounded first order differentiable functions with a bounded derivative; hence, $\Theta_d(t)$ and $\dot{\Theta}_d(t)$ can be assumed to be bounded. In the subsequent tracking control development, the desired signals $\dot{p}_{ed}(t)$ and $\dot{\Theta}_d(t)$ will be used as a feedforward control term.

3.3.2   Control Formulation

To develop the open-loop error system for $e_\omega(t)$, the time derivative of (3.38) is determined as (see Lemma B.1 in Appendix B)

$$\dot{e}_\omega = L_\omega R \omega_e - \dot{\Theta}_d. \tag{3.46}$$

In (3.46), the Jacobian-like matrix $L_\omega(t) \in \mathbb{R}^{3\times3}$ is defined as

$$L_\omega = I_3 - \frac{\theta}{2} [u]_\times + \left(1 - \frac{sinc(\theta)}{sinc^2\left(\dfrac{\theta}{2}\right)} \right) [u]_\times^2, \tag{3.47}$$

where

$$sinc(\theta(t)) \triangleq \frac{\sin \theta(t)}{\theta(t)},$$

and $\omega_e(t) \in \mathbb{R}^3$ denotes the angular velocity of the object expressed in $\mathcal{F}$. By exploiting the fact that $u(t)$ is a unit vector (i.e., $\|u\|^2 = 1$), the determinant of $L_\omega(t)$ can be calculated as [83]

$$\det(L_\omega) = \frac{1}{sinc^2\left(\dfrac{\theta}{2}\right)}, \tag{3.48}$$

where $\det(\cdot)$ signifies the determinant operator. From (3.48), it is clear that $L_\omega(t)$ is only singular for multiples of 2π (i.e., out of the assumed workspace); therefore, $L_\omega(t)$ is invertible in the assumed workspace.

To develop the open-loop error system for $e_v(t)$, the time derivative of (3.36) is determined as (see Lemma B.2 in Appendix B)

$$z_1^* \dot{e}_v = \alpha_1 A_e L_v R \left[v_e + [\omega_e]_\times s_1 \right] - z_1^* \dot{p}_{ed}, \tag{3.49}$$

where $v_e(t) \in \mathbb{R}^3$ denotes the linear velocity of the object expressed in $\mathcal{F}$. In (3.49), $A_e \in \mathbb{R}^{3 \times 3}$ is defined as

$$A_e = A - \begin{bmatrix} 0 & 0 & u_0 \\ 0 & 0 & v_0 \\ 0 & 0 & 0 \end{bmatrix}, \tag{3.50}$$

where u_0 and v_0 were introduced in (3.15), and the auxiliary Jacobian-like matrix $L_v(t) \in \mathbb{R}^{3 \times 3}$ is defined as

$$L_v = \begin{bmatrix} 1 & 0 & -\dfrac{x_1}{z_1} \\ 0 & 1 & -\dfrac{y_1}{z_1} \\ 0 & 0 & 1 \end{bmatrix}. \tag{3.51}$$

The product $A_e L_v(t)$ is an invertible upper triangular matrix from (3.50) and (3.51).

Based on the structure of the open-loop error systems and subsequent stability analysis, the angular and linear camera velocity control inputs for the object are defined as

$$\omega_e = R^T L_\omega^{-1} (\dot{\Theta}_d - K_\omega e_\omega) \tag{3.52}$$

$$v_e = -\frac{1}{\alpha_1} R^T \left(A_e L_v \right)^{-1} (K_v e_v - \hat{z}_1^* \dot{p}_{ed}) - [\omega_e]_\times \hat{s}_1. \tag{3.53}$$

In (3.52) and (3.53), K_ω, $K_v \in \mathbb{R}^{3 \times 3}$ denote diagonal matrices of positive constant control gains, and $\hat{z}_1^*(t) \in \mathbb{R}$, $\hat{s}_1(t) \in \mathbb{R}^3$ denote parameter estimates that are generated according to the following adaptive update laws

$$\dot{\hat{z}}_1^*(t) = -\gamma_1 e_v^T \dot{p}_{ed} \tag{3.54}$$

$$\dot{\hat{s}}_1 = -\alpha_1 \Gamma_2 [\omega_e]_\times R^T L_v^T A_e^T e_v, \tag{3.55}$$

where $\gamma_1 \in \mathbb{R}$ denotes a positive constant adaptation gain, and $\Gamma_2 \in \mathbb{R}^{3 \times 3}$ denotes a positive constant diagonal adaptation gain matrix. After substituting (3.52) into (3.46), the following closed-loop error dynamics can be obtained

$$\dot{e}_\omega = -K_\omega e_\omega. \tag{3.56}$$

After substituting (3.53) into (3.49), the closed-loop translation error dynamics can be determined as

$$z_1^* \dot{e}_v = -K_v e_v + \alpha_1 A_e L_v R \left[\omega_e\right]_\times \tilde{s}_1 - \tilde{z}_1^* \dot{p}_{ed}, \tag{3.57}$$

where the parameter estimation error signals $\tilde{z}_1^*(t) \in \mathbb{R}$ and $\tilde{s}_1(t) \in \mathbb{R}^3$ are defined as

$$\tilde{z}_1^* = z_1^* - \hat{z}_1^* \qquad\qquad \tilde{s}_1 = s_1 - \hat{s}_1 . \tag{3.58}$$

From (3.56) it is clear that the angular velocity control input given in (3.52) is designed to yield an exponentially stable rotational error system. The linear velocity control input given in (3.53) and the adaptive update laws given in (3.54) and (3.55) are motivated to yield a negative feedback term in the translational error system with additional terms included to cancel out cross-product terms involving the parameter estimation errors in the subsequent stability analysis.

3.3.3   Stability Analysis

Theorem 3.1 *The control inputs designed in (3.52) and (3.53), along with the adaptive update laws defined in (3.54) and (3.55), ensure that $e_\omega(t)$ and $e_v(t)$ are asymptotically driven to zero in the sense that*

$$\lim_{t \to \infty} \|e_\omega(t)\| , \; \|e_v(t)\| = 0 . \tag{3.59}$$

Proof: To prove Theorem 3.1, a non-negative function $V(t) \in \mathbb{R}$ is defined as

$$V \triangleq \frac{1}{2} e_\omega^T e_\omega + \frac{z_1^*}{2} e_v^T e_v + \frac{1}{2\gamma_1} \tilde{z}_1^{*2} + \frac{1}{2} \tilde{s}_1^T \Gamma_2^{-1} \tilde{s}_1. \tag{3.60}$$

After taking the time derivative of (3.60) and then substituting for the closed-loop error systems developed in (3.56) and (3.57), the following expression can be obtained

$$\begin{aligned} \dot{V} &= -e_\omega^T K_\omega e_\omega + e_v^T \left(-K_v e_v + \alpha_1 A_e L_v R \left[\omega_e\right]_\times \tilde{s}_1 - \tilde{z}_1^* \dot{p}_{ed} \right) \\ &\quad - \frac{1}{\gamma_1} \tilde{z}_1^* \dot{\hat{z}}_1^* - \tilde{s}_1^T \Gamma_2^{-1} \dot{\hat{s}}_1, \end{aligned} \tag{3.61}$$

where the time derivative of (3.58) was utilized. After substituting the adaptive update laws designed in (3.54) and (3.55) into (3.61), the following simplified expression can be obtained

$$\dot{V} = -e_\omega^T K_\omega e_\omega - e_v^T K_v e_v, \tag{3.62}$$

where the fact that $\left[\omega_e\right]_\times^T = -\left[\omega_e\right]_\times$ was utilized. Based on (3.58), (3.60), and (3.62), it can be determined that $e_\omega(t), e_v(t), \tilde{z}_1^*(t), \hat{z}_1^*(t), \tilde{s}_1(t), \hat{s}_1(t) \in$

$\mathcal{L}_\infty$ and that $e_\omega(t)$, $e_v(t) \in \mathcal{L}_2$. Based on the assumption that $\dot{\Theta}_d(t)$ is designed as a bounded function, the expressions given in (3.38), (3.47), (3.48), and (3.52) can be used to conclude that $\omega_e(t) \in \mathcal{L}_\infty$. Since $e_v(t) \in \mathcal{L}_\infty$, (3.7), (3.13), (3.14), (3.36), (3.37), and (3.51) can be used to prove that $m_1(t)$, $L_v(t) \in \mathcal{L}_\infty$. Given that $\dot{p}_{ed}(t)$ is assumed to be bounded function, the expressions in (3.53)–(3.57) can be used to conclude that $\dot{z}_1^*(t)$, $\dot{s}_1(t)$, $v_e(t)$, $\dot{e}_v(t)$, $\dot{e}_\omega(t) \in \mathcal{L}_\infty$. Since $e_\omega(t)$, $e_v(t) \in \mathcal{L}_2$ and $e_\omega(t)$, $\dot{e}_\omega(t)$, $e_v(t)$, $\dot{e}_v(t) \in \mathcal{L}_\infty$, Barbalat's Lemma [102] can be used to prove the result given in (3.59). ∎

Remark 3.1 *The result in (3.59) is practically global in the sense that it is valid over the entire domain with the exception of the singularity introduced by the exponential parameterization of the rotation matrix (see (3.40)) and the physical restriction that $z_i(t)$, $z_i^*(t)$, and $z_{di}(t)$ must remain positive. Although the result stated in Theorem 3.1 indicates asymptotic convergence for the rotation error $e_\omega(t)$, it is evident from (3.56) that*

$$e_\omega(t) \le e_\omega(0) \exp(-\lambda_{\min}(K_\omega)t)$$

where $\lambda_{\min}(K_\omega)$ denotes the minimum eigenvalue of the constant matrix K_ω. However, the fact that $e_\omega(t) \le e_\omega(0) \exp(-\lambda_{\min}(K_\omega)t)$ does not simplify the control development or stability analysis and the overall resulting control objective of tracking a desired set of prerecorded images is still asymptotically achieved.

3.3.4 Camera-in-Hand Extension

The open-loop error dynamics for the rotation system can be derived as

$$\dot{e}_\omega = -L_\omega \omega_c - \dot{\Theta}_d \tag{3.63}$$

where the fact that

$$[\omega_c]_\times = -\dot{R}R^T \tag{3.64}$$

is used, and $\omega_c(t)$ denotes the camera angular velocity expressed in $\mathcal{F}$. After taking the time derivative of (3.17), the following expression for $\dot{\bar{m}}_1(t)$ can be derived for the camera-in-hand [43]

$$\dot{\bar{m}}_1 = -v_c + [\bar{m}_1]_\times \omega_c, \tag{3.65}$$

where $v_c(t)$ denotes the linear velocity of the camera expressed in terms of $\mathcal{F}$. After utilizing (3.65), the open-loop dynamics for $e_v(t)$ can be determined as

$$z_1^* \dot{e}_v = -\alpha_1 A_e L_v v_c + \left(A_e L_v [m_1]_\times \omega_c - \dot{p}_{ed} \right) z_1^*, \tag{3.66}$$

where $e_v(t)$, $p_e(t)$, $p_{ed}(t)$ are defined in (3.36) and (3.37).

Based on the open-loop error systems in (3.63) and (3.66), the following control inputs and adaptive update law are designed

$$\omega_c \triangleq L_\omega^{-1}\left(K_\omega e_\omega - \dot{\Theta}_d\right) \tag{3.67}$$

$$v_c \triangleq \frac{1}{\alpha_1}(A_e L_v)^{-1}\left(K_v e_v - \hat{z}_1^* \dot{p}_{ed}\right) + \frac{1}{\alpha_1}[m_1]_\times \omega_c \hat{z}_1^* \tag{3.68}$$

$$\dot{\hat{z}}_1^* \triangleq \gamma_1 e_v^T\left(A_e L_v [m_1]_\times \omega_c - \dot{p}_{ed}\right) \tag{3.69}$$

resulting in the following closed-loop error systems

$$\dot{e}_\omega = -K_\omega e_\omega \tag{3.70}$$

$$z_1^* \dot{e}_v = -K_v e_v + \left(A_e L_v [m_1]_\times \omega_c - \dot{p}_{ed}\right)\tilde{z}_1^* . \tag{3.71}$$

The result in (3.59) can now be proven for the camera-in-hand problem using the same analysis techniques and the same nonnegative function as defined in (3.60) with the term containing $\tilde{s}_1(t)$ eliminated.

3.3.5 *Simulation Results*

Simulation studies were performed to illustrate the performance of the controller given in (3.52)–(3.55). For the simulation, the intrinsic camera calibration matrix is given as in (3.15) where $u_0 = 257$ [pixels], $v_0 = 253$ [pixels], $k_u = 101.4$ [pixels·mm^{-1}] and $k_v = 101.4$ [pixels·mm^{-1}] represent camera scaling factors, $\phi = 90$ [Deg] is the angle between the camera axes, and $f = 12.5$ [mm] denotes the camera focal length. The control objective is defined in terms of tracking a desired image sequence. For the simulation, the desired image sequence was required to be artificially generated. To generate an artificial image sequence for the simulation, the Euclidean coordinates of four target points were defined as

$$s_1 = \begin{bmatrix} 0.1 \\ -0.1 \\ 0 \end{bmatrix} \quad s_2 = \begin{bmatrix} 0.1 \\ 0.1 \\ 0 \end{bmatrix} \quad s_3 = \begin{bmatrix} -0.1 \\ 0.1 \\ 0 \end{bmatrix} \quad s_4 = \begin{bmatrix} -0.1 \\ -0.1 \\ 0 \end{bmatrix} \tag{3.72}$$

and the initial translation and rotation between the current, desired, and reference image feature planes were defined as

$$x_f(0) = \begin{bmatrix} -0.3 \\ -0.1 \\ 3.7 \end{bmatrix} \quad x_{fd}(0) = \begin{bmatrix} 0.2 \\ 0.1 \\ 4 \end{bmatrix} \quad x_f^* = \begin{bmatrix} 0.2 \\ 0.1 \\ 4 \end{bmatrix} \tag{3.73}$$

$$R(0) = \begin{bmatrix} -0.4698 & -0.8660 & -0.1710 \\ -0.6477 & 0.4698 & -0.5997 \\ 0.5997 & -0.1710 & -0.7817 \end{bmatrix} \tag{3.74}$$

$$R_d(0) = \begin{bmatrix} 0.9568 & -0.2555 & -0.1386 \\ -0.2700 & -0.9578 & -0.0984 \\ -0.1077 & 0.1316 & -0.9854 \end{bmatrix} \tag{3.75}$$

$$R^* = \begin{bmatrix} 0.9865 & 0.0872 & -0.1386 \\ 0.0738 & -0.9924 & -0.0984 \\ -0.1462 & 0.0868 & -0.9854 \end{bmatrix}. \tag{3.76}$$

The initial pixel coordinates can be computed from (3.72)–(3.76) as

$$p_1(0) = \begin{bmatrix} 170 & 182 & 1 \end{bmatrix}^T \quad p_2(0) = \begin{bmatrix} 110 & 213 & 1 \end{bmatrix}^T$$
$$p_3(0) = \begin{bmatrix} 138 & 257 & 1 \end{bmatrix}^T \quad p_4(0) = \begin{bmatrix} 199 & 224 & 1 \end{bmatrix}^T$$

$$p_{d1}(0) = \begin{bmatrix} 359 & 307 & 1 \end{bmatrix}^T \quad p_{d2}(0) = \begin{bmatrix} 343 & 246 & 1 \end{bmatrix}^T$$
$$p_{d3}(0) = \begin{bmatrix} 282 & 263 & 1 \end{bmatrix}^T \quad p_{d4}(0) = \begin{bmatrix} 298 & 324 & 1 \end{bmatrix}^T$$

$$p_1^* = \begin{bmatrix} 349 & 319 & 1 \end{bmatrix}^T \quad p_2^* = \begin{bmatrix} 355 & 256 & 1 \end{bmatrix}^T$$
$$p_3^* = \begin{bmatrix} 292 & 251 & 1 \end{bmatrix}^T \quad p_4^* = \begin{bmatrix} 286 & 314 & 1 \end{bmatrix}^T.$$

The time-varying desired image trajectory was then generated by the kinematics of the target plane where the desired linear and angular velocities were selected as

$$v_{ed}(t) = \begin{bmatrix} 0.2\sin(t) & 0.3\sin(t) & 0 \end{bmatrix} [\text{m}/\text{sec}] \tag{3.77}$$
$$\omega_{ed}(t) = \begin{bmatrix} 0 & 0 & 0.52\sin(t) \end{bmatrix} [\text{rad}/\text{sec}].$$

The desired translational trajectory is given in Figure 3.5, and the desired rotational trajectory is depicted in Figure 3.6. The generated desired image trajectory is a continuous function; however, in practice, the image trajectory would be discretely represented by a sequence of prerecorded images and would require a data interpolation scheme; hence, a spline function (i.e., the MATLAB spline routine) was utilized to generate a continuous curve to fit the desired image trajectory. For the top two subplots in Figure 3.5, the pixel values obtained from the prerecorded image sequence are denoted by an asterisk (only select data points were included for clarity of illustration), and a cubic spline interpolation that was used to fit the data points is illustrated by a solid line. For the bottom subplot in Figure 3.5 and all the subplots in Figure 3.6, a plus sign denotes reconstructed Euclidean values computed using the prerecorded pixel data, and the spline function is illustrated by a solid line.

The control gains K_v and K_ω and the adaptation gains γ_1 and Γ_2 were adjusted through trial and error to the following values

$$K_v = \text{diag}\{6, 8, 5\} \quad K_\omega = \text{diag}\{0.6, 0.8, 0.7\} \qquad (3.78)$$
$$\gamma_1 = 3 \times 10^{-6} \quad \Gamma_2 = 10^{-5} \times \text{diag}\{4.2, 5.6, 2.8\}.$$

The resulting errors between the actual relative translational and rotational of the target with respect to the reference target and the desired translational and rotational of the target with respect to the reference target are depicted in Figure 3.7 and Figure 3.8, respectively. The parameter estimate signals are depicted in Figure 3.9 and Figure 3.10. The angular and linear control input velocities (i.e., $\omega_e(t)$ and $v_e(t)$) defined in (3.52) and (3.53) are depicted in Figure 3.11 and Figure 3.12.

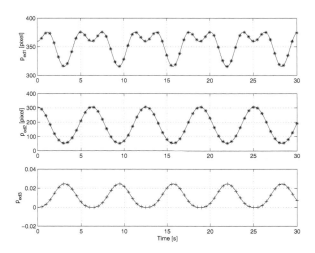

FIGURE 3.5. Desired Translational Trajectory of the Manipulator End-Effector Generated by a Spline Function to Fit Prerecorded Image Data.

While the results in Figure 3.7–Figure 3.12 provide an example of the performance of the tracking controller under ideal conditions, several issues must be considered for a practical implementation. For example, the performance of the tracking control algorithm is influenced by the accuracy of the image-space feedback signals and the accuracy of the reconstructed Euclidean information obtained from constructing and decomposing the homography. That is, inaccuracies in determining the location of a feature from one frame to the next frame (i.e., feature tracking) will lead to errors in the construction and decomposition of the homography matrix, leading

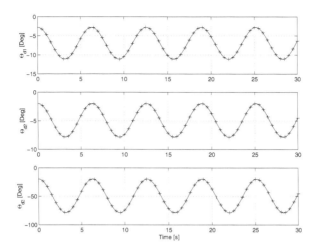

FIGURE 3.6. Desired Rotational Trajectory of the Manipulator End-Effector Generated by a Spline Function to Fit Prerecorded Image Data.

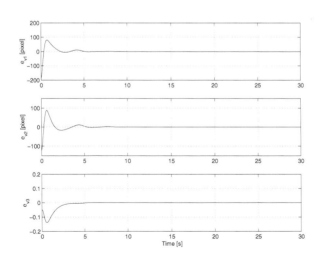

FIGURE 3.7. Error between the Actual Translation Trajectory and the Desired Translation Trajectory given in Figure 3.5 for the Noise-Free Case.

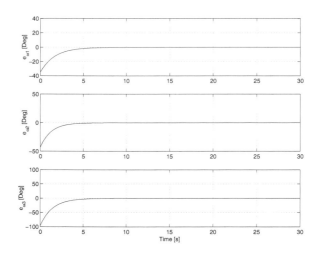

FIGURE 3.8. Error between the Actual Rotation Trajectory and the Desired Rotation Trajectory given in Figure 3.6 for the Noise-Free Case.

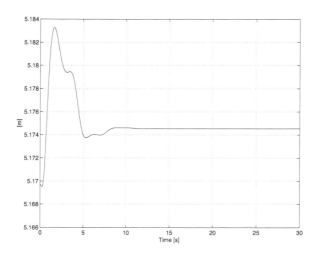

FIGURE 3.9. Parameter Estimate for z_1^* for the Noise-Free Case.

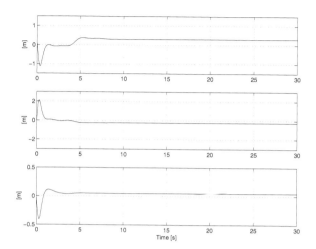

FIGURE 3.10. Parameter Estimates for s_1 for the Noise-Free Case.

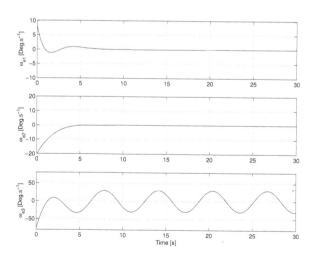

FIGURE 3.11. Angular Velocity Control Input for the Noise-Free Case.

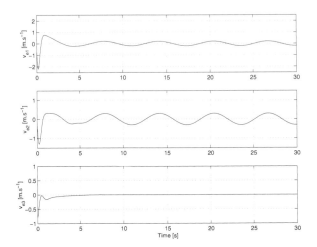

FIGURE 3.12. Linear Velocity Control Input for the Noise-Free Case.

to errors in the feedback control signal. Inaccuracies in determining the feature point coordinates in an image is a similar problem faced in numerous sensor-based feedback applications (e.g., noise associated with a force/torque sensor). Practically, errors related to sensor inaccuracies can often be addressed with an *ad hoc* filter scheme or other mechanisms (e.g., an intelligent image processing and feature tracking algorithm, redundant feature points and an optimal homography computation algorithm).

In light of these practical issues, another simulation was performed where random noise was injected with a standard deviation of 1 pixel (i.e., the measured feature coordinate was subject to ± 4 pixels of measurement error) as in [84]. As in any practical feedback control application in the presence of sensor noise, a filter was employed. Specifically, *ad hoc* third order Butterworth low pass filters with a cutoff frequency of 10 rad/sec were utilized to preprocess the corrupted image data. The control gains K_v and K_ω and the adaptation gains γ_1 and Γ_2 were tuned through trial and error to the following values

$$K_v = \mathrm{diag}\,\{17, 11, 9\} \quad K_\omega = \mathrm{diag}\,\{0.4, 0.4, 0.4\} \qquad (3.79)$$
$$\gamma_1 = 5 \times 10^{-7} \quad \Gamma_2 = 10^{-5} \times \mathrm{diag}\,\{2.4, 3.2, 1.6\}\,.$$

The resulting translational and rotational errors of the target are depicted in Figure 3.13 and Figure 3.14, respectively. The parameter estimate signals are depicted in Figure 3.15 and Figure 3.16. The control input velocities

$\omega_e(t)$ and $v_e(t)$ defined in (3.52) and (3.53) are depicted in Figure 3.17 and Figure 3.18.

Another simulation was also performed to test the robustness of the controller with respect to the constant rotation matrix R^*. The constant rotation matrix R^* in (3.5) is coarsely calibrated as $\mathrm{diag}\{1, -1, -1\}$. The resulting translational and rotational errors of the target are depicted in Figure 3.19 and Figure 3.20, respectively.

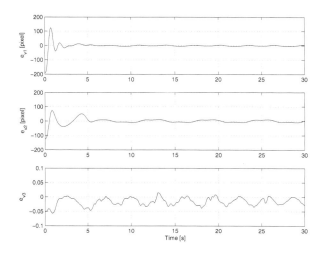

FIGURE 3.13. Error between the Actual Translation Trajectory and the Desired Translation Trajectory given in Figure 3.5 for the Noise-Injected Case.

3.4   Continuum Robots

This section explores the problem of measuring the shape of a continuum robot manipulator using visual information from a fixed camera. The motion of a set of fictitious planes can be captured by an image of four or more feature points defined at various strategic locations along the body of the robot. Using expressions for the robot forward kinematics and the decomposition of a homography relating a reference image of the robot to the actual robot image, the three dimensional shape information can be continuously determined. This information can be used to demonstrate the development of a kinematic controller to regulate the manipulator end-effector to a constant desired position and orientation.

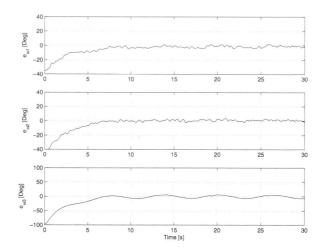

FIGURE 3.14. Error between the Actual Rotation Trajectory and the Desired Rotation Trajectory given in Figure 3.6 for the Noise-Injected Case.

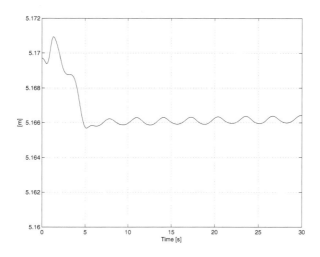

FIGURE 3.15. Parameter Estimate for z_1^* for the Noise-Injected Case.

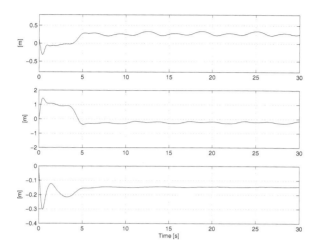

FIGURE 3.16. Parameter Estimates for s_1 for the Noise-Injected Case.

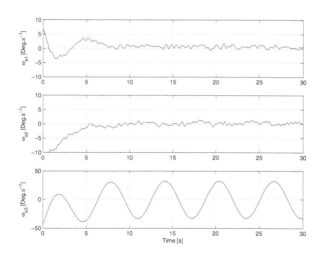

FIGURE 3.17. Angular Velocity Control Input for the Noise-Injected Case.

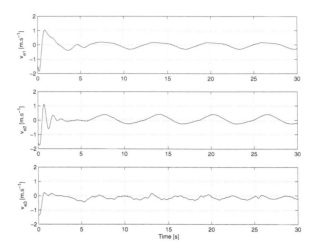

FIGURE 3.18. Linear Velocity Control Input for the Noise-Injected Case.

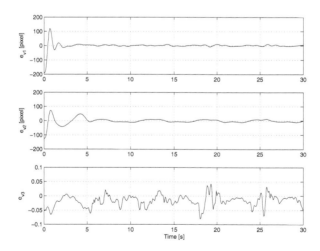

FIGURE 3.19. Error between the Actual Translation Trajectory and the Desired Translation Trajectory given in Figure 3.5 for the Noise-Injected Case with a Coarse Calibration of R^*.

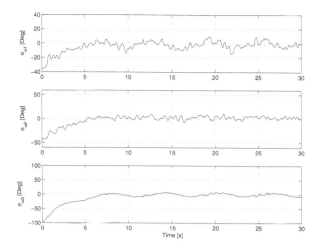

FIGURE 3.20. Error between the Actual Rotation Trajectory and the Desired Rotation Trajectory given in Figure 3.6 for the Noise-Injected Case with a Coarse Calibration of R^*.

3.4.1   Continuum Robot Kinematics

The kinematics of a conventional, rigid-link, industrial robot can be conveniently described as a function of joint angles and link lengths using the standard Denavit-Hartenberg convention [105]. This is a systematic method of assigning orthogonal coordinate frames to the joints of the robot such that the relative position and orientation between frames along the kinematic chain can be obtained as a product of homogeneous transformation matrices. In comparison, continuum robots resemble snakes or tentacles in their physical structure, and due to their continuous and curving shape, there is no intuitive way to define links and joints on them. The concept of curvature [20, 53, 58, 59] is a natural way to describe the kinematics of a continuum robot. One such continuum robot is the Clemson Elephant Trunk [58] which is composed of sixteen two degrees-of-freedom joints divided into four sections, each section designed to bend with a constant planar curvature. Every section is cable driven, and can be actuated such that it defines a different orientation of the plane of its curvature relative to its preceding section. Due to the rigid nature of the joints, torsion is not possible within a section.

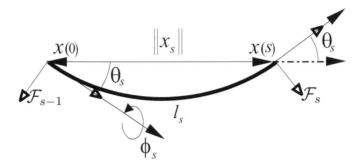

FIGURE 3.21. A Planar Curve.

Based on the concept pioneered in [58], and further refined in [68], the fundamental idea behind development of kinematics for an individual section of this robot is to fit a virtual conventional rigid-link manipulator to its continuous curvature, and develop relationships utilizing the well established Denavit-Hartenberg procedure. Consider the sth section of the robot. Using basic geometry, the kinematics of a 2D planar curve of arc length l_s and curvature k_s can be described by three coupled movements — rotation by an angle θ_s, followed by a translation x_s, and a further rotation by angle θ_s as shown in Figure 3.21. In Figure 3.21, $x_s \in \mathbb{R}^3$ is the position vector of the endpoint of the curve relative to its initial point, and

$$\theta_s = \frac{k_s l_s}{2} \tag{3.80}$$

$$\|x_s\| = \frac{l_s}{\theta_s} \sin(\theta_s). \tag{3.81}$$

After treating the two rotations in the curve as discrete rotational joints and the translation as a coupled discrete prismatic joint, the standard Denavit-Hartenberg procedure [105] can be applied to obtain the forward kinematics for the curve. Thus, the homogeneous transformation matrix for the planar curve, denoted by $A_{sp} \in \mathbb{R}^{4 \times 4}$, can be obtained as [58]

$$A_{sp} = \begin{bmatrix} \cos(k_s l_s) & -\sin(k_s l_s) & 0 & \frac{1}{k_s}\{\cos(k_s l_s) - 1\} \\ \sin(k_s l_s) & \cos(k_s l_s) & 0 & \frac{1}{k_s}\sin(k_s l_s) \\ 0 & 0 & 1 & 0 \\ 0 & 0 & 0 & 1 \end{bmatrix}. \tag{3.82}$$

The out-of-plane rotation of the section relative to the preceding plane can be modelled as an additional rotational joint with rotation of angle ϕ_s

about the initial tangent of the curve (see Figure 3.21). As discovered in [68], this results in incorrect orientation of the frame defined at the other end of the curve, since the body of the robot cannot experience torsion. Therefore, in order to cancel out this torsion, the frame defined at the distal end of the curve is finally rotated by $-\phi_s$. Hence, for the 3D case, the forward kinematics for the sth section of the continuum robot can be obtained from the following homogeneous transformation matrix

$$A_s = \begin{bmatrix} R^s_{s-1} & t^s_{s-1} \\ 0 & 1 \end{bmatrix}, \tag{3.83}$$

where

$$R^s_{s-1} = \begin{bmatrix} 1+\cos^2(\phi_s)\mathrm{ck}_s & \sin(\phi_s)\cos(\phi_s)\mathrm{ck}_s & -\cos(\phi_s)\sin(k_sl_s) \\ \sin(\phi_s)\cos(\phi_s)\mathrm{ck}_s & \cos(k_sl_s)-\cos^2(\phi_s)\mathrm{ck}_s & -\sin(\phi_s)\sin(k_sl_s) \\ \sin(k_sl_s)\cos(\phi_s) & \sin(k_sl_s)\sin(\phi_s) & \cos(k_sl_s) \end{bmatrix}, \tag{3.84}$$

$$t^s_{s-1} = \begin{bmatrix} \dfrac{1}{k_s}\cos(\phi_s)\mathrm{ck}_s \\ \dfrac{1}{k_s}\sin(\phi_s)\mathrm{ck}_s \\ \dfrac{1}{k_s}\sin(k_sl_s) \end{bmatrix}, \tag{3.85}$$

and $\mathrm{ck}_s(t) = \cos(k_sl_s) - 1$. The matrix A_s in (3.83) transforms the coordinates of a point defined in the coordinate frame $\mathcal{F}_s$ at the end of the sth curved section to the coordinate frame $\mathcal{F}_{s-1}$ defined at the end of the $(s-1)^{th}$ section. In the above equations, $R^s_{s-1} \in SO(3)$ and $t^s_{s-1} \in \mathbb{R}^3$ define, respectively, the rotation matrix and translation vector between the frames $\mathcal{F}_s$ and $\mathcal{F}_{s-1}$. Thus, for the entire robot with four sections, the homogeneous transformation matrix can be calculated as

$$T^4_0 = A_1A_2A_3A_4. \tag{3.86}$$

From (3.86) the end-effector position and orientation in the task-space of the robot, denoted by $p(t) \in \mathbb{R}^6$, can be written as

$$\chi = f(q), \tag{3.87}$$

where $f(q) \in \mathbb{R}^6$ denotes the forward kinematics, and $q(t) \in \mathbb{R}^8$ denotes the joint space variables for the robot defined as

$$q(t) = \begin{bmatrix} \phi_1 & k_1 & \phi_2 & k_2 & \phi_3 & k_3 & \phi_4 & k_4 \end{bmatrix}^T, \tag{3.88}$$

where $\phi_i(t)$ and $k_i(t)$ are the out-of-plane rotation and the curvature, respectively, for the ith section.

Based on (3.87), a differential relationship between the end-effector position and the joint space variables can be defined as [68]

$$\dot{\chi} = J(q)\dot{q}, \tag{3.89}$$

where $J(q) \triangleq \dfrac{\partial f(q)}{\partial q} \in \mathbb{R}^{6 \times 8}$ is called a Jacobian matrix, and $\dot{q}(t) \in \mathbb{R}^8$ denotes the joint space velocity vector. Note here that the determination of the Jacobian matrix requires knowledge of the joint space vector $q(t)$. In the following section, we describe how $q(t)$ can be constructed from images of feature points along the manipulator as obtained from the fixed camera.

3.4.2 Joint Variables Extraction

The development in Section 3.2 can now be used to relate the feedback from the video camera to the Euclidean coordinates of the sections of the robot. An inertial coordinate system $\mathcal{I}$ is defined with an origin that coincides with the center of the fixed camera (see Figure 3.22). For the sake of simplicity, the origin of the inertial frame $\mathcal{I}$ is also assumed to coincide with the origin of the robot base frame. At the end of sth section of the robot, consider a transverse plane π_s defined by four non-collinear target points denoted by O_{si} $\forall i = 1, 2, 3, 4$ such that the origin of the previously defined coordinate system $\mathcal{F}_s$ lies in π_s. Also consider a fixed transverse plane denoted by π_s^* with four non-collinear target points denoted by O_{si}^* $\forall i = 1, 2, 3, 4$, and a coordinate system $\mathcal{F}_s^*$, which are defined when the end of the sth section is at a reference position and orientation relative to the fixed camera (i.e., π_s^*'s and $\mathcal{F}_s^*$'s are defined by a reference image of the robot). This reference image doesn't necessarily represent the desired position to which we want to regulate the end-effector of the robot. The 3D coordinates of the target points O_{si}, O_{si}^*, denoted by $\bar{m}_{si}(t)$, $\bar{m}_{si}^* \in \mathbb{R}^3$ in π_s and π_s^*, respectively, can be expressed in the inertial coordinate system $\mathcal{I}$ as in (3.1). The points O_{si} and O_{si}^* represent the same features at different geometric locations, and when expressed in the object reference frames $\mathcal{F}_s$ and $\mathcal{F}_s^*$, they have the same coordinates. Likewise, the normalized coordinates, denoted by $m_{si}(t)$, $m_{si}^* \in \mathbb{R}^3$ in π_s and π_s^*, respectively, can be defined as in (3.7) and (3.9). The pixel coordinates expressed in terms of $\mathcal{I}$ are denoted by $u_{si}(t)$, $v_{si}(t) \in \mathbb{R}$ and $u_{si}^*, v_{si}^* \in \mathbb{R}$, and are respectively defined as elements of $p_{si}(t), p_{si}^* \in \mathbb{R}^3$ as in (3.13), where the pixel coordinates are related to their normalized Euclidean coordinates by the pinhole camera model given in (3.14).

To develop a relationship between the coordinate system $\mathcal{I}$ and the coordinate system $\mathcal{F}_s$ defined at the end of the sth section of the robot,

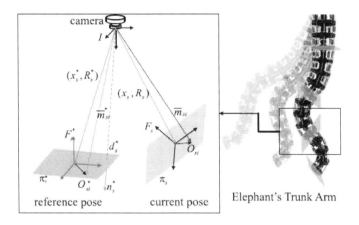

FIGURE 3.22. Coordinate Frame Relationships.

the rotation matrix between $\mathcal{F}_s$ and $\mathcal{I}$ is defined as $R_s(t) \in SO(3)$, and $x_s(t) \in \mathbb{R}^3$ as the translation vector between $\mathcal{F}_s$ and $\mathcal{I}$ for $s = 1, 2, 3, 4$. Similarly, let $x_s^* \in \mathbb{R}^3$ be a constant translation vector between $\mathcal{F}_s^*$ and $\mathcal{I}$, and $R_s^* \in SO(3)$ be the constant rotation matrix between $\mathcal{F}_s^*$ and $\mathcal{I}$. The constant rotation matrix R_s^* is assumed to be known, since the robot can be set to a known reference configuration, or that the constant rotation can be obtained *a priori* using various methods such as a second camera or calibrated Euclidean measurements. As also illustrated in Figure 3.22, $n_s^* \in \mathbb{R}^3$ denotes a constant normal to the reference plane π_s^* expressed in the coordinates of $\mathcal{I}$, and the constant distance $d_s^* \in \mathbb{R}$ from $\mathcal{I}$ to plane π_s^* along the unit normal as in (3.2). By using the fact that O_{si} and O_{si}^* represent the same feature point at different geometric locations and the geometry between the coordinate frames $\mathcal{F}_s$, $\mathcal{F}_s^*$ and $\mathcal{I}$ depicted in Figure 3.22, the following relationships can be developed

$$\bar{m}_{si} = x_s + R_s O_{si} \qquad (3.90)$$

$$\bar{m}_{si}^* = x_s^* + R_s^* O_{si}. \qquad (3.91)$$

After solving (3.91) for O_{si} and substituting the resulting expression into (3.90), the following relationships can be obtained

$$\bar{m}_{si} = \bar{x}_s + \bar{R}_s \bar{m}_{si}^*, \qquad (3.92)$$

where $\bar{R}_s(t) \in SO(3)$ and $\bar{x}_s(t) \in \mathbb{R}^3$ are the new rotational and translational variables, respectively, defined as

$$\bar{R}_s = R_s \left(R_s^*\right)^T \qquad \bar{x}_s = x_s - \bar{R}_s x_s^*. \qquad (3.93)$$

After utilizing (3.2), the relationship in (3.92) can be expressed as in (3.10) as

$$m_{si} = \underbrace{\frac{z_{si}^*}{z_{si}}}_{\alpha_{si}} \underbrace{\left(\bar{R}_s + \frac{\bar{x}_s}{d_s^*} n_s^{*T} \right)}_{H_s} m_{si}^*, \tag{3.94}$$

where $\alpha_{si}(t) \in \mathbb{R}$ is the depth ratio, and $H_s(t) \in \mathbb{R}^{3 \times 3}$ denotes the Euclidean homography between the coordinate systems $\mathcal{F}_s$ and $\mathcal{F}_s^*$. The relationships in (3.14) can be used to rewrite (3.94) in terms of the projective homography as in (3.16).

By utilizing the methods described in Section 3.2 (see also [47, 116]), $H_s(t)$ can be decomposed into rotational and translational components as in (3.94). Specifically, the rotation matrix $\bar{R}_s(t)$ can be computed from the decomposition of $H_s(t)$. The rotation matrix $R_s(t)$, defining the orientation of the end of the sth section of the robot relative to the camera fixed frame $\mathcal{I}$, can then be computed from $\bar{R}_s(t)$ by using (3.93) and the fact that R_s^* is known *a priori*. Since $R_s(t)$ is a rotation matrix between $\mathcal{I}$ and $\mathcal{F}_s$, it can be viewed as a composition of two rotational transformations; a rotational transformation from frame $\mathcal{I}$ to $\mathcal{F}_{s-1}$ followed by a second rotational transformation from $\mathcal{F}_{s-1}$ to $\mathcal{F}_s$. Hence, $R_{s-1}^s(t)$ in (3.84) can be progressively computed (i.e., the rotation matrix from one section of the robot to the next) as [105]

$$R_{s-1}^s = (R_{s-1})^T R_s \qquad \forall s = 1, 2, 3, 4. \tag{3.95}$$

From (3.83), the joint space variables for the sth section can hence be determined as

$$\begin{aligned} k_s &= \frac{1}{l_s} \cos^{-1}([R_{s-1}^s]_{33}) \\ \phi_s &= \sin^{-1}\left(\frac{[R_{s-1}^s]_{32}}{\sin(k_s l_s)} \right), \end{aligned} \tag{3.96}$$

where $l_s \in \mathbb{R}$ is the known arc length of the section and the notation $[\cdot]_{xy}$ denotes a matrix element at row x and column y. With the knowledge of all the joint variables $q(t)$ as computed from (3.96), T_0^4 of (3.86), and consequently, the Jacobian $J(q)$ of (3.89) can be calculated online.

3.4.3   Task-Space Kinematic Controller

The control objective is the regulation of the end-effector of the manipulator to a desired position and orientation denoted by $\chi_d \in \mathbb{R}^6$. This desired configuration of the robot may be available as an image, and the technique described in the previous sections may be applied to compute χ_d. The mismatch between the desired and actual end-effector Cartesian coordinates is the task-space position error, denoted by $e(t) \in \mathbb{R}^6$, as

$$e \triangleq \chi - \chi_d. \tag{3.97}$$

Utilizing the velocity kinematics in (3.89), and the fact that $\dot{\chi}_d = 0$, the open loop error dynamics for $e(t)$ can be expressed as

$$\dot{e} = J\dot{q}. \tag{3.98}$$

The kinematic control input $\dot{q}(t)$ can be designed [115] as

$$\dot{q} = -J^+\beta e + (I_8 - J^+J)g, \tag{3.99}$$

where $\beta \in \mathbb{R}^{6\times6}$ is a diagonal, positive definite gain matrix, $I_n \in \mathbb{R}^{n\times n}$ denotes the $n \times n$ identity matrix, and $J^+(q)$ denotes the pseudo-inverse [6] of $J(q)$, defined as

$$J^+ \triangleq J^T(JJ^T)^{-1}. \tag{3.100}$$

In (3.99), $g(t) \in \mathbb{R}^8$ is a bounded auxiliary signal that is constructed according to a sub-task control objective such as obstacle avoidance. For example, if the joint-space configuration that avoids an obstacle in the manipulator's work-space is known to be q_r, then $g(t)$ can be designed as

$$g \triangleq \gamma(q_r - q), \tag{3.101}$$

where $\gamma \in \mathbb{R}$ is a positive gain constant. In designing $\dot{q}(t)$ as in (3.99), an inherent assumption is that the minimum singular value of the Jacobian, denoted by σ_m, is greater than a known small positive constant $\delta > 0$, such that $max\{\|J^+(q)\|\}$ is known a priori, and all kinematic singularities are avoided. Note that $J^+(q)$ satisfies the following equalities

$$JJ^+ = I_n \tag{3.102}$$
$$J(I_8 - J^+J) = 0. \tag{3.103}$$

Substituting the control input of (3.99) into (3.98) yields

$$\dot{e} = -\beta e, \tag{3.104}$$

where (3.102) and (3.103) have been used. Hence, $e(t)$ is bounded by the following exponentially decreasing envelope

$$\|e(t)\| \le \|e(0)\| \exp(-\lambda t), \tag{3.105}$$

where $\lambda \in \mathbb{R}$ is the minimum eigenvalue of β.

From (3.97) and (3.105), it is clear that $\chi(t) \in \mathcal{L}_\infty$. Based on the assumption that kinematic singularities are avoided, $J(t)$ is always defined and bounded. Hence, the control input $\dot{q}(t)$ is bounded since $J^+(q)$ is bounded for all possible $q(t)$, and $g(t)$ is bounded by assumption. We make the assumption that if $\chi(t) \in \mathcal{L}_\infty$, then $q(t) \in \mathcal{L}_\infty$. From (3.98), $\dot{e}(t), \dot{\chi}(t) \in \mathcal{L}_\infty$.

3.4.4 Simulations and Discussion

The primary contribution in this section is the development of a vision-based technique for the measurement of shape of a continuum robot, given the expressions for the forward kinematics. For the sake of demonstration, a simple task-space kinematic controller is formulated as

$$\dot{q} = -J^T \beta e. \tag{3.106}$$

Substituting (3.106) into the error dynamics of (3.98) results in the same exponential stability result as (3.105), except that λ is now the minimum eigenvalue of $JJ^T\beta$. The desired task-space position of the end-effector was selected as

$$\chi_d = \begin{bmatrix} 0.30 & 0.01 & 0.769 & 0.0 & 0.4 & 0.0 \end{bmatrix}^T. \tag{3.107}$$

The initial configuration of the manipulator, denoted by $q(t_0)$ was selected as

$$q(t_0) = \begin{bmatrix} 0.01 & 0.1 & 0.01 & 0.1 & 0.01 & 0.1 & 0.01 & 0.1 \end{bmatrix}^T, \tag{3.108}$$

which is close to the relaxed configuration of the manipulator where all sections lie extended along the principal axis. The diagonal elements of feedback gain matrix β were set to 30. Based on calibration parameters from an actual camera, the internal camera calibration matrix A was set to

$$A = \begin{bmatrix} 1268.16 & 0 & 257.49 \\ 0 & 1267.51 & 253.10 \\ 0 & 0 & 1 \end{bmatrix}. \tag{3.109}$$

The reference image of the robot was constructed from a configuration where the robot is fully extended along its backbone. The position error and joint variable trajectories from the resulting simulations are shown in Figures 3.23 and 3.24.

In a physical implementation, multiple cameras at known positions relative to the base frame of the robot will be required to successfully track all visual markers on the robot and avoid problems of occlusion. Utilizing the technique in this section, it is possible to accomplish more than just

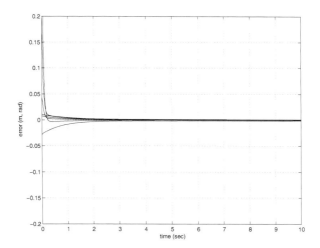

FIGURE 3.23. End Effector Position Error.

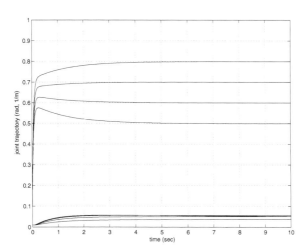

FIGURE 3.24. Time Evolution of Joint Trajectories.

end-effector regulation. Since all joint variables are recovered from processing the images, a joint level controller may also be implemented which will enable complete shape control of the robot (i.e., the manipulator may be servoed to any desired shape given an image of the manipulator at that configuration). The result may be further extended to shape tracking, if a video sequence of the desired trajectory of the robot body is available.

3.5 Mobile Robot Regulation and Tracking

A monocular camera-based vision system attached to a mobile robot (i.e., the camera-in-hand configuration) is considered in this section. By comparing corresponding target points of an object from two different camera images, geometric relationships are exploited to derive a transformation that relates the actual position and orientation of the mobile robot to a reference position and orientation for the regulation problem. For the tracking problem, a prerecorded image sequence (e.g., a video) of three target points is used to define a desired trajectory for the mobile robot. By comparing the target points from a stationary reference image with the corresponding target points in the live image (for the regulation problem) and also with the prerecorded sequence of images (for the tracking problem), projective geometric relationships are exploited to construct Euclidean homographies. The information obtained by decomposing the Euclidean homography is used to develop kinematic controllers. A Lyapunov-based analysis is used to develop an adaptive update law to actively compensate for the lack of depth information required for the translation error system. Experimental results are provided to illustrate the performance of the controller.

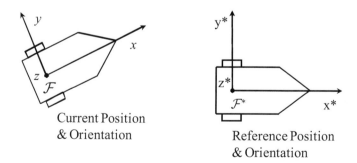

FIGURE 3.25. Mobile Robot Coordinate Systems.

3.5.1   Regulation Control

The objective of this section is to regulate the position/orientation of a mobile robot based on image-feedback of a fixed target. As illustrated in Figure 3.25, the origin of the orthogonal coordinate system $\mathcal{F}$ attached to the camera is coincident with the center of mass of the mobile robot. As also illustrated in Figure 3.25, the xy-axes of $\mathcal{F}$ define the mobile robot plane of motion where the x-axis of $\mathcal{F}$ is aligned with the front of the mobile robot, and the y-axis is parallel to the wheel axis. The z-axis of $\mathcal{F}$ is perpendicular to the mobile robot plane of motion and is located at the center of the wheel axis. The linear velocity of the mobile robot along the x-axis is denoted by $v_c(t)$, and the angular velocity $\omega_c(t)$ is about the z-axis. In addition to $\mathcal{F}$, another fixed orthogonal coordinate system, denoted by $\mathcal{F}^*$, is defined to represent the desired fixed position and orientation of the camera relative to a target. Hence, the goal is to develop a controller that will regulate the position and orientation of $\mathcal{F}$ to $\mathcal{F}^*$.

Camera Model

A target viewed by a camera attached to the mobile robot is assumed to be distinguished by three points O_i, $i = 1, 2, 3$ that compose a plane, denoted by π. The Euclidean position of point O_i expressed in the coordinate frames $\mathcal{F}$ and $\mathcal{F}^*$ is denoted by $\bar{m}_i(t)$, $\bar{m}_i^* \in \mathbb{R}^3$, respectively, and is defined as in (3.1) (see Figure 3.26). The normalized position vectors are defined as

$$
\begin{aligned}
m_i(t) &\triangleq \begin{bmatrix} 1 & m_{iy}(t) & m_{iz}(t) \end{bmatrix}^T = \frac{\bar{m}_i(t)}{x_i(t)} \triangleq \begin{bmatrix} 1 & \dfrac{y_i(t)}{x_i(t)} & \dfrac{z_i(t)}{x_i(t)} \end{bmatrix}^T \\
m_i^* &\triangleq \begin{bmatrix} 1 & m_{iy}^* & m_{iz}^* \end{bmatrix}^T = \frac{\bar{m}_i^*}{x_i^*} \triangleq \begin{bmatrix} 1 & \dfrac{y_i^*}{x_i^*} & \dfrac{z_i^*}{x_i^*} \end{bmatrix}^T,
\end{aligned}
$$

(3.110)

where the standard assumption is made that $x_i(t)$ and x_i^* are positive [85] (i.e., the target is always in front of the camera). In addition to the normalized Euclidean position, each point has an image-space representation, denoted by $p_i(t)$, $p_i^* \in \mathbb{R}^3$

$$
p_i(t) \triangleq \begin{bmatrix} 1 & u_i(t) & v_i(t) \end{bmatrix}^T \qquad p_i^* \triangleq \begin{bmatrix} 1 & u_i^* & v_i^* \end{bmatrix}^T \tag{3.111}
$$

where $u_i(t)$, $v_i(t) \in \mathbb{R}$ denote the pixel coordinates of the point O_i. The image-space coordinates given in (3.111) are related to the normalized coordinates given in (3.110) by the pinhole camera model given in (3.14). Since the camera is assumed to be calibrated (i.e., the matrix A is assumed to be known), $m_i(t)$ and m_i^* can be calculated using (3.14) from the known camera pixel-space vectors $p_i(t)$ and p_i^*.

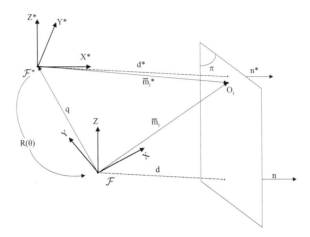

FIGURE 3.26. Geometric Relationship of the Mobile Robot System

In Figure 3.26, $\theta(t) \in \mathbb{R}$ is the angle between the axes x^* and x, the unit vectors $n(t)$, $n^* \in \mathbb{R}^3$ are normal to the plane π expressed in $\mathcal{F}$ and $\mathcal{F}^*$, respectively, and $d(t)$, $d^* \in \mathbb{R}$ are the unknown, positive distances from the origin of $\mathcal{F}$ and $\mathcal{F}^*$ to the plane π along n and n^*, respectively. Based on Figure 3.26, the following relationship can be determined

$$\bar{m}_i = R\bar{m}_i^* + q \ . \tag{3.112}$$

In (3.112), $R(t) \in SO(3)$ denotes the rotation matrix from $\mathcal{F}^*$ to $\mathcal{F}$ as

$$R \triangleq \begin{bmatrix} \cos(\theta) & -\sin(\theta) & 0 \\ \sin(\theta) & \cos(\theta) & 0 \\ 0 & 0 & 1 \end{bmatrix}, \tag{3.113}$$

and $q(t) \in \mathbb{R}^3$ is the translation vector from $\mathcal{F}$ to $\mathcal{F}^*$ given by

$$q(t) \triangleq \begin{bmatrix} q_x(t) & q_y(t) & 0 \end{bmatrix}^T . \tag{3.114}$$

Since d^* is the projection of $\bar{m}_i^*$ along n^*, the relationship given in (3.2) can be used to write (3.112) as

$$\bar{m}_i = H\bar{m}_i^* \tag{3.115}$$

where the Euclidean homography $H(t) \in \mathbb{R}^{3 \times 3}$ is defined as

$$H \triangleq R + \frac{qn^{*T}}{d^*}. \tag{3.116}$$

By using (3.113)-(3.116), the Euclidean homography can be rewritten as

$$
H = [H_{jk}] =
\begin{bmatrix}
\cos\theta + \dfrac{q_x n_x^*}{d^*} & -\sin\theta + \dfrac{q_x n_y^*}{d^*} & \dfrac{q_x n_z^*}{d^*} \\[2ex]
\sin\theta + \dfrac{q_y n_x^*}{d^*} & \cos\theta + \dfrac{q_y n_y^*}{d^*} & \dfrac{q_y n_z^*}{d^*} \\[2ex]
0 & 0 & 1
\end{bmatrix},
\qquad (3.117)
$$

where $n^* = \begin{bmatrix} n_x^* & n_y^* & n_z^* \end{bmatrix}^T$. By examining the terms in (3.117), it is clear that $H(t)$ contains signals that are not directly obtained from the vision system (e.g., $\theta(t)$, $q(t)$, and d^* are not directly available from the camera image). However, the six unknown elements of $H_{jk}(t)$ $\forall j = 1, 2$, $k = 1, 2, 3$ can be determined indirectly from the image coordinates by solving a set of linear equations. Specifically, by using the definition given in (3.110), the expression given in (3.115) can be rewritten as

$$
m_i = \underbrace{\left(\frac{x_i^*}{x_i} \right)}_{\alpha_i} H m_i^*,
\qquad (3.118)
$$

where $\alpha_i(t) \in \mathbb{R}$ denotes a depth ratio. By expanding (3.118), the following expressions can be obtained

$$
\begin{aligned}
1 &= \alpha_i \left(H_{11} + H_{12} m_{iy}^* + H_{13} m_{iz}^* \right) & (3.119) \\
m_{iy} &= \alpha_i \left(H_{21} + H_{22} m_{iy}^* + H_{23} m_{iz}^* \right) & (3.120) \\
m_{iz} &= \alpha_i m_{iz}^*. & (3.121)
\end{aligned}
$$

Given that (3.119)–(3.121) will be generated for each of the three target points, a total of nine independent equations will result. Given the nine independent equations, the nine unknown parameters (i.e., $H_{jk}(t)$ $\forall j = 1, 2$, $k = 1, 2, 3$ and $\alpha_i(t)$ $\forall i = 1, 2, 3$) can be determined. Based on the fact that the elements of the homography matrix and the depth ratio can be determined, the methods described in Section 3.2 (see also [47, 116]) can be used to decompose $H(t)$ to obtain $R(t)$, $\alpha_i(t)$, and $\dfrac{q(t)n^*}{d^*}$; hence, $\theta(t)$ and $\alpha_i(t)$ can be calculated and used in the subsequent control development. To compute $\theta(t)$ from $R(t)$ the following expression can be utilized [105]

$$
\theta = \cos^{-1} \left(\frac{1}{2} \left(\operatorname{tr}(R) - 1 \right) \right),
$$

where

$$
0 \le \theta(t) \le \pi.
$$

In practice, caution has to be given to determine a unique solution for $\theta(t)$ from the homography decomposition. To determine the unique solution for $\theta(t)$ from the set of possible solutions generated by the homography decomposition, a best-guess estimate of the constant normal n^* can be selected from the physical relationship between the camera and the plane defined by the object feature points. Of the possible solutions generated for n^* by the decomposition algorithm, the solution that yields the minimum norm difference with the initial best-guess can be determined as the correct solution. The solution that most closely matches the best-guess estimate can then be used to determine the correct solutions for $\theta(t)$. The robustness of the system is not affected by the *a priori* best-guess estimate of n^* since the estimate is only used to resolve the ambiguity in the solutions generated by the decomposition algorithm.

Problem Formulation

The control objective is to ensure that the coordinate frame attached to the mobile robot is regulated to the fixed coordinate frame $\mathcal{F}^*$. This objective is naturally defined in terms of the Euclidean position/orientation of the mobile robot. Yet, the position and orientation of the mobile robot is not required to be known; rather, only relative translation and orientation information between two corresponding images is required to be computed as previously described. The two required images consist of the current image and an *a priori* acquired image (i.e., the desired image). The requirement for an *a priori* desired image of a target is mild. For example, a mobile robot could be guided (e.g., via a teach pendent) to a desired relative position and orientation with respect to a (indoor or outdoor) target where the desired image is then taken. For future tasks, the mobile robot can compare the current image to the previously acquired image to autonomously return to the desired relative position and orientation, based on the subsequent control development.

To quantify the control objective in terms of the Euclidean position and orientation, the translation error between $\mathcal{F}$ and $\mathcal{F}^*$, denoted by $e_t(t) \in \mathbb{R}^2$, can be written for any target point O_i, $i = 1, 2, 3$ as

$$e_t \triangleq \begin{bmatrix} e_{tx} \\ e_{ty} \end{bmatrix} = \begin{bmatrix} q_x \\ q_y \end{bmatrix} = \begin{bmatrix} x_i \\ y_i \end{bmatrix} - \begin{bmatrix} \cos(\theta) & -\sin(\theta) \\ \sin(\theta) & \cos(\theta) \end{bmatrix} \begin{bmatrix} x_i^* \\ y_i^* \end{bmatrix}, \quad (3.122)$$

where (3.112)–(3.114) have been used. The orientation error between $\mathcal{F}$ and $\mathcal{F}^*$, denoted by $e_o(t) \in \mathbb{R}$, can be written as

$$e_o(t) \triangleq \theta(t), \quad (3.123)$$

where θ was defined in (3.113). Based on the definitions of (3.122) and (3.123), the control objective is to regulate $e_t(t)$ and $e_o(t)$ to zero. The open-loop error system for $e_t(t)$ and $e_o(t)$ can be determined by taking the time derivative of (3.122) and (3.123) and then utilizing the fact that the time derivative of the Euclidean position given in (3.1) can be determined as [34, 85]

$$\dot{\bar{m}}_i = -v - \omega \times \bar{m}_i, \qquad (3.124)$$

where $v(t), \omega(t) \in \mathbb{R}$ denote the linear and angular velocity of the mobile robot expressed in $\mathcal{F}$ as

$$v(t) \triangleq \begin{bmatrix} v_c(t) & 0 & 0 \end{bmatrix}^T \qquad \omega(t) \triangleq \begin{bmatrix} 0 & 0 & \omega_c(t) \end{bmatrix}^T = \begin{bmatrix} 0 & 0 & -\dot{\theta}(t) \end{bmatrix}^T,$$
$$(3.125)$$

respectively. From the expression given in (3.1), (3.124) and (3.125), the Euclidean mobile robot velocity can be written in terms of the linear and angular velocity as

$$\begin{aligned} \dot{x}_i &= -v_c + y_i \omega_c \\ \dot{y}_i &= -x_i \omega_c. \end{aligned} \qquad (3.126)$$

After utilizing (3.122), (3.125), and (3.126) the open-loop error system can be obtained as

$$\begin{aligned} \dot{e}_{tx} &= -v_c + \omega_c e_{ty} \\ \dot{e}_{ty} &= -\omega_c e_{tx} \\ \dot{e}_o &= -\omega_c. \end{aligned} \qquad (3.127)$$

Control Development

The structure of the resulting open-loop error system developed in (3.127) has been extensively examined in mobile robot control literature. However, unlike the typical mobile robot control problem, the Euclidean translation error signals $e_{tx}(t)$ and $e_{ty}(t)$ are unmeasurable, and hence, new analytical development is required. To address this issue, an adaptive controller is developed in this section that actively compensates for the unknown depth information through a gradient-based adaptive update law.

To facilitate the subsequent control design, a composite translation and rotation error signal, denoted by $r(t) \in \mathbb{R}^3$, is defined as

$$r \triangleq \begin{bmatrix} r_1(t) & r_2(t) & r_3(t) \end{bmatrix} = \begin{bmatrix} -e_o & -\dfrac{e_{tx}}{x_i^*} & \dfrac{e_{ty}}{x_i^*} \end{bmatrix}. \qquad (3.128)$$

By utilizing the relationship introduced in (3.122), the following expressions can be developed for $r_2(t)$ and $r_3(t)$

$$r_2 = -\left[\frac{1}{\alpha_i} - \cos(\theta) + m_{iy}^* \sin(\theta) \right] \qquad (3.129)$$

$$r_3 = \left[\frac{1}{\alpha_i} m_{iy} - \sin(\theta) - m_{iy}^* \cos(\theta) \right]. \qquad (3.130)$$

From the expressions given in (3.128)-(3.130), it is clear that $r_1(t)$, $r_2(t)$, and $r_3(t)$ can be computed from (3.119)-(3.121) and the decomposition of the homography matrix. After taking the time derivative of (3.128) and utilizing (3.127), the resulting simplified open-loop dynamics for $r(t)$ can be determined as

$$
\begin{aligned}
\dot{r}_1 &= \omega_c \\
\gamma \dot{r}_2 &= v_c - \gamma \omega_c r_3 \\
\dot{r}_3 &= \omega_c r_2,
\end{aligned}
\qquad (3.131)
$$

where (3.128) has been utilized, and $\gamma \in \mathbb{R}$ denotes the following positive constant

$$\gamma \triangleq x_i^*. \qquad (3.132)$$

To further facilitate the subsequent control design and analysis, an auxiliary signal $\eta(t) \in \mathbb{R}$ is designed as

$$\eta \triangleq r_2 - r_3 \sin(t), \qquad (3.133)$$

where the following open-loop dynamics for $\eta(t)$ can be determined by using (3.131)

$$\gamma \dot{\eta} = v_c - \gamma r_3 \omega_c - \gamma \sin(t) \omega_c r_2 - \gamma r_3 \cos(t). \qquad (3.134)$$

Based on the open-loop dynamics of (3.131), (3.134), and the subsequently stability analysis, an adaptive kinematic controller can be designed as

$$v_c = -k_2 \eta + \hat{\gamma} r_3 \cos(t) + \hat{\gamma} \omega_c r_3 \qquad (3.135)$$

$$\omega_c = -k_1 (r_1 + \chi), \qquad (3.136)$$

where k_1, $k_2 \in \mathbb{R}$ denote positive control gains, and $\chi(t) \in \mathbb{R}$ is an auxiliary signal defined as

$$\chi = (\eta + r_3 \sin(t)) (r_3 - \eta \sin(t)). \qquad (3.137)$$

In (3.135), $\hat{\gamma}(t) \in \mathbb{R}$ denotes a dynamic estimate of γ generated by the following differential expression

$$\dot{\hat{\gamma}} = \Gamma \left(-r_3 \eta \cos(t) - \omega_c \eta r_3 \right), \qquad (3.138)$$

where $\Gamma \in \mathbb{R}$ denotes a positive adaptation gain. After substituting the control inputs given in (3.135) and (3.136) into (3.131) and (3.134), respectively, the following closed-loop error system is obtained

$$
\begin{aligned}
\dot{r}_1 &= -k_1 (r_1 + \chi) & (3.139) \\
\gamma \dot{\eta} &= -k_2 \eta - \tilde{\gamma} (r_3 \cos (t) + w_c r_3) + \gamma k_1 r_2 \sin (t) (r_1 + \chi) \\
\dot{r}_3 &= -k_1 r_2 (r_1 + \chi),
\end{aligned}
$$

where $\tilde{\gamma}(t) \in \mathbb{R}$ denotes the following parameter estimation error

$$
\tilde{\gamma} = \gamma - \hat{\gamma}. \tag{3.140}
$$

Stability Analysis

Theorem 3.2 *The control law given in (3.135) and (3.136) ensures that the position and orientation of the mobile robot coordinate frame $\mathcal{F}$ is regulated to the desired position/orientation described by $\mathcal{F}^*$ in the sense that*

$$
\lim_{t \to \infty} e_t(t), e_o(t) = 0. \tag{3.141}
$$

Proof. To prove (3.141), a non-negative function $V(t)$ is defined as

$$
V \triangleq \frac{1}{2} \gamma \left(r_1^2 + \eta^2 + r_3^2 \right) + \frac{1}{2} \Gamma^{-1} \tilde{\gamma}^2. \tag{3.142}
$$

After taking the time derivative of (3.142) and substituting for the closed-loop system of (3.139), the following expression is obtained

$$
\begin{aligned}
\dot{V} &= -k_1 \gamma r_1 (r_1 + \chi) - k_1 \gamma r_2 r_3 (r_1 + \chi) - \Gamma^{-1} \tilde{\gamma} \dot{\hat{\gamma}} & (3.143) \\
&\quad + \eta [-k_2 \eta - \tilde{\gamma} (r_3 \cos (t) + w_c r_3) + \gamma k_1 r_2 \sin (t) (r_1 + \chi)].
\end{aligned}
$$

Substituting (3.138) into (3.143) and cancelling common terms yields

$$
\dot{V} = -k_2 \eta^2 - \gamma k_1 (r_1 + \chi)^2, \tag{3.144}
$$

where (3.133) and (3.137) have been used. From (3.142) and (3.144), $r_1(t)$, $r_3(t)$, $\eta(t)$, $\tilde{\alpha}(t) \in \mathcal{L}_\infty$ and $\eta(t), [r_1(t) + \chi(t)] \in \mathcal{L}_2$. Based on the previous facts, (3.133)–(3.137), (3.139), and (3.140) can be used to determine that $\chi(t)$, $\hat{\alpha}(t)$, $w_c(t)$, $v_c(t)$, $r_2(t)$, $\dot{r}_1(t)$, $\dot{r}_3(t)$, $\dot{\eta}(t) \in \mathcal{L}_\infty$. Based on the facts that $r_1(t)$, $r_2(t)$, $r_3(t) \in \mathcal{L}_\infty$ and that x_i^* is a positive constant, (3.128) can be used to prove that $e_o(t)$, $e_{tx}(t)$, $e_{ty}(t) \in \mathcal{L}_\infty$. The expressions in (3.138) and (3.140) can be used to prove that $\dot{\tilde{\gamma}}(t)$, $\dot{\hat{\gamma}}(t) \in \mathcal{L}_\infty$. Since $\dot{r}_3(t)$ and $\dot{\tilde{\gamma}}(t) \in \mathcal{L}_\infty$, then $r_3(t)$, $\tilde{\gamma}(t)$ are uniformly continuous (UC). After taking the time derivative of (3.137), the following expression can be obtained

$$
\begin{aligned}
\dot{\chi} &= (\dot{\eta} + \dot{r}_3 \sin (t) + r_3 \cos(t)) (r_3 - \eta \sin (t)) & (3.145) \\
&\quad + (\eta + r_3 \sin (t)) (\dot{r}_3 - \dot{\eta} \sin (t) - \eta \cos (t)).
\end{aligned}
$$

From the previous facts and (3.145), $\dot{\chi}(t) \in \mathcal{L}_\infty$. Based on the facts that $\eta(t)$, $\dot{\eta}(t)$, $[r_1(t) + \chi(t)]$, $[\dot{r}_1(t) + \dot{\chi}(t)] \in \mathcal{L}_\infty$ and that $\eta(t)$, $[r_1(t) + \chi(t)] \in \mathcal{L}_2$, Barbalat's lemma [102] can be employed to conclude that

$$\lim_{t\to\infty} \eta(t), [r_1(t) + \chi(t)] = 0. \tag{3.146}$$

The result in (3.146) can be used in conjunction with the closed-loop dynamics for $r_1(t)$ and $r_3(t)$ given in (3.139) and the control input of (3.136), to determine that

$$\lim_{t\to\infty} \dot{r}_1(t), \omega_c(t), \dot{r}_3(t) = 0. \tag{3.147}$$

By utilizing (3.136), the second equation of (3.139) can be rewritten as

$$\gamma\dot{\eta} = [-k_2\eta - \tilde{\gamma}\omega_c r_3 - \gamma r_2 \sin(t)\,\omega_c] - \tilde{\gamma}r_3\cos(t). \tag{3.148}$$

The results in (3.146) and (3.147) can be used to determine that the bracketed term of (3.148) goes to zero as $t \to \infty$; therefore, since $\tilde{\alpha}(t)$, $r_3(t)$ are UC and $\eta(t)$ has a finite limit as $t \to \infty$, the Extended Barbalat's Lemma (see Lemma A.2 of Appendix A) can be invoked to prove that

$$\lim_{t\to\infty} \dot{\eta}(t) = 0. \tag{3.149}$$

After taking the time derivative of $[r_1(t) + \chi(t)]$, substituting (3.139) and (3.145) for $\dot{r}_1(t)$ and $\dot{\chi}(t)$, respectively, the following resulting expression can be obtained

$$\frac{d}{dt}[r_1(t) + \chi(t)] = [-k_1(r_1 + \chi) + \vartheta(t)] + r_3^2\cos(t), \tag{3.150}$$

where the auxiliary signal $\vartheta(t) \in \mathbb{R}$ is defined as

$$\vartheta(t) \triangleq (\eta + r_3\sin(t))(\dot{r}_3 - \dot{\eta}\sin(t) - \eta\cos(t)) + r_3(\dot{r}_3\sin(t) \\ +\dot{\eta}) - \eta\sin(t)(\dot{\eta} + \dot{r}_3\sin(t) + r_3\cos(t)). \tag{3.151}$$

Based on (3.146), (3.147), and (3.149), it can be shown that

$$\lim_{t\to\infty} \vartheta(t) = 0. \tag{3.152}$$

From (3.146) and (3.152), the bracketed term of (3.150) also goes to zero as $t \to \infty$. Since $r_3(t)$ is UC and $[r_1(t) + \chi(t)]$ has a finite limit as $t \to \infty$, the Extended Barbalat's Lemma (see Lemma A.2 of Appendix A) can be utilized to conclude that

$$\lim_{t\to\infty} r_3^2\cos(t) = 0. \tag{3.153}$$

The result in (3.153) implies that

$$\lim_{t \to \infty} r_3(t) = 0. \tag{3.154}$$

Based on the previous facts, (3.137) and (3.146) can now be utilized to prove that

$$\lim_{t \to \infty} r_1(t) = 0. \tag{3.155}$$

By utilizing (3.146), (3.154), (3.155) and the definitions introduced in (3.128) and (3.133), the result in (3.141) can be obtained. Specifically, given that

$$\lim_{t \to \infty} r_1(t), r_3(t), \eta(t) = 0,$$

then it can be determined that

$$\lim_{t \to \infty} e_t(t), e_o(t) = 0. \quad \blacksquare$$

Experimental Verification

FIGURE 3.27. Mobile Robot Testbed.

The testbed depicted in Figure 3.27 was constructed to implement the adaptive regulation controller given by (3.135), (3.136), and (3.138). The mobile robot testbed consists of the following components: a modified K2A mobile robot (with an inclusive Pentium 133 MHz personal computer (PC)) manufactured by Cybermotion Inc., a Dalsa CAD-6 camera that captures 955 frames per second with 8-bit gray scale at a 260×260 resolution, a

Road Runner Model 24 video capture board, and two Pentium-based PCs. In addition to the mobile robot modifications described in detail in [38], additional modifications particular to this experiment included mounting a camera and the associated image processing Pentium IV 800 MHz PC (operating under QNX, a real-time micro-kernel based operating system) on the top of the mobile robot as depicted in Figure 3.27. The internal mobile robot computer (also operating under QNX) hosts the control algorithm that was written in "C/C++," and implemented using Qmotor 3.0 [76]. In addition to the image processing PC, a second PC (operating under the MS Windows 2000 operating system) was used to remotely log in to the internal mobile robot PC via the QNX Phindows application. The remote PC was used to access the graphical user interface of Qmotor for execution of the control program, gain adjustment, and data management, plotting, and storage. Light-emitting diodes (LEDs) were rigidly attached to a rigid structure that was used as the target, where the intensity of the LEDs contrasted sharply with the background. Due to the contrast in intensity, a simple thresholding algorithm was used to determine the coordinates of each LED.

The mobile robot is controlled by a torque input applied to the drive and steer motors. As subsequently described, to facilitate a torque controller the actual linear and angular velocity of the mobile robot is required. To acquire these signals a backwards difference algorithm was applied to the drive and steering motor encoders. Encoder data acquisition and the control implementation were performed at a frequency of 1.0kHz using the Quanser MultiQ I/O board. For simplicity the electrical and mechanical dynamics of the system were not incorporated in the control design (i.e., the emphasis of this experiment is to illustrate the visual servo controller). However, since the developed kinematic controller is differentiable, standard backstepping techniques could be used to incorporate the mechanical and electrical dynamics. See [37] and [38] for several examples that incorporate the mechanical dynamics. Permanent magnet DC motors provide steering and drive actuation through a 106:1 and a 96:1 gear coupling, respectively. The dynamics for the modified K2A mobile robot are given as follows

$$
\frac{1}{r_o} \begin{bmatrix} 1 & 0 \\ 0 & \frac{L_o}{2} \end{bmatrix} \begin{bmatrix} \tau_1 \\ \tau_2 \end{bmatrix} = \begin{bmatrix} m_o & 0 \\ 0 & I_o \end{bmatrix} \begin{bmatrix} \dot{v}_1 \\ \dot{v}_2 \end{bmatrix}
\tag{3.156}
$$

where $\tau_1(t)$, $\tau_2(t) \in \mathbb{R}$ denote the drive and steering motor torques, respectively, $m_o = 165$ [kg] denotes the mass of the robot, $I_o = 4.643$ [kg·m^2] denotes the inertia of the robot, $r_o = 0.010$ [m] denotes the radius of the wheels, and $L_o = 0.667$ [m] denotes the length of the axis between the wheels. Using the Camera Calibration Toolbox for MATLAB (Zhengyou Zhang's data) [5] the intrinsic calibration parameters of the camera were determined. The pixel coordinates of the principal point (i.e., the image center that is defined as the frame buffer coordinates of the intersection of the optical axis with the image plane) were determined to be $u_0 = v_0 = 130$ [pixels], the focal length and camera scaling factors were determined to be $fk_u = 1229.72$ [pixels] and $fk_v = 1235.29$ [pixels].

Based on (3.128)–(3.130), (3.133), (3.135), (3.136), and (3.138), the signals required to implement the controller include m_{1y}^*, $\alpha_1(t)$, $\theta(t)$, and $m_{1y}(t)$. As previously described, to obtain these signals, an image is required to be obtained at the desired relative position and orientation of the camera with respect to a target. As also previously described, the subscript $i = 1$, is used to indicate that the signal corresponds to the first target point (without loss of generality). The mobile robot was driven by a joystick to a desired position and orientation relative to the target, the desired image was acquired, and the coordinates of the target features were saved on the image processing PC. From the coordinates of the target features and knowledge of the intrinsic calibration parameters, (3.15) was used to determine m_{1y}^*. The constant value for m_{1y}^* was included in the control code hosted by the internal mobile robot PC.

After obtaining the desired image, the mobile robot was driven away from the target by a joystick approximately 6 [m] along the x-axis, with some small offset along the y-axis, and with approximately 34 [deg] of orientation error. Before the control program was executed, the image processing PC was set to acquire the live camera images at 955 frames/sec, to determine the pixel coordinates of the target points, to construct and decompose the homography, and to transmit the signals $\alpha_1(t)$, $\theta(t)$, and $m_{1y}(t)$ that are computed from the homography decomposition via a server program over a dedicated 100Mb/sec network connection to the internal mobile robot computer. A camera with an image capture rate of 955 frames/sec is not required for the experiment. The high speed camera was utilized to enable a higher closed-loop control frequency.

A client program was executed on the internal mobile robot computer to receive $\alpha_1(t)$, $\theta(t)$, and $m_{1y}(t)$ from the server program and write the information into a shared memory location. When the control program was executed, the values for $\alpha_1(t)$, $\theta(t)$, and $m_{1y}(t)$ were acquired from the shared memory location (rather than directly from the network connection to maintain a near deterministic response and for program stability). The values for m_{1y}^*, $\alpha_1(t)$, $\theta(t)$, and $m_{1y}(t)$ were utilized to determine $r(t)$ and

$\eta(t)$ as described by (3.128)–(3.130) and (3.133), and to compute the control signals.

To execute a torque level controller, a feedback loop was implemented as

$$\tau = K_h \bar{\eta}, \tag{3.157}$$

where $\tau = [\tau_1(t),\ \tau_2(t)]^T \in \mathbb{R}^2$ denotes a vector of the drive and steering motor torques, respectively, $K_h \in \mathbb{R}^{2\times 2}$ is a diagonal scaling term, and $\bar{\eta}(t) \in \mathbb{R}^2$ is a velocity mismatch signal defined as

$$\bar{\eta} = \left[\begin{array}{cc} v_c & \omega_c \end{array}\right]^T - \left[\begin{array}{cc} v_a & \omega_a \end{array}\right]^T, \tag{3.158}$$

where $v_c(t)$ and $\omega_c(t)$ denote the linear and angular velocity inputs computed in (3.135) and (3.136), and $v_a(t)$ and $\omega_a(t)$ denote actual linear and angular velocity of the mobile robot computed from the time derivative of the motor encoders.

The control and adaptation gains were adjusted to reduce the position/orientation error with the initial adaptive estimate set to zero. In practice, the adaptive estimate would be initialized to a best-guess value. In this experiment, the adaptive estimate was initialized to zero to illustrate the ability of the estimate to converge in the presence of a large initial error. The final feedback and adaptation gain values were recorded as

$$k_1 = 55.35, \quad k_2 = 21.25, \quad \Gamma = 0.15, \quad K_h = diag\{8,\ 0.185\}.$$

The resulting orientation error is provided in Figure 3.28, and the unitless planar position regulation errors $r_2(t)$ and $r_3(t)$, are depicted in Figure 3.29. Figure 3.30 illustrates that the adaptive estimate for the depth parameter d^* approaches a constant. From Figures 3.28 and 3.29, it is clear that some steady state errors exist in the orientation and the translation along the lateral mobile robot axis, previously defined as $e_o(t)$ and $e_{ty}(t)$, respectively. The steady-state error in $e_o(t)$ is due, in large part, to the fact that as the mobile robot approaches the target, changes in the image-space orientation are magnified (i.e., a one pixel difference from a far distance has less orientation error than a one pixel difference at a close distance). The steady-state error in $e_o(t)$ is propagated in $e_{ty}(t)$. That is, the lateral position of the mobile robot is directly influenced by the orientation error. The computed unitless depth ratio $\gamma_1(t)$ is provided in Figure 3.31. The control torque inputs at the wheels of the mobile robot (i.e., after the 106:1 and 96:1 gear coupling) that is applied by the steer and drive motors is depicted in Figure 3.32.

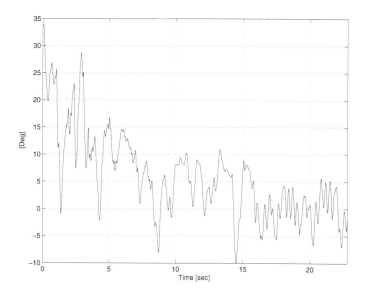

FIGURE 3.28. Orientation Error, $e_o(t) = r_1(t)$.

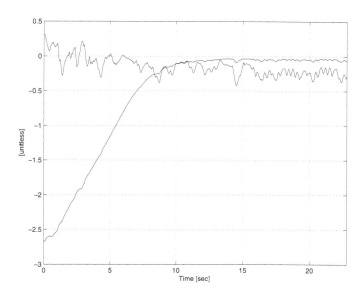

FIGURE 3.29. Position Error, $r_2(t)$ and $r_3(t)$.

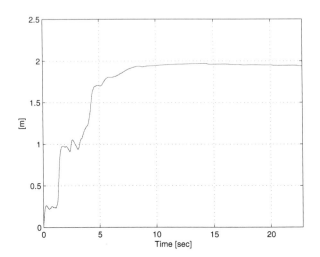

FIGURE 3.30. Adaptive Estimate.

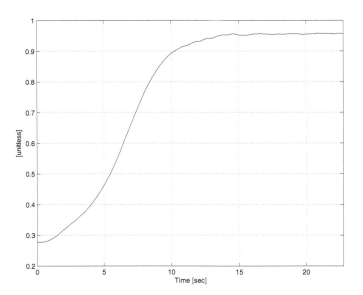

FIGURE 3.31. Computed Depth Ratio.

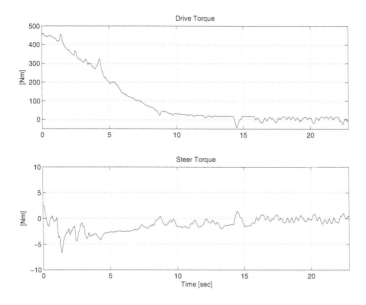

FIGURE 3.32. Computed Torque Inputs.

3.5.2   Tracking Control

The tracking problem can be described as in Figure 3.25 with the addition of a desired "phantom" robot as illustrated in Figure 3.33. The phantom robot is used to generate a prerecorded desired time-varying trajectory of $\mathcal{F}_d$ that is assumed to be second-order differentiable. The desired trajectory is obtained from a prerecorded set of images of a stationary target viewed by the on-board camera as the mobile robot moves. For example, the desired mobile robot motion could be obtained as an operator drives the robot via a teach pendant, with the on-board camera capturing and storing the sequence of images of the stationary target. For this scenario, the fixed orthogonal coordinate system, denoted by $\mathcal{F}^*$, enables the current and desired image trajectories to be compared to a constant reference image. The use of a constant reference image also facilitates the development of a constant parameter that can be related to the time-varying depth from the mobile robot to the target. Relating the time-varying depth information to a depth related parameter facilitates adaptive control methods. Based on the definition of these coordinate frames, the goal in this section is to develop a homography-based visual servo controller that will force $\mathcal{F}$ to track the position and orientation trajectory provided by $\mathcal{F}_d$.

From a practical standpoint, numerous applications can be represented by the described problem formulation. For example, the mobile robot could

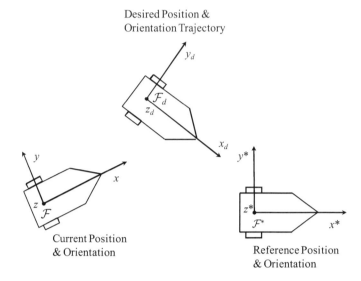

FIGURE 3.33. Mobile Robot Coordinate Systems.

be navigated via a teach pendant, while the camera records a desired set of images that represent the trajectory of the mobile robot relative to the target. Then in subsequent tasks, the mobile robot will be able to track the same relative trajectory independent of the possibility that the target has moved between the time the image sequence was recorded and the autonomous task execution. A simple practical example is if the mobile robot is taught a path (via the set of images) to a docking station to recharge the batteries. The mobile robot will be able to track this path to achieve successful docking with the charging station even if the station has been moved from the original location (or likewise, if the initial position and orientation of the mobile robot is different), provided obstacles have not been placed in the path of the mobile robot that would inhibit the mobile robot trajectory. See [101] for further discussion and motivation for the problem formulation.

Geometric Model

In this section, geometric relationships are developed between the coordinate systems $\mathcal{F}$, $\mathcal{F}_d$, and $\mathcal{F}^*$, and a reference plane π that is defined by three target points O_i $\forall i = 1, 2, 3$ that are not collinear. The 3D Euclidean coordinates of O_i expressed in terms of $\mathcal{F}$, $\mathcal{F}_d$, and $\mathcal{F}^*$ as $\bar{m}_i\left(t\right)$, $\bar{m}_{di}\left(t\right)$, $\bar{m}_i^* \in \mathbb{R}^3$, respectively, are defined as in (3.1). The rotation from $\mathcal{F}^*$ to $\mathcal{F}$ is denoted by $R\left(t\right) \in SO(3)$, and the translation from $\mathcal{F}$ to $\mathcal{F}^*$ is denoted by $x_f(t) \in \mathbb{R}^3$ where $x_f(t)$ is expressed in $\mathcal{F}$. Similarly, $R_d(t) \in SO(3)$ denotes the desired time-varying rotation from $\mathcal{F}^*$ to $\mathcal{F}_d$, and $x_{fd}(t) \in \mathbb{R}^3$ denotes the desired translation from $\mathcal{F}_d$ to $\mathcal{F}^*$ where $x_{fd}(t)$ is expressed in $\mathcal{F}_d$. Since the motion of the mobile robot is constrained to the xy-plane, $x_f(t)$ and $x_{fd}(t)$ are defined as

$$x_f(t) \triangleq \begin{bmatrix} x_{f1} & x_{f2} & 0 \end{bmatrix}^T \tag{3.159}$$

$$x_{fd}(t) \triangleq \begin{bmatrix} x_{fd1} & x_{fd2} & 0 \end{bmatrix}^T .$$

From the geometry between the coordinate frames depicted in Figure 3.34, $\bar{m}_i^*$ can be related to $\bar{m}_i(t)$ and $\bar{m}_{di}(t)$ as

$$\bar{m}_i = x_f + R\bar{m}_i^* \qquad \bar{m}_{di} = x_{fd} + R_d\bar{m}_i^* . \tag{3.160}$$

In (3.160), $R(t)$ and $R_d(t)$ are defined as in (3.113), where $\theta(t) \in \mathbb{R}$ denotes the right-handed rotation angle about $z_i(t)$ that aligns the rotation of $\mathcal{F}$ with $\mathcal{F}^*$, and $\theta_d(t) \in \mathbb{R}$ denotes the right-handed rotation angle about $z_{di}(t)$ that aligns the rotation of $\mathcal{F}_d$ with $\mathcal{F}^*$. From the definition of the coordinate systems

$$\dot{\theta} = -\omega_c \qquad \dot{\theta}_d = -\omega_{cd} \tag{3.161}$$

where $\omega_{cd}(t) \in \mathbb{R}$ denotes the desired angular velocity of the mobile robot expressed in $\mathcal{F}_d$. The rotation angles are assumed to be confined to the following regions

$$-\pi < \theta(t) < \pi \qquad -\pi < \theta_d(t) < \pi . \tag{3.162}$$

Based on the definition of d^* in (3.2) and the fact that n^* and $\bar{m}^*$ do not change, it is clear that d^* is a constant. From (3.2), the relationships in (3.160) can be expressed as

$$\bar{m}_i = \left(R + \frac{x_f}{d^*} n^{*T} \right) \bar{m}_i^* \tag{3.163}$$

$$\bar{m}_{di} = \left(R_d + \frac{x_{fd}}{d^*} n^{*T} \right) \bar{m}_i^* .$$

Euclidean Reconstruction

The relationship given in (3.160) provides a means to quantify the translational and rotational error between $\mathcal{F}$ and $\mathcal{F}^*$ and between $\mathcal{F}_d$ and $\mathcal{F}^*$. Since the position of $\mathcal{F}$, $\mathcal{F}_d$, and $\mathcal{F}^*$ cannot be directly measured, this section illustrates how the normalized Euclidean coordinates of the target

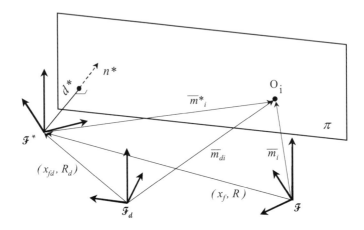

FIGURE 3.34. Coordinate Frame Relationships.

points can be reconstructed by relating multiple images. Specifically, comparisons are made between an image acquired from the camera attached to $\mathcal{F}$, the reference image, and the prerecorded sequence of images that define the trajectory of $\mathcal{F}_d$. To facilitate the subsequent development, the normalized Euclidean coordinates of O_i expressed in terms of $\mathcal{F}$, $\mathcal{F}_d$, and $\mathcal{F}^*$ are denoted by $m_i(t)$, $m_{di}(t)$, $m_i^* \in \mathbb{R}^3$, respectively, and are explicitly defined in (3.110). In addition to having a Euclidean coordinate, each target point O_i will also have a projected pixel coordinate denoted by $u_i(t), v_i(t) \in \mathbb{R}$ for $\mathcal{F}$, $u_i^*, v_i^* \in \mathbb{R}$ for $\mathcal{F}^*$, and $u_{di}(t), v_{di}(t) \in \mathbb{R}$ for $\mathcal{F}_d$, that are defined as elements of $p_i(t) \in \mathbb{R}^3$ (i.e., the actual time-varying image points), $p_{di}(t) \in \mathbb{R}^3$ (i.e., the desired image point trajectory), and $p_i^* \in \mathbb{R}^3$ (i.e., the constant reference image points), respectively, as in (3.111). The normalized Euclidean coordinates of the target points are related to the image data through the pinhole model as in (3.14).

Given that $m_i(t)$, $m_{di}(t)$, and m_i^* can be obtained from (3.14), the rotation and translation between the coordinate systems can be related in terms of the normalized Euclidean coordinates as

$$m_i = \underbrace{\frac{x_i^*}{x_i}}_{\alpha_i} \underbrace{\left(R + x_h n^{*T}\right) m_i^*}_{H} \tag{3.164}$$

$$m_{di} = \underbrace{\frac{x_i^*}{x_{di}}}_{\alpha_{di}} \underbrace{\left(R_d + x_{hd} n^{*T}\right) m_i^*}_{H_d}, \tag{3.165}$$

where $\alpha_i(t)$, $\alpha_{di}(t) \in \mathbb{R}$ denote the depth ratios, $H(t)$, $H_d(t) \in \mathbb{R}^{3 \times 3}$ denote Euclidean homographies, and $x_h(t)$, $x_{hd}(t) \in \mathbb{R}^3$ denote scaled translation vectors that are defined as

$$x_h \triangleq \begin{bmatrix} x_{h1} & x_{h2} & 0 \end{bmatrix}^T = \frac{x_f}{d^*} \qquad (3.166)$$

$$x_{hd} \triangleq \begin{bmatrix} x_{hd1} & x_{hd2} & 0 \end{bmatrix}^T = \frac{x_{fd}}{d^*}.$$

By using (3.113) and (3.12), the Euclidean homography in (3.164) can be rewritten as

$$H = [H_{jk}] \qquad (3.167)$$

$$= \begin{bmatrix} \cos\theta + x_{h1}n_x^* & -\sin\theta + x_{h1}n_y^* & x_{h1}n_z^* \\ \sin\theta + x_{h2}n_x^* & \cos\theta + x_{h2}n_y^* & x_{h2}n_z^* \\ 0 & 0 & 1 \end{bmatrix}.$$

By examining the terms in (3.167), it is clear that $H(t)$ contains signals that are not directly measurable (e.g., $\theta(t)$, $x_h(t)$, and n^*). By expanding $H_{jk}(t)$ $\forall j = 1, 2, k = 1, 2, 3$, the expressions given in (3.119)–(3.121). Following the development for the regulation problem, $R(t)$, $R_d(t)$, $x_h(t)$, and $x_{hd}(t)$ can all be computed from (3.119)–(3.121) and used for the subsequent control synthesis. Since $R(t)$ and $R_d(t)$ are known matrices, then (3.113) can be used to determine $\theta(t)$ and $\theta_d(t)$.

Control Development

The control objective is to ensure that the coordinate frame $\mathcal{F}$ tracks the time-varying trajectory of $\mathcal{F}_d$ (i.e., $\bar{m}_i(t)$ tracks $\bar{m}_{di}(t)$). This objective is naturally defined in terms of the Euclidean position/orientation of the mobile robot. Specifically, based on the previous development, the translation and rotation tracking error, denoted by $e(t) \triangleq \begin{bmatrix} e_1 & e_2 & e_3 \end{bmatrix}^T \in \mathbb{R}^3$, is defined as

$$\begin{aligned} e_1 &\triangleq x_{h1} - x_{hd1} \\ e_2 &\triangleq x_{h2} - x_{hd2} \\ e_3 &\triangleq \theta - \theta_d, \end{aligned} \qquad (3.168)$$

where $x_{h1}(t)$, $x_{h2}(t)$, $x_{hd1}(t)$, and $x_{hd2}(t)$ are introduced in (3.166), and $\theta(t)$ and $\theta_d(t)$ are introduced in (3.113). Based on the definition in (3.168), it can be shown that the control objective is achieved if the tracking error $e(t) \to 0$. Specifically, it is clear from (3.12) that if $e_1(t) \to 0$ and $e_2(t) \to 0$, then $x_f(t) \to x_{fd}(t)$. If $e_3 \to 0$, then it is clear from (3.113) and (3.168) that $R(t) \to R_d(t)$. If $x_f(t) \to x_{fd}(t)$ and $R(t) \to R_d(t)$, then (3.160) can be used to prove that $\bar{m}_i(t) \to \bar{m}_{di}(t)$.

As described in Section 3.3.1, it is typical that the desired trajectory is used as a feedforward component in tracking control designs. As described previously, the functions $\alpha_{di}(t)$, $H_d(t)$, $R_d(t)$, and $x_{hd}(t)$ can be obtained from a sufficiently smooth function derived from the prerecorded sequence of images. Given $\theta_d(t)$ and the time derivative of $R_d(t)$, $\dot{\theta}_d(t)$ can be determined. In the subsequent tracking control development, $\dot{x}_{hd1}(t)$ and $\dot{\theta}_d(t)$ will be used in feedforward control terms.

Open-loop Error System

As a means to develop the open-loop tracking error system, the time derivative of the Euclidean position $x_f(t)$ is determined as [85]

$$\dot{x}_f = -v + [x_f]_\times \omega, \tag{3.169}$$

where $v(t)$, $\omega(t) \in \mathbb{R}^3$ denote the respective linear and angular velocity of the mobile robot expressed in $\mathcal{F}$ as

$$v \triangleq \begin{bmatrix} v_c & 0 & 0 \end{bmatrix}^T \qquad \omega \triangleq \begin{bmatrix} 0 & 0 & \omega_c \end{bmatrix}^T, \tag{3.170}$$

and $[x_f]_\times$ denotes the 3×3 skew-symmetric form of $x_f(t)$. After substituting (3.166) into (3.169), the time derivative of the translation vector $x_h(t)$ can be written in terms of the linear and angular velocity of the mobile robot as

$$\dot{x}_h = -\frac{v}{d^*} + [x_h]_\times \omega. \tag{3.171}$$

After incorporating (3.170) into (3.171), the following expression can be obtained

$$\begin{aligned} \dot{x}_{h1} &= -\frac{v_c}{d^*} + x_{h2}\omega_c \\ \dot{x}_{h2} &= -x_{h1}\omega_c, \end{aligned} \tag{3.172}$$

where (3.166) was utilized. Given that the desired trajectory is generated from a prerecorded set of images taken by the on-board camera as the mobile robot was moving, an expression similar to (3.169) can be developed as

$$\dot{x}_{fd} = -\begin{bmatrix} v_{cd} & 0 & 0 \end{bmatrix}^T + [x_{fd}]_\times \begin{bmatrix} 0 & 0 & \omega_{cd} \end{bmatrix}^T, \tag{3.173}$$

where $v_{cd}(t) \in \mathbb{R}$ denotes the desired linear[1] velocity of the mobile robot expressed in $\mathcal{F}_d$. After substituting (3.166) into (3.173), the time derivative of the translation vector $x_{hd}(t)$ can be written as

$$\begin{aligned} \dot{x}_{hd1} &= -\frac{v_{cd}}{d^*} + x_{hd2}\omega_{cd} \\ \dot{x}_{hd2} &= -x_{hd1}\omega_{cd}. \end{aligned} \tag{3.174}$$

---

[1] Note that $v_{cd}(t)$ is not measurable.

After taking the time derivative of (3.168) and utilizing (3.161) and (3.172), the following open-loop error system can be obtained

$$
\begin{aligned}
d^* \dot{e}_1 &= -v_c + d^* (x_{h2} \omega_c - \dot{x}_{hd1}) \\
\dot{e}_2 &= - \left(x_{h1} \omega_c + x_{hd1} \dot{\theta}_d \right) \\
\dot{e}_3 &= - \left(\omega_c + \dot{\theta}_d \right),
\end{aligned}
\tag{3.175}
$$

where the definition of $e_2(t)$ given in (3.168), and the second equation of (3.174) was utilized. To facilitate the subsequent development, the auxiliary variable $\bar{e}_2 (t) \in \mathbb{R}$ is defined as

$$
\bar{e}_2 \triangleq e_2 - x_{hd1} e_3.
\tag{3.176}
$$

After taking the time derivative of (3.176) and utilizing (3.175), the following expression is obtained

$$
\dot{\bar{e}}_2 = - \left(e_1 \omega_c + \dot{x}_{hd1} e_3 \right).
\tag{3.177}
$$

Based on (3.176), it is clear that if $\bar{e}_2(t)$, $e_3(t) \rightarrow 0$, then $e_2(t) \rightarrow 0$. Based on this observation and the open-loop dynamics given in (3.177), the following control development is based on the desire to prove that $e_1 (t), \bar{e}_2 (t), e_3 (t)$ are asymptotically driven to zero.

Closed-Loop Error System

Based on the open-loop error systems in (3.175) and (3.177), the linear and angular velocity kinematic control inputs for the mobile robot are designed as

$$
v_c \triangleq k_v e_1 - \bar{e}_2 \omega_c + \hat{d}^* (x_{h2} \omega_c - \dot{x}_{hd1})
\tag{3.178}
$$

$$
\omega_c \triangleq k_\omega e_3 - \dot{\theta}_d - \dot{x}_{hd1} \bar{e}_2,
\tag{3.179}
$$

where k_v, $k_\omega \in \mathbb{R}$ denote positive, constant control gains. In (3.178), the parameter update law $\hat{d}^*(t) \in \mathbb{R}$ is generated by the following differential equation

$$
\dot{\hat{d}}^* = \gamma_1 e_1 (x_{h2} \omega_c - \dot{x}_{hd1}),
\tag{3.180}
$$

where $\gamma_1 \in \mathbb{R}$ is a positive, constant adaptation gain. After substituting the kinematic control signals designed in (3.178) and (3.179) into (3.175), the following closed-loop error systems are obtained

$$
\begin{aligned}
d^* \dot{e}_1 &= -k_v e_1 + \bar{e}_2 \omega_c + \tilde{d}^* (x_{h2} \omega_c - \dot{x}_{hd1}) \\
\dot{\bar{e}}_2 &= - (e_1 \omega_c + \dot{x}_{hd1} e_3) \\
\dot{e}_3 &= -k_\omega e_3 + \dot{x}_{hd1} \bar{e}_2,
\end{aligned}
\tag{3.181}
$$

where (3.177) was utilized, and the depth-related parameter estimation error $\tilde{d}^*(t) \in \mathbb{R}$ is defined as

$$\tilde{d}^* \triangleq d^* - \hat{d}^* . \tag{3.182}$$

Stability Analysis

Theorem 3.3 *The adaptive update law defined in (3.180) along with the control input designed in (3.178) and (3.179) ensure that the mobile robot tracking error $e(t)$ is asymptotically driven to zero in the sense that*

$$\lim_{t\to\infty} e(t) = 0 \tag{3.183}$$

provided the time derivative of the desired trajectory satisfies the condition

$$\lim_{t\to\infty} \dot{x}_{hd1} \neq 0. \tag{3.184}$$

Proof: To prove Theorem 3.3, the non-negative function $V(t) \in \mathbb{R}$ is defined as

$$V \triangleq \frac{1}{2}d^* e_1^2 + \frac{1}{2}\bar{e}_2^2 + \frac{1}{2}e_3^2 + \frac{1}{2\gamma_1}\tilde{d}^{*2} . \tag{3.185}$$

The following simplified expression can be obtained by taking the time derivative of (3.185), substituting the closed-loop dynamics in (3.181) into the resulting expression, and then cancelling common terms

$$\dot{V} = -k_v e_1^2 + e_1 \tilde{d}^* (x_{h2}\omega_c - \dot{x}_{hd1}) - k_\omega e_3^2 - \frac{1}{\gamma_1}\tilde{d}^* \dot{\hat{d}}^*. \tag{3.186}$$

Substituting (3.180) into (3.186) yields

$$\dot{V} = -k_v e_1^2 - k_\omega e_3^2 . \tag{3.187}$$

From (3.185) and (3.187), it is clear that $e_1(t)$, $\bar{e}_2(t)$, $e_3(t)$, $\tilde{d}^*(t) \in \mathcal{L}_\infty$ and that $e_1(t)$, $e_3(t) \in \mathcal{L}_2$. Since $\tilde{d}^*(t) \in \mathcal{L}_\infty$ and d^* is a constant, the expression in (3.182) can be used to determine that $\hat{d}^*(t) \in \mathcal{L}_\infty$. From the assumption that $x_{hd1}(t)$, $\dot{x}_{hd1}(t)$, $x_{hd2}(t)$, $\theta_d(t)$, and $\dot{\theta}_d(t)$ are constructed as bounded functions, and the fact that $\bar{e}_2(t)$, $e_3(t) \in \mathcal{L}_\infty$, the expressions in (3.168), (3.176), and (3.179) can be used to prove that $e_2(t)$, $x_{h1}(t)$, $x_{h2}(t)$, $\theta(t)$, $\omega_c(t) \in \mathcal{L}_\infty$. Based on the previous development, the expressions in (3.178), (3.180), and (3.181) can be used to conclude that $v_c(t)$, $\dot{\hat{d}}^*(t)$, $\dot{e}_1(t)$, $\dot{\bar{e}}_2(t)$, $\dot{e}_3(t) \in \mathcal{L}_\infty$. Based on the fact that $e_1(t)$, $e_3(t)$, $\dot{e}_1(t)$, $\dot{e}_3(t) \in \mathcal{L}_\infty$ and that $e_1(t)$, $e_3(t) \in \mathcal{L}_2$, Barbalat's lemma [102] can be employed to prove that

$$\lim_{t\to\infty} e_1(t), e_3(t) = 0 . \tag{3.188}$$

From (3.188) and the fact that the signal $\dot{x}_{hd1}(t)\bar{e}_2(t)$ is uniformly continuous (i.e., $\dot{x}_{hd1}(t)$, $\ddot{x}_{hd1}(t)$, $\bar{e}_2(t)$, $\dot{\bar{e}}_2(t) \in \mathcal{L}_\infty$), Extended Barbalat's Lemma (see Lemma A.2 of Appendix A) can be applied to the last equation in (3.181) to prove that

$$\lim_{t\to\infty} \dot{e}_3(t) = 0 \tag{3.189}$$

and that

$$\lim_{t\to\infty} \dot{x}_{hd1}(t)\bar{e}_2(t) = 0 \ . \tag{3.190}$$

If the desired trajectory satisfies (3.184), then (3.190) can be used to prove that

$$\lim_{t\to\infty} \bar{e}_2(t) = 0 \ . \tag{3.191}$$

Based on the definition of $\bar{e}_2(t)$ given in (3.176), the results in (3.188) and (3.191) can be used to conclude that

$$\lim_{t\to\infty} e_2(t) = 0 \tag{3.192}$$

provided the condition in (3.184) is satisfied. ∎

Remark 3.2 *The condition given in (3.184) is in terms of the time derivative of the desired translation vector. Typically, for WMR tracking problems, this assumption is expressed in terms of the desired linear and angular velocity of the WMR. To this end, (3.174) can be substituted into (3.184) to obtain the following condition*

$$\lim_{t\to\infty} \frac{v_{cd}(t)}{d^*} \neq x_{hd2}(t)\omega_{cd}(t). \tag{3.193}$$

The condition in (3.193) is comparable to typical WMR tracking results that restrict the desired linear and angular velocity. For an in-depth discussion of this type of restriction including related previous results see [38].

Experimental Results

The adaptive tracking controller given by (3.178)–(3.180) was implemented on the same mobile robot testbed as described for the regulation problem. To acquire the desired image trajectory, the mobile robot was driven by a joystick while the image processing PC acquired the camera images at 955 frames/sec, determined the pixel coordinates of the feature points, and saved the pixel data to a file. The last image was also saved as the reference image. The desired image file and the reference image were read into a stand-alone program that computed $x_{hd}(t)$ and $\theta_d(t)$ offline. To determine the unique solution for $x_{hd}(t)$ and $\theta_d(t)$ (and likewise for $x_h(t)$ and $\theta(t)$)

from the set of possible solutions generated by the homography decomposition, a best-guess estimate of the constant normal n^* was selected as $n^* = \begin{bmatrix} 1 & 0 & 0 \end{bmatrix}^T$ (i.e., from the physical relationship between the camera and the plane defined by the object feature points, the focal axis of the camera mounted on the mobile robot was assumed to be roughly perpendicular to π). Of the possible solutions generated for n^* by the decomposition algorithm, the solution that yielded the minimum norm difference with the initial best-guess was determined as the correct solution. The solution that most closely matched the best-guess estimate was then used to determine the correct solutions for $x_{hd}(t)$ and $\theta_d(t)$ (or $x_h(t)$ and $\theta(t)$). The robustness of the system is not affected by the *a priori* estimate of n^* since the estimate is only used to resolve the ambiguity in the solutions generated by the decomposition algorithm, and the n^* generated by the decomposition algorithm is used to further decompose the homography. A Butterworth filter was applied to $x_{hd}(t)$ and $\theta_d(t)$ to reduce noise effects. A filtered backwards difference algorithm was used to compute $\dot{x}_{hd}(t)$ and $\dot{\theta}_d(t)$. Figure 3.35 and Figure 3.36 depict the desired translation and rotation signals, respectively.

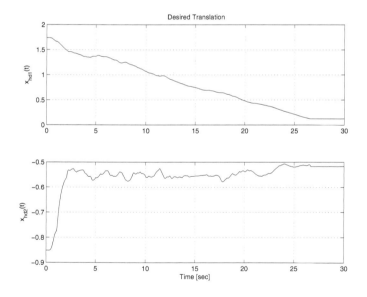

FIGURE 3.35. Desired Translation.

The desired trajectory signals $x_{hd}(t)$, $\dot{x}_{hd1}(t)$, $\theta_d(t)$, and $\dot{\theta}_d(t)$ were stored in a file that was opened by the control algorithm and loaded into memory when the control algorithm was loaded in Qmotor. After determining $x_h(t)$

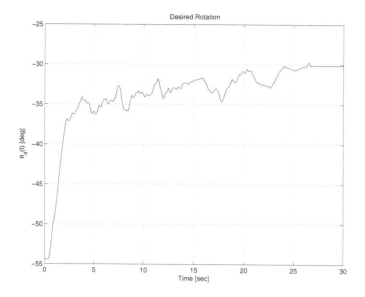

FIGURE 3.36. Desired Rotation.

and $\theta(t)$ from the homography decomposition, comparisons with $x_{hd}(t)$ and $\theta_d(t)$ were made at each time instant to compute the error signal $e_1(t)$, $\bar{e}_2(t)$, and $e_3(t)$, which were subsequently used to compute $v_c(t)$, $\omega_c(t)$, and $\hat{d}^*(t)$ given in (3.178)–(3.180). The torque level controller given in (3.157) was implemented as described in the regulation experimental section.

The control gains were adjusted to reduce the position/orientation tracking error with the adaptation gains set to zero and the initial adaptive estimate set to zero. After some tuning, the position/orientation tracking error response could not be significantly improved by further adjustments of the feedback gains. The adaptation gains were then adjusted to allow the parameter estimation to reduce the position/orientation tracking error. After the tuning process was completed, the final adaptation and feedback gain values were recorded as

$$k_v = 4.15, \quad k_\omega = 0.68, \quad \gamma = 40.1, \quad K_h = diag\{99.7, 23.27\}.$$

The unitless position/orientation tracking errors $e_1(t)$ and $e_2(t)$, are depicted in Figure 3.37 and Figure 3.38, respectively. Figure 3.39 illustrates that the adaptive estimate for the depth parameter d^* approaches a constant. Figure 3.40 illustrates the linear and angular velocity of the mobile robot. The control torque inputs are presented in Figure 3.41 and represent the torques applied after the gearing mechanism.

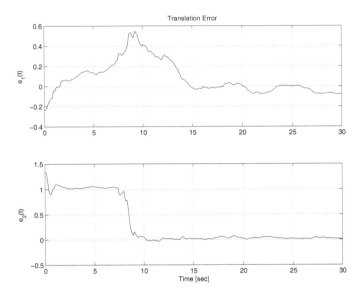

FIGURE 3.37. Translation Error.

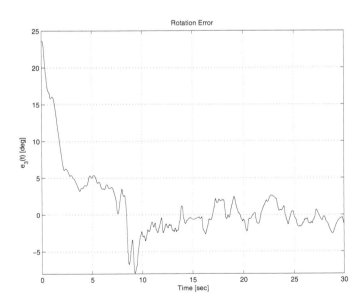

FIGURE 3.38. Rotation Error.

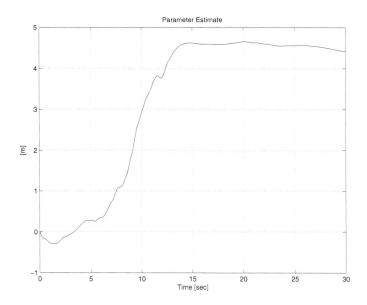

FIGURE 3.39. Parameter Estimate.

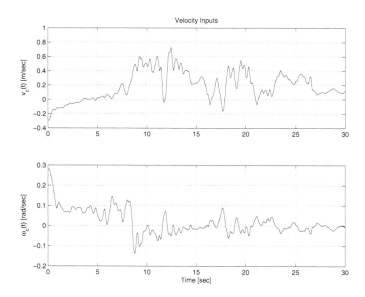

FIGURE 3.40. Linear and Angular Velocity Control Inputs.

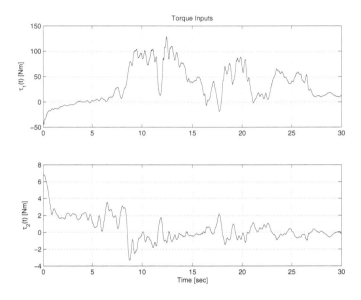

FIGURE 3.41. Drive and Steer Motor Torque Inputs.

From Figure 3.37 and Figure 3.38, it is clear that $e_2(t)$ is relatively un-changing in the first 8 seconds, whereas $e_1(t)$ and $e_3(t)$ are changing signif-icantly. This phenomena is due to the nonholonomic nature of the vehicle. Specifically, since there is an initial position and orientation error, the con-troller moves the vehicle to minimize the error and align the mobile robot with the desired image trajectory. Since the mobile robot can not move along both axes of the Cartesian plane simultaneously while also rotating (i.e., due to the nonholonomic motion constraints), the mobile robot ini-tially moves to minimize $e_1(t)$ and $e_3(t)$. Likewise, when $e_2(t)$ undergoes change between 8 and 10 seconds, $e_1(t)$ remains relatively unchanged. While performing the experiment, slightly different responses were obtained each run due to variations in the initial position and orientation of the mobile robot and variations in the control parameters as the gains were adjusted. With a constant set of control gains, the transient response still exhibited some variations due to differences in the initial conditions; however, the steady state response remained constant for each trial.

Note that $e_1(t)$ and $e_2(t)$ depicted in Figure 3.37 are unitless. From (3.12) and (3.168), it is clear that $e_1(t)$ and $e_2(t)$ are unitless because both the translation $x_f(t)$ and the depth related constant d^* have units of me-ters. That is, $x_h(t)$ and $x_{hd}(t)$ are unitless translation terms computed from the homography decomposition (note that no units are provided in Figure 3.35). In practice, the mobile robot traversed an arc than approximately spanned a 6×1 meter space, with an approximate speed of 0.22 meters/second (i.e., approximately the same speed as the numerous mobile robot experiments presented in [38]).

Based on the outcome of this experiment, several issues for future research and technology integration are evident. For example, the problem formulation has a number of practical applications in environments where the reference object may not be stationary between each task execution. However, the result does not address cases where an obstacle enters the task-space and inhibits the mobile robot from tracking the prerecorded trajectory. To address this issue, there is a clear need for continued research that targets incorporating image-space path planning with the control design as in [28], [29], [50], and [87]. Additionally, the result does not address a method to automatically reselect feature points. For example, methods to automatically determine new feature points if they become nearly aligned, or if a feature point leaves the field-of-view (e.g., becomes occluded), could add robustness to the implemented control system. Of course, an *ad hoc* approach of simply continuously tracking multiple redundant feature points could be utilized, but this approach may excessively restrict the image processing bandwidth.

3.6   Structure from Motion

In this section, an adaptive nonlinear estimator is developed to identify the Euclidean coordinates of feature points on a moving object using a single fixed camera. No explicit model is used to describe the movement of the object. Homography-based techniques are used in the development of the object kinematics, while Lyapunov design methods are utilized in the synthesis of the adaptive estimator. An extension of this development to the dual case of camera-in-hand is also presented. The performance of the estimator is demonstrated by simulation and experimental results.

3.6.1   Object Kinematics

The development in this section is based on the image geometry for a fixed camera viewing a moving target as described in Section 3.2.1. To quantify the translation of $\mathcal{F}$ relative to the fixed coordinate system $\mathcal{F}^*$ (see Figure 3.1), a translation error $e_v(t) \in \mathbb{R}^3$ is defined in terms of the image coordinates of the feature point O_1 as

$$e_v \triangleq \begin{bmatrix} u_1 - u_1^* & v_1 - v_1^* & -\ln(\alpha_1) \end{bmatrix}^T, \tag{3.194}$$

where $\ln(\cdot) \in \mathbb{R}$ denotes the natural logarithm. As stated in previous sections, any point O_i on π could have been utilized; however, O_1 is used to reduce the notational complexity. The signal $e_v(t)$ is measurable, since the first two elements of the vector are obtained from the images and the last element is available from known signals. The translational kinematics for the fixed-camera problem can be expressed as

$$\dot{e}_v = \frac{\alpha_1}{z_1^*} A_e R \left[v_e - [s_1]_\times \omega_e \right], \qquad (3.195)$$

where the notation $[s_1]_\times$ denotes the 3×3 skew symmetric form of s_1, the *unknown* linear and angular velocity of the object expressed in the local coordinate frame $\mathcal{F}$ are defined as $v_e(t)$, $\omega_e(t) \in \mathbb{R}^3$, respectively, and $A_e(t)$ is introduced in (3.50). The rotation error $e_\omega(t) \in \mathbb{R}^3$ is defined to quantify the rotation of $\mathcal{F}$ relative to $\mathcal{F}^*$, using the axis-angle representation [105], where the open-loop error system can be developed as in (3.46) as

$$\dot{e}_\omega = L_\omega R \omega_e . \qquad (3.196)$$

From (3.195) and (3.196), the kinematics of the object under motion can be expressed as

$$\dot{e} = Jv \qquad (3.197)$$

where $e(t) \triangleq \begin{bmatrix} e_v^T & e_\omega^T \end{bmatrix}^T \in \mathbb{R}^6$, $v(t) \triangleq \begin{bmatrix} v_e^T & \omega_e^T \end{bmatrix}^T \in \mathbb{R}^6$, and $J(t) \in \mathbb{R}^{6\times6}$ is a Jacobian-like matrix defined as

$$J = \begin{bmatrix} \dfrac{\alpha_1}{z_1^*} A_e R & -\dfrac{\alpha_1}{z_1^*} A_e R [s_1]_\times \\ 0_3 & L_\omega R \end{bmatrix} \qquad (3.198)$$

where $0_3 \in \mathbb{R}^{3\times3}$ denotes a zero matrix. In the subsequent analysis, it is assumed that a single geometric length $s_1 \in \mathbb{R}^3$ between two feature points is known. With this assumption, each element of $J(t)$ is known with the possible exception of the constant $z_1^* \in \mathbb{R}$; however, z_1^* can also be computed given s_1. In the subsequent development, it is assumed that the object never leaves the field of view of the camera; hence, $e(t) \in L_\infty$. It is also assumed that the object velocity, acceleration, and jerk are bounded, i.e., $v(t), \dot{v}(t), \ddot{v}(t) \in L_\infty$; hence the structure of (3.197) indicates that $\dot{e}(t), \ddot{e}(t), \dddot{e}(t) \in L_\infty$.

3.6.2 *Identification of Velocity*

In [22], an estimator was developed for online asymptotic identification of the signal $\dot{e}(t)$. Designating $\hat{e}(t)$ as the estimate for $e(t)$, the estimator was designed as

$$
\begin{aligned}
\dot{\hat{e}} \quad &\triangleq \quad \int_{t_0}^{t} (K + I_6)\tilde{e}(\tau)d\tau + \int_{t_0}^{t} \rho \mathrm{sgn}\left(\tilde{e}(\tau)\right) d\tau \\
&\quad + (K + I_6)\tilde{e}(t)
\end{aligned} \tag{3.199}
$$

where $\tilde{e}(t) \triangleq e(t) - \hat{e}(t) \in \mathbb{R}^6$ is the estimation error for the signal $e(t)$, $K, \rho \in \mathbb{R}^{6 \times 6}$ are positive definite constant diagonal gain matrices, $I_6 \in \mathbb{R}^{6 \times 6}$ is the 6×6 identity matrix, t_0 is the initial time, and $\mathrm{sgn}(\tilde{e}(t))$ denotes the standard signum function applied to each element of the vector $\tilde{e}(t)$. The reader is referred to [22] and the references therein for analysis pertaining to the development of the above estimator. In essence, it was shown in [22] that the above estimator asymptotically identifies the signal $\dot{e}(t)$ provided the following inequality is satisfied for each diagonal element ρ_i of the gain matrix ρ,

$$
\rho_i \geq \left|\ddot{e}_i\right| + \left|\dddot{e}_i\right| \quad \forall i = 1, 2, ...6. \tag{3.200}
$$

Hence, $\dot{\hat{e}}_i(t) \rightarrow \dot{e}_i(t)$ as $t \rightarrow \infty, \forall i = 1, 2, ...6$. Since $J(t)$ is known and invertible, the six degree-of-freedom velocity of the moving object can be identified as

$$
\hat{v}(t) = J^{-1}(t)\dot{\hat{e}}(t), \text{ and hence } \hat{v}(t) \rightarrow v(t) \text{ as } t \rightarrow \infty. \tag{3.201}
$$

Euclidean Reconstruction of Feature Points

This Section is focused on the identification of Euclidean coordinates of the feature points on a moving object (i.e., the vector s_i relative to the object frame $\mathcal{F}$, $\bar{m}_i(t)$ and $\bar{m}_i^*$ relative to the camera frame $\mathcal{I}$ for all i feature points on the object). To facilitate the development of the estimator, the time derivative of the extended image coordinates introduced in (3.37) are expressed as

$$
\begin{aligned}
\dot{p}_{ei} \quad &= \quad \frac{\alpha_i}{z_i^*} A_e R \left[v_e + [\omega_e]_\times s_i\right] \\
&= \quad W_i V_{vw} \theta_i
\end{aligned} \tag{3.202}
$$

where $W_i(.) \in \mathbb{R}^{3 \times 3}, V_{vw}(t) \in \mathbb{R}^{3 \times 4}$ and $\theta_i \in \mathbb{R}^4$ are defined as

$$
W_i \triangleq \alpha_i A_{ei} R \tag{3.203}
$$

$$
V_{vw} \triangleq \left[\begin{array}{cc} v_e & [\omega_e]_\times \end{array} \right] \tag{3.204}
$$

$$
\theta_i \triangleq \left[\begin{array}{cc} \dfrac{1}{z_i^*} & \dfrac{s_i{}^T}{z_i^*} \end{array} \right]^T. \tag{3.205}
$$

The elements of $W_i(.)$ are known and bounded, and an estimate of $V_{vw}(t)$, denoted by $\hat{V}_{vw}(t)$, is available by appropriately re-ordering the vector $\hat{v}(t)$ given in (3.201).

The objective is to identify the unknown constant θ_i in (3.202). To facilitate this objective, a parameter estimation error signal, denoted by $\tilde{\theta}_i(t) \in \mathbb{R}^4$, is defined as

$$\tilde{\theta}_i(t) \triangleq \theta_i - \hat{\theta}_i(t), \qquad (3.206)$$

where $\hat{\theta}_i(t) \in \mathbb{R}^4$ is a subsequently designed parameter update signal. A measurable filter signal $W_{fi}(t) \in \mathbb{R}^{3\times4}$, and a non-measurable filter signal $\eta_i(t) \in \mathbb{R}^3$ are also introduced as

$$\dot{W}_{fi} = -\beta_i W_{fi} + W_i \hat{V}_{vw} \qquad (3.207)$$

$$\dot{\eta}_i = -\beta_i \eta_i + W_i \tilde{V}_{vw} \theta_i, \qquad (3.208)$$

where $\beta_i \in \mathbb{R}$ is a scalar positive gain, and $\tilde{V}_{vw}(t) \triangleq V_{vw}(t) - \hat{V}_{vw}(t) \in \mathbb{R}^{3\times4}$ is an estimation error signal.

Motivated by the subsequent stability analysis, an estimator, denoted by $\hat{p}_{ei}(t) \in \mathbb{R}^3$, is designed for the extended image coordinates as

$$\dot{\hat{p}}_{ei} = \beta_i \tilde{p}_{ei} + W_{fi}\dot{\hat{\theta}}_i + W_i \hat{V}_{vw}\hat{\theta}_i, \qquad (3.209)$$

where $\tilde{p}_{ei}(t) \triangleq p_{ei}(t) - \hat{p}_{ei}(t) \in \mathbb{R}^3$ denotes the measurable estimation error signal for the extended image coordinates of the feature points. The time derivative of this estimation error signal is computed from (3.202) and (3.209) as

$$\dot{\tilde{p}}_{ei} = -\beta_i \tilde{p}_{ei} - W_{fi}\dot{\hat{\theta}}_i + W_i \tilde{V}_{vw}\theta_i + W_i \hat{V}_{vw}\tilde{\theta}_i. \qquad (3.210)$$

From (3.208) and (3.210), it can be shown that

$$\tilde{p}_{ei} = W_{fi}\tilde{\theta}_i + \eta_i. \qquad (3.211)$$

Based on the subsequent analysis, the following least-squares update law [102] is developed for $\hat{\theta}_i(t)$

$$\dot{\hat{\theta}}_i = L_i W_{fi}^T \tilde{p}_{ei}, \qquad (3.212)$$

where $L_i(t) \in \mathbb{R}^{4\times4}$ is an estimation gain that is recursively computed as

$$\frac{d}{dt}(L_i^{-1}) = W_{fi}^T W_{fi}. \qquad (3.213)$$

The subsequent analysis requires that $L_i^{-1}(0)$ in (3.213) be positive definite. This requirement can be easily satisfied by selecting the appropriate non-zero initial values.

Analysis

Theorem 3.4 *The update law defined in (3.212) ensures that $\tilde{\theta}_i(t) \to 0$ as $t \to \infty$ provided that the following persistent excitation condition [102] holds*

$$\gamma_1 I_4 \leq \int_{t_0}^{t_0+T} W_{fi}^T(\tau) W_{fi}(\tau) d\tau \leq \gamma_2 I_4 \qquad (3.214)$$

and provided that the gains β_i satisfy the following inequality

$$\beta_i > k_{1i} + k_{2i} \|W_i\|_\infty^2 \qquad (3.215)$$

where $t_0, \gamma_1, \gamma_2, T, k_{1i}, k_{2i} \in \mathbb{R}$ are positive constants, $I_4 \in \mathbb{R}^{4\times 4}$ is the 4×4 identity matrix, and k_{1i} is selected such that

$$k_{1i} > 2. \qquad (3.216)$$

Proof: Let $V(t) \in \mathbb{R}$ denote a non-negative scalar function defined as

$$V \triangleq \frac{1}{2}\tilde{\theta}_i^T L_i^{-1} \tilde{\theta}_i + \frac{1}{2}\eta_i^T \eta_i. \qquad (3.217)$$

After taking the time derivative of (3.217), the following expression can be obtained

$$
\begin{aligned}
\dot{V} &= -\frac{1}{2}\left\| W_{fi}\tilde{\theta}_i \right\|^2 - \tilde{\theta}_i^T W_{fi}^T \eta_i - \beta_i \|\eta_i\|^2 + \eta_i^T W_i \tilde{V}_{vw}\theta_i \\
&\leq -\frac{1}{2}\left\| W_{fi}\tilde{\theta}_i \right\|^2 - \beta_i \|\eta_i\|^2 + \|\theta_i\| \|W_i\|_\infty \left\| \tilde{V}_{vw} \right\|_\infty \|\eta_i\| \\
&\quad + \left\| W_{fi}\tilde{\theta}_i \right\| \|\eta_i\| - k_{1i} \|\eta_i\|^2 + k_{1i} \|\eta_i\|^2 \\
&\quad + k_{2i} \|W_i\|_\infty^2 \|\eta_i\|^2 - k_{2i} \|W_i\|_\infty^2 \|\eta_i\|^2.
\end{aligned} \qquad (3.218)
$$

After using the nonlinear damping argument [75], the expression in (3.218) can be further simplified as

$$
\begin{aligned}
\dot{V} &\leq -\left(\frac{1}{2} - \frac{1}{k_{1i}}\right)\left\| W_{fi}\tilde{\theta}_i \right\|^2 - \left(\beta_i - k_{1i} - k_{2i}\|W_i\|_\infty^2\right)\|\eta_i\|^2 \\
&\quad + \frac{1}{k_{2i}}\|\theta_i\|^2 \left\| \tilde{V}_{vw} \right\|_\infty^2,
\end{aligned} \qquad (3.219)
$$

where $k_{1i}, k_{2i} \in \mathbb{R}$ are positive constants as previously mentioned. The gains k_{1i}, k_{2i}, and β_i must be selected to ensure that

$$\frac{1}{2} - \frac{1}{k_{1i}} \geq \mu_{1i} > 0 \qquad (3.220)$$

$$\beta_i - k_{1i} - k_{2i}\|W_i\|_\infty^2 \geq \mu_{2i} > 0 \qquad (3.221)$$

where $\mu_{1i}, \mu_{2i} \in \mathbb{R}$ are positive constants. The gain conditions given by (3.220) and (3.221) allow us to formulate the conditions given by (3.215) and (3.216), as well as allowing us to further upper bound the time derivative of (3.217) as

$$\dot{V} \leq -\mu_{1i} \left\| W_{fi} \tilde{\theta}_i \right\|^2 - \mu_{2i} \left\| \eta_i \right\|^2 + \frac{1}{k_{2i}} \left\| \theta_i \right\|^2 \left\| \tilde{V}_{vw} \right\|_\infty^2 . \tag{3.222}$$

In the analysis provided in [22], it was shown that a filter signal $r(t) \in \mathbb{R}^6$ defined as $r(t) = \tilde{e}(t) + \dot{\tilde{e}}(t) \in L_\infty \cap L_2$. From this result it is easy to show that $\tilde{e}(t), \dot{\tilde{e}}(t) \in L_2$ [35]. Since $J(t) \in L_\infty$ and invertible, it follows that $J^{-1}(t)\dot{\tilde{e}}(t) \in L_2$. Hence $\tilde{v}(t) \triangleq v(t) - \hat{v}(t) \in L_2$, and it is easy to show that $\left\| \tilde{V}_{vw}(t) \right\|_\infty^2 \in L_1$ [71]. Therefore, the last term in (3.222) is L_1, and

$$\int_0^\infty \frac{1}{k_{2i}} \left\| \theta_i(\tau) \right\|^2 \left\| \tilde{V}_{vw}(\tau) \right\|_\infty^2 d\tau \leq \varepsilon, \tag{3.223}$$

where $\varepsilon \in \mathbb{R}$ is a positive constant. The expressions in (3.217), (3.222), and (3.223) can be used to conclude that

$$\int_0^\infty \left(\mu_{1i} \left\| W_{fi}(\tau)\tilde{\theta}_i(\tau) \right\|^2 + \mu_{2i} \left\| \eta_i(\tau) \right\|^2 \right) d\tau \leq V(0) - V(\infty) + \varepsilon. \tag{3.224}$$

The inequality in (3.224) can be used to determine that $W_{fi}(t)\tilde{\theta}_i(t), \eta_i(t) \in L_2$. From (3.224) and the fact that $V(t)$ is non-negative, it can be concluded that $V(t) \leq V(0) + \varepsilon$ for any t, and hence $V(t) \in L_\infty$. Therefore, from (3.217), $\eta_i(t) \in L_\infty$ and $\tilde{\theta}_i^T(t)L_i^{-1}(t)\tilde{\theta}_i(t) \in L_\infty$. Since $L_i^{-1}(0)$ is positive definite, and the persistent excitation condition in (3.214) is assumed to be satisfied, (3.213) can be used to show that $L_i^{-1}(t)$ is always positive definite. It must follow that $\tilde{\theta}_i(t) \in L_\infty$. Since $\hat{v}(t) \in L_\infty$ as shown in [22], it follows from (3.204) that $\hat{V}_{vw}(t) \in L_\infty$. From (3.207), and the fact that $W_i(.)$ defined in (3.203) are composed of bounded terms, $W_{fi}(t), \dot{W}_{fi}(t) \in L_\infty$ [35], and consequently, $W_{fi}(t)\tilde{\theta}_i(t) \in L_\infty$. Therefore, (3.211) can be used to conclude that $\tilde{p}_{ei}(t) \in L_\infty$. It follows from (3.212) that $\dot{\hat{\theta}}_i(t) \in L_\infty$, and hence, $\dot{\tilde{\theta}}_i(t) \in L_\infty$. From the fact that $\dot{W}_{fi}(t), \dot{\tilde{\theta}}_i(t) \in L_\infty$, it is easy to show that $\frac{d}{dt}\left(W_{fi}(t)\tilde{\theta}_i(t) \right) \in L_\infty$. Hence, $W_{fi}(t)\tilde{\theta}_i(t)$ is uniformly continuous. Since $W_{fi}(t)\tilde{\theta}_i(t) \in L_2$, then

$$W_{fi}(t)\tilde{\theta}_i(t) \to 0 \text{ as } t \to \infty. \tag{3.225}$$

It can be shown that the output $W_{fi}(t)$ of the filter defined in (3.207) is persistently exciting if the input $W_i(t)\hat{V}_{vw}^T(t)$ to the filter is persistently exciting [96]. Hence, the condition in (3.214) is satisfied if

$$\gamma_3 I_4 \leq \underbrace{\int_{t_0}^{t_0+T} \hat{V}_{vw}^T(\tau) W_i^T(\tau) W_i(\tau) \hat{V}_{vw}(\tau)\, d\tau}_{W} \leq \gamma_4 I_4, \tag{3.226}$$

where $\gamma_3, \gamma_4 \in \mathbb{R}$ are positive constants. It can be shown upon expansion of the integrand $W(t) \in \mathbb{R}^{4 \times 4}$ of (3.226) that even if only one of the components of translational velocity is non-zero, the first element of $\hat{\theta}_i(t)$ (i.e. $\frac{1}{z_i^*}$), will converge to the correct value. It should be noted that the translational velocity of the object has no bearing on the convergence of the remaining three elements of $\hat{\theta}_i(t)$, and unfortunately, it seems that no inference can be made about the relationship between convergence of the three remaining elements of $\hat{\theta}_i(t)$ and the rotational velocity of the object. If the signal $W_{fi}(t)$ satisfies the persistent excitation condition [102] given in (3.214), then it can be concluded from (3.225) that (see Lemma B.3 in Appendix B)

$$\tilde{\theta}_i(t) \to 0 \text{ as } t \to \infty. \blacksquare \tag{3.227}$$

As previously stated, the estimation of object velocity requires the knowledge of the constant rotation matrix $R^* \in \mathbb{R}^{3 \times 3}$ and a single geometric length $s_1 \in \mathbb{R}^3$ on the object. After utilizing (3.110), (3.14), (3.205) and (3.212), the estimates for $\bar{m}_i^*$, denoted by $\hat{\bar{m}}_i^*(t) \in \mathbb{R}^3$, can be obtained as

$$\hat{\bar{m}}_i^*(t) = \frac{1}{\left[\hat{\theta}_i(t)\right]_1} A^{-1} p_i^*, \tag{3.228}$$

where the term in the denominator denotes the first element of the vector $\hat{\theta}_i(t)$. Similarly, the estimates for the time varying Euclidean position of the feature points on the object relative to the camera frame, denoted by $\hat{\bar{m}}_i(t) \in \mathbb{R}^3$, can be calculated as

$$\hat{\bar{m}}_i(t) = \frac{1}{\alpha_i(t) \left[\hat{\theta}_i(t)\right]_1} A^{-1} p_i(t). \tag{3.229}$$

3.6.3 Camera-in-Hand Extension

The fixed camera problem can be extended to the case where the camera can move relative to the object (i.e., the camera-in-hand problem). For example, as shown in Figure 3.42, a camera could be mounted on the end-effector of a robot and used to scan an object in its workspace to determine its structure, as well as determine the robot's position. Let three feature points on the object, denoted by O_1, O_2 and O_3 define the plane π^* in Figure

3.42. The time derivative of $e(t)$ introduced in (3.197) can be expressed as follows for the camera-in-hand problem [16]

$$\dot{e} = J_c v \tag{3.230}$$

where $J_c(t) \in \mathbb{R}^{6 \times 6}$ is given by

$$J_c = \begin{bmatrix} -\dfrac{\alpha_1}{z_1^*} A_{e1} & A_{e1} \, [m_1]_\times \\ 0_3 & -L_\omega \end{bmatrix} \tag{3.231}$$

where $0_3 \in \mathbb{R}^{3 \times 3}$ is a zero matrix, $L_\omega(t) \in \mathbb{R}^{3 \times 3}$ has exactly the same form as for the fixed camera case in (3.47), and $v(t) \triangleq \begin{bmatrix} v_c^T & \omega_c^T \end{bmatrix}^T \in \mathbb{R}^6$ now denotes the velocity of the camera expressed relative to $\mathcal{I}$. Terms from the rotation matrix $R(t)$ are not present in (3.231), and therefore, unlike the fixed camera case, the estimate of the velocity of the camera, denoted by $\hat{v}(t)$, can be computed without knowledge of the constant rotation matrix R^*.

With the exception of the term $z_1^* \in \mathbb{R}$, all other terms in (3.231) are either measurable or known *a priori*. If the camera can be moved away from its reference position by a known translation vector $\bar{x}_{fk} \in \mathbb{R}^3$, then z_1^* can be computed offline. Decomposition of the Euclidean homography between the normalized Euclidean coordinates of the feature points obtained at the reference position, and at $\bar{x}_{fk}$ away from the reference position, respectively, can be used to calculate the scaled translation vector $\dfrac{\bar{x}_{fk}}{d^*} \in \mathbb{R}^3$, where $d^* \in \mathbb{R}$ is the distance from the initial camera position, denoted by $\mathcal{I}^*$, to the plane π^*. Then, it can be seen that

$$z_1^* = \frac{d^*}{n^{*T} m_1^*} = \frac{d^*}{n^{*T} A^{-1} p_1^*}. \tag{3.232}$$

From (3.230) and (3.231), any feature point O_i can be shown to satisfy

$$
\begin{aligned}
\dot{p}_{ei} &= -\frac{\alpha_i}{z_i^*} A_{ei} v_c + A_{ei} \, [m_i]_\times \omega_c \\
&= W_{1i} v_c \theta_i + W_{2i} \omega_c,
\end{aligned}
\tag{3.233}
$$

where $p_{ei}(t) \in \mathbb{R}^3$ is defined in (3.37), and $W_{1i}(.) \in \mathbb{R}^{3 \times 3}$, $W_{2i}(t) \in \mathbb{R}^{3 \times 3}$ and $\theta_i \in \mathbb{R}$ are given as

$$
\begin{aligned}
W_{1i} &= -\alpha_i A_{ei} & (3.234) \\
W_{2i} &= A_{ei} \, [m_i]_\times & (3.235) \\
\theta_i &= \frac{1}{z_i^*}. & (3.236)
\end{aligned}
$$

The matrices $W_{1i}(.)$ and $W_{2i}(t)$ are measurable and bounded. The objective is to identify the unknown constant θ_i in (3.236). With the knowledge of z_i^* for all feature points on the object, their Euclidean coordinates $\bar{m}_i^*$ relative to the camera reference position, denoted by $\mathcal{I}^*$, can be computed from (3.110)–(3.14). As before, the mismatch between the actual signal θ_i and the estimated signal $\hat{\theta}_i(t)$ is defined as $\tilde{\theta}(t) \in \mathbb{R}$ where $\tilde{\theta}(t) \triangleq \theta - \hat{\theta}(t)$.

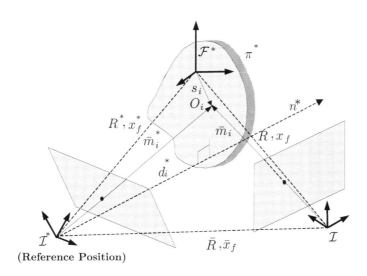

FIGURE 3.42. Coordinate Frame Relationships for the Camera-in-Hand Case.

To facilitate the subsequent development, a measurable filter signal $\zeta_i(t) \in \mathbb{R}^3$ and two non-measurable filter signals $\kappa_i(t), \eta_i(t) \in \mathbb{R}^3$ are introduced as

$$\dot{\zeta}_i = -\beta_i\zeta_i + W_{1i}\hat{v}_c, \qquad (3.237)$$

$$\dot{\kappa}_i = -\beta_i\kappa_i + W_{1i}\tilde{v}_c\theta_i, \qquad (3.238)$$

$$\dot{\eta}_i = -\beta_i\eta_i + W_{2i}\tilde{\omega}_c, \qquad (3.239)$$

where $\beta_i \in \mathbb{R}$ is a scalar positive gain, $\hat{v}_c(t), \hat{\omega}_c(t) \in \mathbb{R}^3$ are the estimates for the translational and rotational velocity, respectively, obtained from the velocity observer in Section 3.6.2, $\tilde{v}_c(t) \triangleq v_c(t) - \hat{v}_c(t)$ is the mismatch in estimated translational velocity, and $\tilde{\omega}_c(t) \triangleq \omega_c(t) - \hat{\omega}_c(t)$ is the mismatch in estimated rotational velocity. The structure of the velocity observer and the proof of its convergence are exactly identical to the fixed camera case. Likewise, an estimator for the $p_{ei}(t)$ can be designed as

$$\dot{\hat{p}}_{ei} = \beta_i\tilde{p}_{ei} + \zeta_i\dot{\hat{\theta}}_i + W_{1i}\hat{v}_c\hat{\theta}_i + W_{2i}\hat{\omega}_c. \qquad (3.240)$$

The following expression can then be developed from (3.233) and (3.240) as

$$\dot{\tilde{p}}_{ei} = -\beta_i \tilde{p}_{ei} - \zeta_i \dot{\tilde{\theta}}_i + W_{1i} \tilde{v}_c \theta_i + W_{1i} \hat{v}_c \tilde{\theta}_i + W_{2i} \tilde{\omega}_c. \tag{3.241}$$

From (3.238), (3.239), and (3.241), it can be shown that

$$\tilde{p}_{ei} = \zeta_i \tilde{\theta}_i + \kappa_i + \eta_i. \tag{3.242}$$

Based on the subsequent analysis, a least-squares update law for $\hat{\theta}_i(t)$ can be designed as

$$\dot{\hat{\theta}}_i = L_i \zeta_i^T \tilde{p}_{ei}, \tag{3.243}$$

where $L_i(t) \in \mathbb{R}$ is an estimation gain that is determined as

$$\frac{d}{dt}(L_i^{-1}) = \zeta_i^T \zeta_i \tag{3.244}$$

and initialized such that $L_i^{-1}(0) > 0$.

Theorem 3.5 *The update law defined in (3.243) ensures that $\tilde{\theta}_i \to 0$ as $t \to \infty$ provided that the following persistent excitation condition [102] holds*

$$\gamma_5 \le \int_{t_0}^{t_0+T} \zeta_i^T(\tau)\zeta_i(\tau)d\tau \le \gamma_6 \tag{3.245}$$

and provided that the gains β_i satisfy the following inequalities

$$\beta_i > k_{3i} + k_{4i} \|W_{1i}\|_\infty^2 \tag{3.246}$$

$$\beta_i > k_{5i} + k_{6i} \|W_{2i}\|_\infty^2 \tag{3.247}$$

where $t_0, \gamma_5, \gamma_6, T, k_{3i}, k_{4i}, k_{5i}, k_{6i} \in \mathbb{R}$ are positive constants, and k_{3i}, k_{5i} are selected such that

$$\frac{1}{k_{3i}} + \frac{1}{k_{5i}} < \frac{1}{2}. \tag{3.248}$$

Proof: Similar to the analysis for the fixed camera case, a non-negative function denoted by $V(t) \in \mathbb{R}$ is defined as

$$V \triangleq \frac{1}{2}\tilde{\theta}_i^T L_i^{-1} \tilde{\theta}_i + \frac{1}{2}\kappa_i^T \kappa_i + \frac{1}{2}\eta_i^T \eta_i. \tag{3.249}$$

After taking the time derivative of (3.249), the following expression can be obtained

$$\begin{aligned}
\dot{V} &= -\frac{1}{2}\left\|\tilde{\theta}_i\right\|^2 \|\zeta_i\|^2 - \beta_i \|\kappa_i\|^2 - \beta_i \|\eta_i\|^2 \\
&\quad -\tilde{\theta}_i^T \zeta_i^T \kappa_i - \tilde{\theta}_i^T \zeta_i^T \eta_i + \kappa_i^T W_{1i}\tilde{v}_c\theta_i + \eta_i^T W_{2i}\tilde{\omega}_c,
\end{aligned} \tag{3.250}$$

where (3.238), (3.239), and (3.243)–(3.242) were used. After some simplification and expansion of some terms, (3.250) can be upper bounded as

$$
\begin{aligned}
\dot{V} \;\leq\; & -\frac{1}{2}\left\|\tilde{\theta}_i\right\|^2 \|\zeta_i\|^2 - \left(\beta_i - k_{3i} - k_{4i}\,\|W_{1i}\|_\infty^2\right)\|\kappa_i\|^2 \\
& -\left(\beta_i - k_{5i} - k_{6i}\,\|W_{2i}\|_\infty^2\right)\|\eta_i\|^2 + \left\|\tilde{\theta}_i\right\|\,\|\zeta_i\|\,\|\kappa_i\| - k_{3i}\,\|\kappa_i\|^2 \\
& + \|\theta_i\|\,\|\tilde{v}_c\|\,\|W_{1i}\|_\infty\,\|\kappa_i\| - k_{4i}\,\|W_{1i}\|_\infty^2\,\|\kappa_i\|^2 \\
& + \|\zeta_i\|\,\left\|\tilde{\theta}_i\right\|\,\|\eta_i\| - k_{5i}\,\|\eta_i\|^2 \\
& + \|\tilde{\omega}_c\|\,\|W_{2i}\|_\infty\,\|\eta_i\| - k_{6i}\,\|W_{2i}\|_\infty^2\,\|\eta_i\|^2 \\
\leq\; & -\left(\frac{1}{2} - \frac{1}{k_{3i}} - \frac{1}{k_{5i}}\right)\left\|\tilde{\theta}_i\right\|^2 \|\zeta_i\|^2 - \left(\beta_i - k_{3i} - k_{4i}\,\|W_{1i}\|_\infty^2\right)\|\kappa_i\|^2 \\
& -\left(\beta_i - k_{5i} - k_{6i}\,\|W_{2i}\|_\infty^2\right)\|\eta_i\|^2 + \frac{1}{k_{4i}}\,\|\theta_i\|^2\,\|\tilde{v}_c\|^2 \\
& + \frac{1}{k_{6i}}\,\|\tilde{\omega}_c\|^2 \,,
\end{aligned}
\tag{3.251}
$$

where $k_{3i}, k_{4i}, k_{5i}, k_{6i} \in \mathbb{R}$ are positive constants. The gain constants are selected to ensure that

$$
\frac{1}{2} - \frac{1}{k_{3i}} - \frac{1}{k_{5i}} \;\geq\; \mu_{3i} > 0 \tag{3.252}
$$

$$
\beta_i - k_{3i} - k_{4i}\,\|W_{1i}\|_\infty^2 \;\geq\; \mu_{4i} > 0 \tag{3.253}
$$

$$
\beta_i - k_{5i} - k_{6i}\,\|W_{2i}\|_\infty^2 \;\geq\; \mu_{5i} > 0, \tag{3.254}
$$

where $\mu_{3i}, \mu_{4i}, \mu_{5i} \in \mathbb{R}$ are positive constants. The gain conditions given by (3.252)–(3.254) allow for (3.249) to be further upper bounded as

$$
\begin{aligned}
\dot{V} \;\leq\; & -\mu_{3i}\left|\tilde{\theta}_i\right|^2 \|\zeta_i\|^2 - \mu_{4i}\,\|\kappa_i\|^2 - \mu_{5i}\,\|\eta_i\|^2 + \frac{1}{k_{4i}}\,|\theta_i|^2\,\|\tilde{v}_c\|^2 + \frac{1}{k_{6i}}\,\|\tilde{\omega}_c\|^2 \\
\leq\; & -\mu_{3i}\left|\tilde{\theta}_i\right|^2 \|\zeta_i\|^2 - \mu_{4i}\,\|\kappa_i\|^2 - \mu_{5i}\,\|\eta_i\|^2 + \mu_{6i}\,\|\tilde{v}\|^2 \,,
\end{aligned}
\tag{3.255}
$$

where $\mu_{6i} = \max\left\{\frac{|\theta_i|^2}{k_{4i}}, \frac{1}{k_{6i}}\right\} \in \mathbb{R}$. Following the argument in the fixed camera case, $\tilde{v}(t) \in L_2$; hence,

$$
\int_{t_0}^{t} \mu_{6i}\,\|\tilde{v}(\tau)\|^2\, d\tau \leq \varepsilon, \tag{3.256}
$$

where $\varepsilon \in \mathbb{R}$ is a positive constant. From (3.249), (3.255) and (3.256), the following inequality can be developed

$$
\int_{t_0}^{t} \mu_{3i}\left|\tilde{\theta}_i(\tau)\right|^2 \|\zeta_i(\tau)\|^2 + \mu_{4i}\,\|\kappa_i(\tau)\|^2 + \mu_{5i}\,\|\eta_i(\tau)\|^2\, d\tau \leq V(0) - V(\infty) + \varepsilon.
\tag{3.257}
$$

From (3.257), it is clear that $\zeta_i(t)\tilde{\theta}_i(t), \kappa_i(t), \eta_i(t) \in L_2$. Applying the same signal chasing arguments as in the fixed camera case, it can be shown that $\tilde{\theta}_i(t), \kappa_i(t), \eta_i(t) \in L_\infty$. It can also be shown that $\dot{\tilde{\theta}}_i(t), \zeta_i(t), \dot{\zeta}_i(t) \in L_\infty$ and therefore $\frac{d}{dt}\zeta_i(t)\tilde{\theta}_i(t) \in L_\infty$. Hence $\zeta_i(t)\tilde{\theta}_i(t)$ is uniformly continuous [39], and since $\zeta_i(t)\tilde{\theta}_i(t) \in L_\infty$, then [39]

$$\zeta_i(t)\tilde{\theta}_i(t) \to 0 \text{ as } t \to \infty. \tag{3.258}$$

Applying the same argument as in the fixed camera case, convergence of $\hat{\theta}_i(t)$ to true parameters is guaranteed (i.e., $\tilde{\theta}_i(t) \to 0$ as $t \to \infty$), if the signal $\zeta_i(t)$ satisfies the persistent excitation condition in (3.245). As in the case of the fixed camera, the persistent excitation condition of (3.245) is satisfied if the translational velocity of the camera is non-zero.

Utilizing (3.110), (3.14) and the update law in (3.243), the estimates for Euclidean coordinates of all i feature points on the object relative to the camera at the reference position, denoted by $\hat{m}_i^*(t) \in \mathbb{R}^3$, can be determined as

$$\hat{m}_i^*(t) = \frac{1}{\hat{\theta}_i(t)} A^{-1} p_i^*. \tag{3.259}$$

The knowledge of z_1^* allows us to resolve the scale factor ambiguity inherent in the Euclidean reconstruction algorithm. However, z_1^* may not always be available, or may not be measurable using the technique described previously in (3.232) due to practical considerations (e.g., a video sequence may be the only available input). With a minor modification to the estimator design, the scale ambiguity can be resolved, and Euclidean coordinates of feature points recovered, if the Euclidean distance between two of the many feature points in the scene is available. Assuming z_1^* is not directly measurable, terms in the velocity kinematics of (3.230) can be re-arranged in the following manner

$$\dot{e} = \bar{J}_c \bar{v}, \tag{3.260}$$

where the Jacobian $\bar{J}_c(t) \in \mathbb{R}^{6 \times 6}$ and a scaled velocity vector $\bar{v}(t) \in \mathbb{R}^6$ are defined as

$$\bar{J}_c = \begin{bmatrix} -\alpha_1 A_{e1} & A_{e1}[m_1]_\times \\ 0_3 & -L_\omega \end{bmatrix} \tag{3.261}$$

$$\bar{v} = \begin{bmatrix} \bar{v}_c^T & \omega_c^T \end{bmatrix}^T, \text{and } \bar{v}_c = \frac{v_c^T}{z_1^*}. \tag{3.262}$$

The velocity observer of Section 3.6.2 can now be utilized to identify the scaled velocity $\bar{v}(t)$. Likewise, the time derivative of the extended image coordinates presented in (3.233) can be re-written in terms of the scaled velocity as

$$\dot{p}_{ei} \;=\; W_{1i}\bar{v}_c\bar{\theta}_i + W_{2i}\omega_c \tag{3.263}$$

$$\text{where } \bar{\theta}_i \;=\; \frac{z_1^*}{z_i^*}. \tag{3.264}$$

The rest of the development is identical. Hence, it is clear from (3.264) that the adaptive estimator of (3.243) now identifies the depth of all feature points scaled by a constant scalar z_1^*. If the Euclidean distance between any two features i and j $(i \neq j)$ are known, then from (3.243), (3.259) and (3.264) it is clear that

$$\lim_{t\to\infty} \left\| \hat{\bar{m}}_i^*(t) - \hat{\bar{m}}_j^*(t) \right\| = z_i^* \left\| \bar{m}_i^* - \bar{m}_j^* \right\|. \tag{3.265}$$

Hence, the unknown scalar z_1^* can be recovered and the Euclidean coordinates of all feature points obtained. Hence, the six degrees-of-freedom position of the moving camera relative to its reference position can be computed online.

3.6.4   Simulations and Experimental Results

A practical implementation of the estimator consists of at least four subcomponents: (a) hardware for image acquisition, (b) implementation of an algorithm to extract and track feature points between image frames, (c) the implementation of the depth estimation algorithm itself, and (d) a method to display the reconstructed scene (3D models, depth maps, etc.). This section presents simulation results for the fixed camera case, and the camera-in-hand camera case. Experimental results are provided for the camera-in-hand system.

Experimental testbed: As shown in Figure 3.43, the camera-in-hand system consists of a calibrated monochrome CCD camera (Sony XC-ST50) mounted on the end-effector of a Puma 560 industrial robotic manipulator whose end-effector was commanded via a PC to move along a smooth trajectory. The camera was interfaced to a second PC dedicated to image processing, and equipped with an Imagenation PXC200AF framegrabber board capable of acquiring images in real time (30 fps) over the PCI bus. A 20 Hz digital signal from the robot control PC triggered the framegrabber to acquire images. The same trigger signal also recorded the actual robot end-effector velocity (which is same as the camera velocity) into a file, which was utilized as ground truth to validate the performance of the estimator in identifying the camera velocity. The computational complexity of feature tracking and depth estimator algorithms made it unfeasible

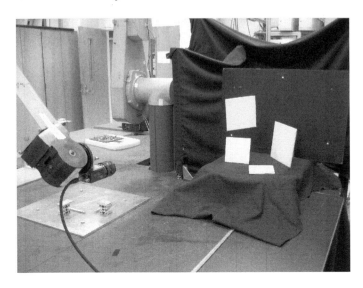

FIGURE 3.43. The Experimental Testbed.

to acquire and process images at the frame rate of the camera. Hence the sequence of images acquired from the camera were encoded into a video file (AVI format, utilizing the lossless Huffman codec for compression, where possible) to be processed offline later.

Feature tracking: The image sequences were at least a minute long, and due to thousands of images that must be processed in every sequence, an automatic feature detector and tracker was necessary. We utilized an implementation of the Kanade-Lucas-Tomasi feature tracking algorithm [77] available at [3] for detection and tracking of feature points from one frame to the next. The libavformat and libavcodec libraries from the FFmpeg project [49] were utilized to extract individual frames from the video files. The output of the feature point tracking stage was a data file containing image space trajectories of all successfully tracked feature points from the video sequence. This data served as the input to the depth estimation algorithm.

Depth Estimator: The adaptive estimation algorithm described in Section 3.6.2 was implemented in C++ and ran at a sampling frequency of 1 kHz to guarantee sufficient accuracy from the numerical integrators in the estimator. A linear interpolator, followed by 2nd order low pass filtering was used to interpolate 20 Hz image data to 1 kHz required as input to the estimator.

Display: A 3D graphical display based on the OpenGL API was developed in order to render the reconstructed scene. A surface model of objects in the scene can be created by generating a "mesh" from the reconstructed feature points. Using OpenGL libraries, it is easy to develop a display that allows the user to view the objects in the scene from any vantage point by virtually navigating around the objects in a scene using an input device such as a computer mouse pointer.

Simulation Results: For simulations, the image acquisition hardware and the image processing step were both replaced with a software component that generates feature trajectories utilizing object kinematics. For the fixed camera simulation, a planar object was selected to be the body undergoing motion – four feature points initially 2 meters away along the axis of the camera were selected to visually track the object. The velocity of the object along each of the six degrees of freedom were set to $0.2\sin(t)$. The coordinates of the object feature points in the object's coordinate frame $\mathcal{F}$ were arbitrarily chosen as

$$
\begin{aligned}
s_1 &= \begin{bmatrix} 1.0 & 0.5 & 0.1 \end{bmatrix}^T \\
s_2 &= \begin{bmatrix} 1.2 & -0.75 & 0.1 \end{bmatrix}^T \\
s_3 &= \begin{bmatrix} 0.0 & -1.0 & 0.1 \end{bmatrix}^T \\
s_4 &= \begin{bmatrix} -1.0 & 0.5 & 0.1 \end{bmatrix}^T.
\end{aligned}
\tag{3.266}
$$

The object's reference orientation R^* relative to the camera was selected as $diag(1, -1, -1)$, where $diag(.)$ denotes a diagonal matrix with arguments as the diagonal entries. The estimator gain β_i was set to 20 for all i feature points. As an example, Figure 3.44 depicts the convergence of $\hat{\theta}_2(t)$ from which the Euclidean coordinates of the second feature point s_2 could be computed as shown in (3.205). Similar graphs were obtained for convergence of the estimates for the remaining feature points. In the simulation for the camera-in-hand system, a non-planar stationary object was selected that could be identified by 12 feature points. Figure 3.45 shows the convergence of the inverse depth estimates, and Figure 3.46 shows the estimation errors.

Experimental Results: An experimental scene with a dollhouse as shown in Figure 3.47 was utilized to verify the practical performance of the camera-in-hand Euclidean estimator. Figure 3.48 shows the reconstructed wire-frame view of the dollhouse in the scene, displayed using the OpenGL-based viewer. Table 3.1 shows a comparison between the actual and the estimated lengths in the scene. The time evolution of the inverse depth estimates is shown in Figure 3.49.

Discussion: The simulation and experimental results in Figures 3.44, 3.45, 3.46, and Table 3.1 clearly demonstrate good performance of the estimator. All estimates converge to their expected values in a span of a few

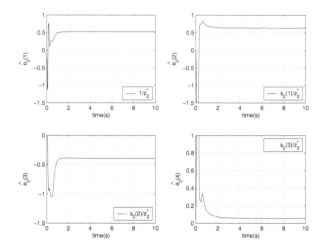

FIGURE 3.44. The Estimated Parameters for the Second Feature Point on the Moving Object (Fixed Camera Simulation).

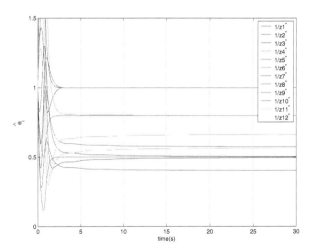

FIGURE 3.45. Inverse Depth Estimates for all Feature Points on the Stationary Object (Camera-in-Hand Simulation).

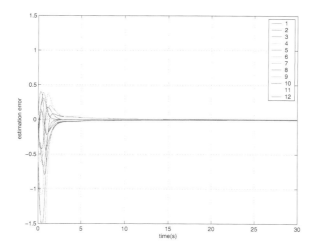

FIGURE 3.46. Error in Depth Estimation Reduce to Zero with Time (Camera-in-Hand Simulation).

FIGURE 3.47. One of the Frames from the Dollhouse Video Sequence. The White Dots Overlayed on the Image are the Tracked Feature Points.

FIGURE 3.48. A Wire-Frame Reconstruction of the Doll-House Scene (Camera-in-Hand Experiment).

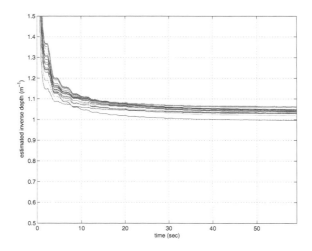

FIGURE 3.49. The Time Evolution of Inverse Depth Estimates from the Camera-in-Hand Experiment.

Object	Actual (cm.)	Estimated (cm.)
Length I	23.6	23.0
Length II	39.7	39.7
Length III	1.0	1.04
Length IV	13.0	13.2
Length V	100.0	99.0
Length VI	19.8	19.6
Length VII	30.3	30.0

TABLE 3.1. Estimated Dimensions from the Scene (Camera-in-Hand Experiment).

seconds. The only difference between the simulations and the experiment is the source of input pixel data. As mentioned previously, the image space feature point trajectories for the simulations were generated based on rigid body kinematics and known dimensions of the object. In the case of experiments, the image space trajectories were obtained from the feature tracker that was run on the image sequence previously recorded from the camera. One of the most challenging aspects in a real-world implementation of the estimator is accurate feature tracking. Features that can be tracked well by a tracking algorithm may not adequately describe the geometrical structure of the object, and vice versa. Quality of feature tracking also depends on factors such as lighting, shadows, magnitude of inter-frame motion, and texturedness, to name a few. Additionally, any compression (such as MPEG) employed in the input video sequence will introduce additional artifacts that can potentially degrade performance of the estimator. There is no one common solution to this problem, and the parameters of the tracker must be tuned to the specific image dataset in hand to obtain best results. See [99] for a discussion of issues related to feature selection and tracking. Apart from the accuracy in feature tracking that directly affects the accuracy in online estimation of the homography matrix relating corresponding feature points, the performance of the estimator also depends on accurate camera calibration. Since the implementation is the same for simulations and the experiment, the errors in estimation (Table 3.1) can mostly be attributed to the jitter in the output of the feature tracker.

3.7   Notes

Two mainstream visual servo control methods include image-based and position-based approaches. Overviews of these methods are provided in

[66, 61, 11]. A third mainstream class of visual servo controllers uses some image-space information combined with some reconstructed information as a means to combine the advantages of image and position-based approaches while avoiding their disadvantages (e.g., [12, 17, 15, 26, 34, 44, 51, 64, 84, 83, 85]). A particular subset of this approach was coined 2.5D visual servo control in [12, 84, 83, 85] because this class of controllers exploits some two dimensional image feedback and reconstructed three-dimensional feedback. This class of controllers is also called homography-based visual servo control in [17, 15, 44, 64] because of the underlying reliance of the construction and decomposition of a homography. As stated in [85], some advantages of this methodology over the aforementioned approaches are that an accurate Euclidean model of the environment (or target image) is not required and potential singularities in the image-Jacobian are eliminated (i.e., the image-Jacobian for homography-based visual servo controllers is typically triangular). Based on the observation that interaction between the translation and rotation of images can result in slower transient performance due to inefficient camera motions, Deguchi proposed two algorithms in [34] for a robot manipulator application that decouple the rotation and translation components using a homography and an epipolar condition. More recently, Corke and Hutchinson [26] also developed a method for decoupling the rotation and translation components from the remaining degrees of freedom.

Motivated by the desire to compensate for the uncertain depth information, homography-based control methods have been developed that can actively adapt for the unknown depth parameter (cf. [17, 15, 24, 25, 43, 44, 63, 64]) along with methods based on direct estimation and approximation (cf. [42, 66]). For example, [24] developed an adaptive kinematic controller for a robot manipulator application to ensure uniformly ultimately bounded (UUB) set-point regulation of the image point errors while compensating for the unknown depth information, provided conditions on the translational velocity and the bounds on uncertain depth parameters are satisfied. In [43] and [44], Fang et al. developed homography-based visual servo controllers to asymptotically regulate a manipulator end-effector and a mobile robot, respectively, by developing an adaptive update law that actively compensates for an unknown depth parameter. In [45], Fang et al. also developed a camera-in-hand regulation controller that incorporated a robust control structure to compensate for uncertainty in the extrinsic calibration parameters. Adaptive homography-based visual servo controllers have been developed in [17, 15] to achieve asymptotic tracking control of a manipulator end-effector and a mobile robot, respectively. Adaptive homography-based visual servo controllers based on the quaternion rotation parameterization were developed in [63, 64] to achieve asymptotic regulation and tracking control of a manipulator end-effector, respectively.

Motivated by the need for new advancements to meet visual servo tracking applications, previous research has concentrated on developing different types of path planning techniques in the image-space (e.g., see [27], [91], [100]). More recently, Mezouar and Chaumette developed a path-following image-based visual servo algorithm in [87] where the path to a goal point is generated via a potential function that incorporates motion constraints. In [29], Cowan et al. develop a hybrid position/image-space controller that forces a manipulator to a desired setpoint while avoiding obstacles and ensuring the object remains in the field-of-view by avoiding pitfalls such as self-occlusion. Related research that focuses on vision-based controllers that focus on tracking a prerecorded sequence of images or reference path is provided in [86], [101], and [114].

Wheeled mobile robots (mobile robots) are often required to execute tasks in environments that are unstructured. Due to the uncertainty in the environment, numerous researchers have investigated different sensing modalities as a means to enable improved autonomous response by the system. Given this motivation, researchers initially targeted the use of a variety of sonar and laser-based sensors. Some initial work also targeted the use of a fusion of various sensors to build a map of the environment for mobile robot navigation (see [62], [74], [108], [111], [113], and the references within; other early innovative mobile robot control research is given in [70]). While this is still an active area of research, various shortcomings associated with these technologies and recent advances in image extraction/interpretation technology and advances in control theory have motivated researchers to investigate the sole use of camera-based vision systems for autonomous navigation. For example, using consecutive image frames and an object database, the authors of [72] recently proposed a monocular visual servo tracking controller for mobile robots based on a linearized system of equations and Extended Kalman Filtering (EKF) techniques. Also using EKF techniques on the linearized kinematic model, the authors of [30] used feedback from a monocular omnidirectional camera system (similar to [2]) to enable wall following, follow-the-leader, and position regulation tasks. In [55], Hager et al. used a monocular vision system mounted on a pan-tilt-unit to generate image-Jacobian and geometry-based controllers by using different snapshots of the target and an epipolar constraint. As stated in [10], a drawback of the method developed in [55] is that the system equations became numerically ill-conditioned for large pan angles. Given this shortcoming, Burschka and Hager [10] used a spherical image projection of a monocular vision system that relied on teaching and replay phases to facilitate the estimation of the unknown object height parameter in the image-Jacobian by solving a least-squares problem. Spatiotemporal apparent velocities obtained from an optical flow of successive images of an object were used in [104] to estimate the depth and time-to-contact to develop a monocular vision "guide robot" that is used as a guide for blind users. A similar optical flow technique was also used in [78]. In [40], Dixon et al. used feedback from an uncalibrated, fixed (ceiling-mounted) camera to develop an adaptive tracking controller for a mobile robot that compensated for the parametric uncertainty in the camera and the mobile robot dynamics. An image-based visual servo controller that exploits an object model was proposed in [112] to solve the mobile robot tracking controller (the regulation problem was not solved due to restrictions on the reference trajectory) that adapted for the constant, unknown height of an object moving in a plane through Lyapunov-based techniques. In [110], an image-based visual servo controller was proposed for a mobile manipulator application; however, the result requires geometric distances associated with the object to be known, and relies on an image-Jacobian that contains singularities for some configurations. Moreover, the result in [110] requires the additional degrees-of-freedom from the manipulator to regulate the orientation of the camera. In [81] and [117], visual servo controllers were recently developed for systems with similar underactuated kinematics as mobile robots. Specifically, Mahony and Hamel [81] developed a semi-global asymptotic visual servoing result for unmanned aerial vehicles that tracked parallel coplanar linear visual features while Zhang and Ostrowski [117] used a vision system to navigate a blimp.

In addition to the visual servo control problem, image-feedback can also be used for state estimation (e.g., the structure-from-motion problem). Although the problem is inherently nonlinear, typical structure-from-motion results use linearization-based methods such as extended Kalman filtering (EKF) [8, 23, 89]. Some of these approaches combine epipolar geometry with the EKF to refine the estimate of an Essential Matrix to determine the position and orientation (pose) and/or velocity [103]. Other methods use external pose estimation schemes and use image features as inputs to the Kalman filter with pose and/or velocity as an output [1, 9, 23]. Some researchers have recast the problem as the state estimation of a continuous-time perspective dynamic system, and have employed nonlinear system analysis tools in the development of state observers that identify motion and structure parameters [65, 67]. These papers show that if the velocity of the moving object (or camera) is known, and satisfies certain observability conditions, an estimator for the unknown Euclidean position of the feature points can be developed. In [19], an observer for the estimation of camera motion was presented based on perspective observations of a single feature point from the (single) moving camera. The observer development was based on sliding mode and adaptive control techniques, and it was shown that upon satisfaction of a persistent excitation condition [102], the rotational velocity could be fully recovered; furthermore, the translational velocity could be recovered up to a scale factor. Other nonlinear estimators/observers have been developed in [18, 22, 41, 67, 79]. Several researches have developed nonlinear observers for state estimation with such omnidirectional cameras (e.g., [54] and [80]).

References

[1] A. Azarbayejani and A. P. Pentland, "Recursive estimation of motion, structure, and focal length," *IEEE Transactions on Pattern Analysis and Machine Intelligence*, Vol. 17, No. 6, pp. 562–575, 1995.

[2] S. Baker and S. Nayar, "A Theory of Catadioptric Image Formation," *Proceedings of the ICCV*, pp. 35–42, Jan. 1998.

[3] S. Birchfield, *KLT: An Implementation of the Kanade-Lucas-Tomasi Feature Tracker*, http://www.ces.clemson.edu/~stb/klt.

[4] C. M. Bishop, *Pattern Recognition and Machine Learning*, Springer, 2007.

[5] J.-Y. Bouguet, *Camera Calibration Toolbox for MATLAB*, Public Domain Internet Software.

[6] T. L. Boullion, and P. L. Odell, *Generalized Inverse Matrices*, Wiley, New York, 1971.

[7] R. Brockett, S. Millman and H. J. Sussmann, *Differential Geometric Control Theory*, Birkhauser, Boston, 1990.

[8] T. J. Broida, S. Chandrashekhar, and R. Chellappa, "Recursive 3-D Motion Estimation From a Monocular Image Sequence," *IEEE Transactions on Aerospace and Electronic Systems*, Vol. 26, No. 4, pp. 639–656, 1990.

[9] T. Broida and R. Chellappa, "Estimating the kinematics and structure of a rigid object from a sequence of monocular images," *IEEE Transactions on Pattern Analysis and Machine Intelligence*, Vol. 13, No. 6, pp. 497–513, 1991.

[10] D. Burschka and G. Hager, "Vision-Based Control of Mobile Robots," *Proceedings of the IEEE International Conference on Robotics and Automation*, pp. 1707–1713, 2001.

[11] F. Chaumette, and S. Hutchinson, "Visual servo control part I: Basic approaches," *IEEE Robotics and Automation Magazine*, Vol. 13, pp. 82–90, 2006.

[12] F. Chaumette and E. Malis, "2 1/2 D Visual Servoing: A Possible Solution to Improve Image-based and Position-based Visual Servoings," *Proceedings of the IEEE International Conference on Robotics and Automation*, pp. 630–635, 2000.

[13] F. Chaumette, E. Malis, and S. Boudet, "2D 1/2 Visual Servoing with Respect to a Planar Object," *Proceedings of the Workshop on New Trends in Image-Based Robot Servoing*, pp. 45–52, 1997.

[14] J. Chen, A. Behal, D. Dawson, and Y. Fang, "2.5D Visual Servoing with a Fixed Camera," *Proceedings of the American Control Conference*, Denver, Colorado, June 2003, pp. 3442–3447.

[15] J. Chen, W. E. Dixon, D. M. Dawson, and M. McIntire, "Homography-based Visual Servo Tracking Control of a Wheeled Mobile Robot," *IEEE Transactions on Robotics*, Vol 22, pp. 406–415, 2006.

[16] J. Chen, W. E. Dixon, D. M. Dawson, and V. Chitrakaran, "Navigation Function Based Visual Servo Control," *Proceedings of the American Control Conference*, pp. 3682–3687, Portland, USA, 2005.

[17] J. Chen, D. Dawson, W. E. Dixon, and A. Behal, "Adaptive Homography-based Visual Servo Tracking for a Fixed Camera Configuration with a Camera-in-Hand Extension," *IEEE Transactions on Control Systems Technology*, Vol. 13, pp. 814–825, 2005.

[18] X. Chen and H. Kano, "A New State Observer for Perspective Systems," *IEEE Transactions on Automatic Control*, Vol. 47, pp. 658–663, 2002.

[19] X. Chen, and H. Kano, "State Observer for a Class of Nonlinear Systems and Its Application to Machine Vision," *IEEE Transactions on Automatic Control*, Vol. 49, No. 11, pp. 2085–2091, 2004.

[20] G. S. Chirikjian, "Theory and Applications of Hyper-Redundant Robotic Manipulators," Ph.D. Thesis, Department of Applied Mechanics, California Institute of Technology, June 1992.

[21] V. K. Chitrakaran, D. M. Dawson, J. Chen, and W. E. Dixon, "Euclidean Position Estimation of Features on a Moving Object Using a Single Camera: A Lyapunov-Based Approach," *Proceedings of the American Control Conference*, pp. 4601–4606, June 2005.

[22] V. Chitrakaran, D. M. Dawson, W. E. Dixon, and J. Chen, "Identification of a Moving Object's Velocity with a Fixed Camera," *Automatica*, Vol. 41, No. 3, pp. 553–562, 2005.

[23] A. Chiuso, P. Favaro, H. Jin, and S. Soatto, "Structure from Motion Causally Integrated Over Time," *IEEE Transactions on Pattern Analysis and Machine Intelligence*, Vol. 24, pp. 523–535, 2002.

[24] F. Conticelli and B. Allotta, "Nonlinear Controllability and Stability Analysis of Adaptive Image-Based Systems," *IEEE Transactions on Robotics and Automation*, Vol. 17, No. 2, April 2001.

[25] F. Conticelli and B. Allota, "Discrete-Time Robot Visual Feedback in 3-D Positioning Tasks with Depth Adaptation," *IEEE/ASME Transactions on Mechatronics*, Vol. 6, No. 3, pp. 356–363, Sept. 2001.

[26] P. Corke and S. Hutchinson, "A new partitioned approach to image-based visual servo control," *IEEE Transactions on Robotics and Automation*, Vol. 17, pp. 507–515, 2001.

[27] N. J. Cowan and D. Koditschek, "Planar Image-based Visual Servoing as a Navigation Problem," *Proceedings of the IEEE International Conference on Robotics and Automation*, pp. 1720–1725, San Francisco, CA, Apr. 2000.

[28] N. J. Cowan, O. Shakernia, R. Vidal, and S. Sastry, "Vision-Based Follow-the-Leader," *Proceedings of the International Conference on Intelligent Robots and Systems*, Las Vegas, Nevada, pp. 1796–1801, October 2003.

[29] N. J. Cowan, J. D. Weingarten, and D. E. Koditschek, "Visual servoing via Navigation Functions," *IEEE Transactions on Robotics and Automation*, Vol. 18, pp. 521–533, Aug. 2002.

[30] A. K. Das, et al., "Real-Time Vision-Based Control of a Nonholonomic Mobile Robot," *Proceedings of the IEEE International Conference on Robotics and Automation*, pp. 1714–1719, 2001.

[31] J. B. C. Davies, "A Flexible Motion Generator," PhD Thesis, Heriot-Watt University, Edinburgh, 1996.

[32] E. R. Davies, *Machine Vision: Theory, Algorithms, Practicalities*, Morgan Kaufmann, 2004.

[33] M. de Queiroz, D. Dawson, S. Nagarkatti, and F. Zhang, *Lyapunov-based Control of Mechanical Systems*, Birkhauser, New York, 2000.

[34] K. Deguchi, "Optimal Motion Control for Image-Based Visual Servoing by Decoupling Translation and Rotation," *Proceedings of the International Conference on Intelligent Robots and Systems*, pp. 705–711, Oct. 1998.

[35] C. A. Desoer, and M. Vidyasagar, *Feedback Systems: Input-Output Properties*, Academic Press, 1975.

[36] G. Dissanayake, P. Newman, S. Clark, H. F. Durrant-Whyte, and M. Csorba, "A Solution to the Simultaneous Localisation and Map Building (SLAM) Problem," *IEEE Transactions in Robotics and Automation*, Vol. 17, No. 3, pp. 229–241, 2001.

[37] W. E. Dixon, M. S. de Queiroz, D. M. Dawson, and T. J. Flynn, "Adaptive Tracking and Regulation Control of a Wheeled Mobile Robot with Controller/Update Law Modularity," *IEEE Transactions on Control Systems Technology*, Vol. 12, No. 1, pp. 138–147, 2004.

[38] W. E. Dixon, D. M. Dawson, E. Zergeroglu and A. Behal, *Nonlinear Control of Wheeled Mobile Robots*, Springer-Verlag London Limited, 2001.

[39] W. E. Dixon, A. Behal, D. M. Dawson, and S. Nagarkatti, *Nonlinear Control of Engineering Systems: A Lyapunov-Based Approach*, Birkhäuser Boston, 2003.

[40] W. E. Dixon, D. M. Dawson, E. Zergeroglu, and A. Behal, "Adaptive Tracking Control of a Wheeled Mobile Robot via an Uncalibrated Camera System," *IEEE Transactions on Systems, Man, and Cybernetics -Part B: Cybernetics*, Vol. 31, No. 3, pp. 341–352, 2001.

[41] W. E. Dixon, Y. Fang, D. M. Dawson, and T. J. Flynn, "Range Identification for Perspective Vision Systems," *IEEE Transactions on Automatic Control*, Vol. 48, pp. 2232–2238, 2003.

[42] B. Espiau, F. Chaumette, and P. Rives, "A New Approach to Visual Servoing in Robotics," *IEEE Transactions on Robotics and Automation*, Vol. 8, pp. 313–326, 1992.

[43] Y. Fang, A. Behal, W. E. Dixon and D. M. Dawson, "Adaptive 2.5D Visual Servoing of Kinematically Redundant Robot Manipulators," *Proceedings of the IEEE Conference on Decision and Control*, Las Vegas, NV, December 2002, pp. 2860–2865.

[44] Y. Fang, W. E. Dixon, D. M. Dawson, and P. Chawda, "Homography-based visual servoing of wheeled mobile robots," *IEEE Transactions on Systems, Man, and Cybernetics -Part B: Cybernetics*, Vol. 35, pp. 1041–1050, 2005.

[45] Y. Fang, W. E. Dixon, D. M. Dawson, and J. Chen, "Robust 2.5D Visual Servoing for Robot Manipulators," *Proceedings of the American Control Conference*, Denver, Colorado, pp. 3311–3316, June 2003.

[46] O. Faugeras, *Three-Dimensional Computer Vision*, The MIT Press, Cambridge Massachusetts, 2001.

[47] O. Faugeras and F. Lustman, "Motion and Structure From Motion in a Piecewise Planar Environment," *International Journal of Pattern Recognition and Artificial Intelligence*, Vol. 2, No. 3, pp. 485–508, 1988.

[48] C. A. Felippa, *A Systematic Approach to the Element-Independent Corotational Dynamics of Finite Elements*, Center for Aerospace Structures Document Number CU-CAS-00-03, College of Engineering, University of Colorado, January 2000.

[49] FFmpeg Project, http://ffmpeg.sourceforge.net/index.php.

[50] N. R. Gans and S. A. Hutchinson, "An Asymptotically Stable Switched System Visual Controller for Eye in Hand Robots," *Proceedings of the IEEE/RSJ International Conference on Intelligent Robots and Systems*, Las Vegas, Nevada, pp. 735–742, Oct. 2003.

[51] N. Gans, and S. Hutchinson, "Stable visual servoing through hybrid switched-system control," *IEEE Transactions on Robotics*, Vol. 23, pp. 530–540, 2007.

[52] R. C. Gonzalez and R. E. Woods, *Digital Image Processing*, Prentice Hall, 3rd edition, 2007.

[53] I. Gravagne, I. D. Walker, "On the Kinematics of Remotely-Actuated Continuum Robots," *Proceedings of the IEEE International Conference on Robotics and Automation*, San Francisco, pp. 2544–2550, May 2000.

[54] S. Gupta, D. Aiken, G. Hu, and W. E. Dixon, "Lyapunov-Based Range and Motion Identification For A Nonaffine Perspective Dynamic System," *Proceedings of the American Control Conference*, pp. 4471–4476, 2006.

[55] G. D. Hagar, D. J. Kriegman, A. S. Georghiades, and O. Ben-Shahar, "Toward Domain-Independent Navigation: Dynamic Vision and Control," *Proceedings of the IEEE Conference on Decision and Control*, pp. 3257–3262, 1998.

[56] M. W. Hannan, I. D. Walker, "Analysis and Experiments with an Elephant's Trunk Robot," *International Journal of the Robotics Society of Japan*, Vol. 15, No. 8, pp. 847–858, 2001.

[57] M.W. Hannan, I. D. Walker, "Real-time Shape Estimation for Continuum Robots Using Vision," *Robotica*, Vol. 23, No. 5, pp. 645–651, Sept. 2005.

[58] M. W. Hannan, "Theory and Experiments with an 'Elephant's Trunk' Robotic Manipulator," PhD Thesis, Department of Electrical and Computer Engineering, Clemson University, 2002.

[59] M. Hannan, I. D. Walker, "Novel Kinematics for Continuum Robots," *Proceedings of the International Symposium on Advances in Robot Kinematics*, Piran, Slovenia, pp. 227–238, June 2000.

[60] R. I. Hartley, "In Defense of the Eight-Point Algorithm," *IEEE Transactions on Pattern Analysis and Machine Intelligence*, Vol. 19, No. 6, pp. 580–593, 1997.

[61] K. Hashimoto, "A review on vision-based control of robot manipulators," *Advanced Robotics*, Vol. 17, pp. 969–991, 2003.

[62] M. Hebert, "3-D Vision for Outdoor Navigation by an Autonomous Vehicle," *Proceedings of the Image Understanding Workshop*, Cambridge, UK, 1998.

[63] G. Hu, W. E. Dixon, S. Gupta, and N. Fitz-coy, "A quaternion formulation for homography-based visual servo control," *Proceedings of the IEEE International Conference Robotics and Automation*, pp. 2391–2396, 2006.

[64] G. Hu, S. Gupta, N. Fitz-coy, and W. E. Dixon, "Lyapunov-based visual servo tracking control via a quaternion formulation," *Proceedings of the IEEE Conference Decision and Control*, pp. 3861–3866, 2006.

[65] X. Hu, and T. Ersson, "Active State Estimation of Nonlinear Systems," *Automatica*, Vol. 40, pp. 2075–2082, 2004.

[66] S. Hutchinson, G. Hager, and P. Corke, "A tutorial on visual servo control," *IEEE Transactions on Robotics and Automation*, Vol. 12, pp. 651–670, 1996.

[67] M. Jankovic, and B. K. Ghosh, "Visually Guided Ranging from Observations of Points, Lines and Curves via an Identifier Based Nonlinear Observer," *Systems and Control Letters*, Vol. 25, pp. 63–73, 1995.

[68] B. A. Jones, "Kinematics and Implementation of Continuum Manipulators," PhD Thesis, Department of Electrical and Computer Engineering, Clemson University, 2005.

[69] D. Jung, J. Heinzmann, and A. Zelinsky, "Range and Pose Estimation for Visual Servoing of a Mobile Robot," *Proceedings of the IEEE International Conference on Robotics and Automation*, pp. 1226–1231, 1998.

[70] Y. Kanayama, Y. Kimura, F. Miyazaki, T. Noguchi, "A Stable Tracking Control Method for an Autonomous Mobile Robot," *Proceedings of the IEEE International Conference on Robotics and Automation*, Cincinnati, Ohio, May 1990, pp. 253–258.

[71] H. K. Khalil, *Nonlinear Systems*, 3rd edition, Prentice Hall, 2002.

[72] B. H. Kim, et al., "Localization of a Mobile Robot using Images of a Moving Target," *Proceedings of the IEEE International Conference on Robotics and Automation*, pp. 253–258, 2001.

[73] M. Kircanski, and M. Vukobratovic, "Contribution to Control of Redundant Robotic Manipulators in an Environment with Obstacles," *International Journal of Robotics Research*, Vol. 5, No. 4. pp. 112–119, 1986.

[74] D. J. Kriegman, E. Triendl, and T. O. Binford, "Stereo Vision Navigation in Buildings for Mobile Robots," *IEEE Transactions on Robotics and Automation*, Vol. 5, No. 6, pp. 792–803, Dec. 1989.

[75] M. Krstić, I. Kanellakopoulos, and P. Kokotović, *Nonlinear and Adaptive Control Design*, New York, NY: John Wiley and Sons, 1995.

[76] M. S. Loffler, N. P. Costescu, and D. M. Dawson, "QMotor 3.0 and the QMotor Robotic Toolkit: a PC-Based Control Platform," *IEEE Control Systems Magazine*, vol. 22, no. 3, pp. 12–26, 2002.

[77] B. D. Lucas and T. Kanade, "An Iterative Image Registration Technique with an Application to Stereo Vision," *International Joint Conference on Artificial Intelligence*, pp. 674–679, 1981.

[78] Y. Ma, J. Kosecka, and S. Sastry, "Vision Guided Navigation for Nonholonomic Mobile Robot," *IEEE Transactions on Robotics and Automation*, Vol. 15, No. 3, pp. 521–536, June 1999.

[79] L. Ma, Y. Chen, and K. L. Moore, "Range Identification for Perspective Dynamic System with Single Homogeneous Observation," *Proceedings of the IEEE International Conference Robotics and Automation*, pp. 5207–5212, 2004.

[80] L. Ma, Y. Chen, and K. L. Moore, "Range Identification for Perspective Dynamic Systems with 3D Imaging Surfaces," *Proceedings of the American Control Conference*, pp. 3671–3675, 2005.

[81] R. Mahony and T. Hamel, "Visual Servoing Using Linear Features for Under-actuated Rigid Body Dynamics," *Proceedings of the IEEE/RJS International Conference on Intelligent Robots and Systems*, pp. 1153–1158, 2001.

[82] E. Malis, *Contributions à la modélisation et à la commande en asservissement visuel*, Ph.D. Dissertation, University of Rennes I, IRISA, France, Nov. 1998.

[83] E. Malis and F. Chaumette, "Theoretical Improvements in the Stability Analysis of a New Class of Model-Free Visual Servoing Methods," *IEEE Transactions on Robotics and Automation*, Vol. 18, No. 2, pp. 176–186, April 2002.

[84] E. Malis and F. Chaumette, "2 1/2 D Visual Servoing with Respect to Unknown Objects Through a New Estimation Scheme of Camera Displacement," *International Journal of Computer Vision*, Vol. 37, No. 1, pp. 79–97, June 2000.

[85] E. Malis, F. Chaumette, and S. Bodet, "2 1/2 D Visual Servoing," *IEEE Transactions on Robotics and Automation*, Vol. 15, No. 2, pp. 238–250, April 1999.

[86] Y. Matsutmoto, M. Inaba, and H. Inoue, "Visual Navigation using View-Sequenced Route Representation," *Proceedings of IEEE International Conference on Robotics and Automation*, pp. 83–88, 1996.

[87] Y. Mezouar and F. Chaumette, "Path Planning for Robust Image-based Control," *IEEE Transactions on Robotics and Automation*, Vol. 18, No. 4, pp. 534–549, August 2002.

[88] M. Nixon and A. S Aguado, *Feature Extraction in Computer Vision and Image Processing*, Newnes, 2002.

[89] J. Oliensis, "A Critique of Structure From Motion Algorithms," *Computer Vision and Image Understanding*, Vol. 80, No. 2, pp. 172–214, 2000.

[90] J.R. Parker, *Algorithms for Image Processing and Computer Vision*, Wiley, 1996.

[91] E. Rimon and D. E. Koditschek, "Exact Robot Navigation Using Artificial Potential Functions," *IEEE Transactions on Robotics and Automation,* Vol. 8, pp. 501–518, Oct. 1992.

[92] G. Robinson, J. B. C. Davies, "Continuum Robots — A State of the Art," *Proceedings of the IEEE Conference on Robotics and Automation*, pp. 2849–2854, 1999.

[93] A. Ruf and R. Horaud, "Visual Trajectories From Uncalibrated Stereo," *IEEE International Conference Intelligent Robots Systems*, pp. 83–91, Sept. 1997.

[94] J. C. Russ, *The Image Processing Handbook*, CRC, 5th edition, 2006.

[95] C. Samson, "Velocity and Torque Feedback Control of a Nonholonomic Cart," *Proceedings of the International Workshop in Adaptive and Nonlinear Control: Issues in Robotics*, Grenoble, France, 1990.

[96] S. Sastry, and M. Bodson, *Adaptive Control: Stability, Convergence, and Robustness*, Prentice Hall, Inc: Englewood Cliffs, NJ, 1989.

[97] B. Siciliano, "Kinematic Control of Redundant Robot Manipulators: A Tutorial," *Journal of Intelligent and Robotic Systems*, Vol. 3, pp. 201–212, 1990.

[98] L. G. Shapiro and G. C. Stockman, *Computer Vision*, Prentice Hall, 2001.

[99] J. Shi, and C. Tomasi, "Good Features to Track," *Proceedings of the IEEE Conference on Computer Vision and Pattern Recognition*, pp. 593–600, 1994.

[100] R. Singh, R. M. Voyle, D. Littau, and N. P. Papanikolopoulos, "Alignment of an Eye-In-Hand System to Real Objects using Virtual Images," *Proceedings of the Workshop Robust Vision, Vision-based Control of Motion, IEEE International Conference Robotics Automation*, May 1998.

[101] S. B. Skaar and J-D. Yoder, "Extending Teach-Repeat to Nonholonomic Robots," *Structronic Systems: Smart Structures, Devices & Systems, Part II*, pp. 316–342, *Series on Stability, Vibration and Control of Systems*, Series B: Vol. 4. 1998.

[102] J. J. E. Slotine and W. Li, *Applied Nonlinear Control*, Prentice Hall, Inc: Englewood Cliffs, NJ, 1991.

[103] S. Soatto, R. Frezza, and P. Perona, "Motion estimation via dynamic vision," *IEEE Transactions on Automatic Control*, Vol. 41, No. 3, pp. 393–413, 1996.

[104] K.-T. Song and J.-H. Huang, "Fast Optical Flow Estimation and Its Application to Real-time Obstacle Avoidance," *Proceedings of the IEEE International Conference on Robotics and Automation*, pp. 2891–2896, 2001.

[105] M. W. Spong and M. Vidyasagar, *Robot Dynamics and Control*, John Wiley and Sons, Inc: New York, NY, 1989.

[106] R. Sukthankar, R. Stockton, and M. Mullin, "Smarter Presentations: Exploiting Homography in Camera-Projector Systems," *Proceedings of the International Conference on Computer Vision*, 2001.

[107] A. R. Teel, R. M. Murray, and C. G. Walsh, "Non-holonomic Control Systems: From Steering to Stabilization with Sinusoids," *International Journal of Control*, Vol. 62, No. 4, pp. 849–870, 1995.

[108] C. E. Thorpe, M. Hebert, T. Kanade, and S. Shafer, "Vision and Navigation for the Carnegie-Mellon Navlab," *IEEE Transactions on Pattern Analysis and Machine Intelligence*, Vol. 10, No. 3, pp. 362–373, May 1988.

[109] Trucco, A. Verri, *Introductory Techniques for 3-D Computer Vision*, Prentice Hall, 1998.

[110] D. Tsakiris, P. Rives, and C. Samson, "Extending Visual Servoing Techniques to Nonholonomic Mobile Robots" pp. 107–117 in *The Confluence of Vision and Control, Lecture Notes in Control and Information System*, D. Kriegman (Ed.), Springer Verlag, 1998.

[111] M. A. Turk, D. G. Morgenthaler, K. D. Gremban, and M. Marra, "VITS-A Vision System for Autonomous Land Vehicle Navigation," *IEEE Transactions on Pattern Analysis and Machine Intelligence*, Vol. 10, No. 3, pp. 342–361, May 1988.

[112] H. Y. Wang, S. Itani, T. Fukao, and N. Adachi, "Image-Based Visual Adaptive Tracking Control of Nonholonomic Mobile Robots," *Proceedings of the IEEE/RJS International Conference on Intelligent Robots and Systems*, pp. 1–6, 2001.

[113] A. M. Waxman, et al., "A Visual Navigation System for Autonomous Land Vehicles," *IEEE Journal of Robotics Automation*, Vol. RA-3, No. 2, pp. 124–141, 1987.

[114] J.-D. Yoder, E. T. Baumgartner, and S. B. Skaar, "Initial Results in the Development of a Guidance System for a Powered Wheelchair," *IEEE Transactions on Rehabilitation Engineering*, Vol. 4, No. 3, pp. 143–151, 1996.

[115] T. Yoshikawa, "Analysis and Control of Robot Manipulators with Redundancy," *Robotics Research - The First International Symposium*, MIT Press, pp. 735–747, 1984.

[116] Z. Zhang and A. R. Hanson, "Scaled Euclidean 3D Reconstruction Based on Externally Uncalibrated Cameras," *IEEE Symposium on Computer Vision*, pp. 37–42, 1995.

[117] H. Zhang and J. P. Ostrowski, "Visual Servoing with Dynamics: Control of an Unmanned Blimp," *Proceedings of the IEEE International Conference on Robotics and Automation*, pp. 618–623, 1999.

4

Path Planning and Control

4.1 Introduction

In this chapter, we discuss the important problems of path planning and control for manipulator arms and wheeled mobile robots (WMRs). In the context of the path planning problem for robots, the basic challenge relates to the issue of finding a solution around obstacles in the robot's workspace whose locations are known *a priori*. The next obvious problem that one is naturally drawn to relates to determining and following a path (possibly to a setpoint) in an unstructured environment that may be cluttered with obstacles whose *a priori* locations are unknown. Furthermore, the introduction of a dynamic environment leads to a more challenging problem that may impose a time constraint on the path being followed. This chapter deals with integrated path planning and control design when the obstacle locations are known, as well as the more interesting case of an unstructured environment where fixed or in-hand/onboard vision is used as an active feedback element in addition to standard feedback sensors such as encoders and tachometers.

Traditionally, robot control research has focused on the position tracking problem where the objective is to force the robot's end-effector to follow an *a priori* known desired time-varying trajectory. Since the objective is encoded in terms of a time dependent trajectory, the robot may be forced to follow an unknown course to catch up with the desired trajectory in the presence of a large initial error. As an example, a radial reduction phenomenon has been reported by several researchers [47], [48] in which the actual path followed has a smaller radius than the specified trajectory. In light of this phenomenon, the control objective for many robotic tasks are more appropriately encoded as a contour following problem in which the objective is to force the robot to follow a state-dependent function that describes the contour — these are problems where strict time parameterization of the contour is not a critical factor. One example of a control strategy aimed at the contour following problem is velocity field control (VFC) where the desired contour is described by a velocity tangent vector [49]. The advantages of the VFC approach are: (a) more effective penalization of the robot for leaving the desired contour through the velocity field error, (b) specification of the control task invariant of the task execution speed, and (c) more explicit task coordination and synchronization. In the first part of this chapter, we illustrate how an example adaptive controller (e.g., the benchmark adaptive tracking controller presented in [67]) for a robot manipulator can be modified to incorporate trajectory planning techniques with the controller. Specifically, the benchmark adaptive controller given in [67] is modified to yield VFC in the presence of parametric uncertainty. Velocity field tracking is achieved by incorporating a norm squared gradient term in the control design that is used to prove that the link positions are bounded through the use of a Lyapunov-analysis. In addition to VFC, some task objectives are motivated by the need to follow a trajectory to a desired goal configuration while avoiding known obstacles in the configuration space. For this class of problems, it is more important for the robot to follow an obstacle free path to the desired goal point than it is to either follow a contour or meet a time-based requirement. In the second part of Section 4.2, this class of problems is addressed via the use of a navigation function (NF) which is a special kind of potential function with a refined mathematical structure guaranteeing the existence of a unique minimum [39, 64]. The NF is then used to modify the benchmark adaptive controller in [67] to track a reference trajectory that yields a collision free path to a constant goal point in an obstacle cluttered environment with known obstacles. In Section 4.3, we address the obstacle avoidance problem for WMRs using both VFC and NF based methods. The focus in this part is to demonstrate how these techniques can be applied in a nonholonomic setting such as the one provided by a WMR.

After discussing path planning and control in *a priori* mapped environments, the chapter will then shift focus to problems where the robot finds itself in an unstructured environment but has the ability to see and recognize features in its environment — these features are exploited for path planning as well as for real-time execution of the planned trajectory, thereby allowing the robot the twin benefits of being able to work in multiple settings as well as being robust to modeling errors and uncertainties, i.e., the robot has reliably autonomous functionality. The need for improved robot autonomy has led researchers to investigate the basic science challenges leading to the development of visual servo controllers. In general, visual servo controllers can be divided into position-based visual servo (PBVS) control, image-based visual servo (IBVS), and hybrid approaches. PBVS is based on the idea of using a vision system to reconstruct the Euclidean-space and then developing the servo controller on the reconstructed information. An issue with this strategy is that the target object may exit the camera field-of-view (FOV). Another issue is that a 3D model of the object is generally required for its implementation. IBVS control is based on the idea of directly servoing on the image-space information, with reported advantages of increased robustness to camera calibration and improved capabilities to ensure the target remains visible. Even for IBVS controllers that are formulated as regulation controllers, excessive control action and transient response resulting from a large initial error can cause the target to leave the FOV, and may lead to trajectories that are not physically valid or optimal due to the nonlinearities and potential singularities associated with the transformation between the image space and the Euclidean-space [6]. In light of the characteristics of IBVS and PBVS, several researchers have recently explored hybrid approaches. Homography-based visual servo control techniques (coined 2.5D controllers) have been recently developed that exploit a robust combination of reconstructed Euclidean information and image-space information in the control design. The Euclidean information is reconstructed by decoupling the interaction between translational and rotational components of a homography matrix. Some advantages of this methodology over the aforementioned IBVS and PBVS approaches are that an accurate Euclidean model of the environment (or target object) is not required, and potential singularities in the image-Jacobian are eliminated (i.e., the image-Jacobian for homography-based visual servo controllers is typically triangular) [55].

While homography-based approaches exploit the advantages of IBVS and PBVS, a common problem with all the aforementioned approaches is the inability to achieve the control objective while ensuring the target features remain visible. Section 4.4 of this chapter presents a solution to the problem of navigating the position and orientation of a camera held by the end-effector of a robot manipulator to a goal position and orientation along the desired image-space trajectory while ensuring the target points remain visible (i.e., the target points avoid self-occlusion and remain in the field-of-view (FOV)) under certain technical restrictions. The desired trajectory is generated in the image space based on a measurable image Jacobian-like matrix and an image space NF while satisfying rigid body constraints. The final part of this chapter (Section 4.5) deals with the design of a PBVS controller that does not require a 3D object model and is able to tackle nonlinear radial distortion and uncertainty in the camera calibration. This strategy is motivated by the need to circumvent the singularity issues associated with IBVS due to the use of the non-square image Jacobian. The idea is to utilize an image-based optimization algorithm that searches for the unknown desired task-space setpoint. The control strategy designed leads to exponential stability of the desired setpoint.

4.2 Velocity Field and Navigation Function Control for Manipulators

Motivated by manipulator task objectives that are more effectively described by on-line, state-dependent trajectories, two adaptive tracking controllers that accommodate on-line path planning objectives are developed in this section. First, an example adaptive controller is modified to achieve velocity field tracking in the presence of parametric uncertainty in the robot dynamics. From a review of VFC literature, it can be determined that previous research efforts have focused on ensuring the robot tracks the velocity field, but no development has been provided to ensure the link position remains bounded. A proportional-integral controller developed in [4] achieves semiglobal practical stabilization of the velocity field tracking errors despite uncertainty in the robot dynamics; however, the link position boundedness issue is addressed by an assumption that the boundedness of the norm

$$\left\| q(0) + \int_0^t \vartheta(q(\sigma))d\sigma \right\| \tag{4.1}$$

yields globally bounded trajectories, where $q(t)$ denotes the position, and $\vartheta(\cdot)$ denotes the velocity field. In lieu of the assumption in (4.1), the VFC development described here is based on the selection of a velocity field that is first order differentiable, and the existence of a first order differentiable, nonnegative function $V(q) \in \mathbb{R}$ such that the following inequality holds

$$\frac{\partial V(q)}{\partial q}\vartheta(q) \leq -\gamma_3(\|q\|) + \zeta_0 \tag{4.2}$$

where $\frac{\partial V(q)}{\partial q}$ denotes the partial derivative of $V(q)$ with respect to $q(t)$, $\gamma_3(\cdot) \in \mathbb{R}$ is a class $\mathcal{K}$ function[1], and $\zeta_0 \in \mathbb{R}$ is a nonnegative constant. That is, the assumption in (4.1) is replaced by a stability-based condition on the velocity field. It is interesting to note that the velocity field described in the experimental results provided in [4] can be shown to satisfy the stability-based condition in (4.2) (see Section B.2.1 in Appendix B for proof). A two part stability analysis is provided to demonstrate the global uniform boundedness (GUB) of the link position as well as the convergence of the velocity field tracking error to zero despite parametric uncertainty in the dynamic model. Experimental results based on the velocity field presented in [4] are provided to demonstrate validation of the VFC approach. An extension is then provided that targets the trajectory planning problem where the task objective is to move to a goal configuration while avoiding known obstacles. Specifically, an adaptive navigation function based controller is designed to provide a path from an initial condition inside the free configuration space of the robot manipulator to the goal configuration. This analysis proves that all the system states are bounded, and that the robot manipulator will track an obstacle-free path to a goal point despite parametric uncertainty in the dynamic model. Experimental results for the adaptive navigation function controller are provided to demonstrate the validity of the approach.

4.2.1   System Model

The mathematical model for an n-DOF robotic manipulator is assumed to have the following form

$$M(q)\ddot{q} + V_m(q, \dot{q})\dot{q} + G(q) = \tau. \tag{4.3}$$

In (4.3), $q(t)$, $\dot{q}(t)$, $\ddot{q}(t) \in \mathbb{R}^n$ denote the link position, velocity, and acceleration, respectively, $M(q) \in \mathbb{R}^{n \times n}$ represents the positive-definite, symmetric inertia matrix, $V_m(q, \dot{q}) \in \mathbb{R}^{n \times n}$ represents the centripetal-Coriolis terms, $G(q) \in \mathbb{R}^n$ represents the known gravitational vector, and $\tau(t) \in \mathbb{R}^n$ represents the torque input vector. The system states, $q(t)$ and $\dot{q}(t)$ are assumed to be measurable. It is also assumed that $M(q)$, $V_m(q, \dot{q})$, and $G(q) \in \mathcal{L}_\infty$ provided $q(t)$, $\dot{q}(t) \in \mathcal{L}_\infty$. The dynamic model in (4.3) exhibits the following properties that are utilized in the subsequent control development and stability analysis.

---

[1]A continuous function $\alpha : [0, \alpha) \to [0, \infty)$ is said to belong to class $\mathcal{K}$ if it is strictly increasing and $\alpha(0) = 0$ [35].

Property 4.2.1: The inertia matrix can be upper and lower bounded by
the following inequalities [45]

$$m_1 \|\xi\|^2 \leq \xi^T M(q)\xi \leq m_2(q)\|\xi\|^2 \quad \forall \xi \in \mathbb{R}^n \tag{4.4}$$

where m_1 is a positive constant, $m_2(\cdot)$ is a positive function, and $\|\cdot\|$
denotes the Euclidean norm.

Property 4.2.2: The inertia and the centripetal-Coriolis matrices satisfy
the following relationship [45]

$$\xi^T \left(\frac{1}{2}\dot{M}(q) - V_m(q,\dot{q}) \right) \xi = 0 \quad \forall \xi \in \mathbb{R}^n \tag{4.5}$$

where $\dot{M}(q)$ represents the time derivative of the inertia matrix.

Property 4.2.3: The robot dynamics given in (4.3) can be linearly parameterized as follows [45]

$$Y(q,\dot{q},\ddot{q})\theta \triangleq M(q)\ddot{q} + V_m(q,\dot{q})\dot{q} + G(q) \tag{4.6}$$

where $\theta \in \mathbb{R}^p$ contains constant system parameters, and $Y(q,\dot{q},\ddot{q}) \in \mathbb{R}^{n \times p}$ denotes a regression matrix composed of $q(t)$, $\dot{q}(t)$, and $\ddot{q}(t)$.

4.2.2 Adaptive VFC Control Objective

As described previously, many robotic tasks can be effectively encapsulated as a velocity field. That is, the velocity field control objective can be
described as commanding the robot manipulator to track a velocity field
which is defined as a function of the current link position. To quantify this
objective, a velocity field tracking error, denoted by $\eta_1(t) \in \mathbb{R}^n$, is defined
as follows

$$\eta_1(t) \triangleq \dot{q}(t) - \vartheta(q) \tag{4.7}$$

where $\vartheta(\cdot) \in \mathbb{R}^n$ denotes the velocity field. To achieve the control objective,
the subsequent development is based on the assumption that $q(t)$ and $\dot{q}(t)$
are measurable, and that $\vartheta(q)$ and its partial derivative $\dfrac{\partial \vartheta(q)}{\partial q} \in \mathbb{R}^n$ are
assumed to be bounded provided $q(t) \in \mathcal{L}_\infty$.

Benchmark Control Modification

To develop the open-loop error dynamics for $\eta_1(t)$, we take the time derivative of (4.7) and pre-multiply the resulting expression by the inertia matrix

as follows

$$M(q)\dot{\eta}_1 = -V_m(q,\dot{q})\dot{q} - G(q) + \tau + V_m(q,\dot{q})\vartheta(q) \qquad (4.8)$$
$$-V_m(q,\dot{q})\vartheta(q) - M(q)\frac{\partial\vartheta(q)}{\partial q}\dot{q}$$

where (4.3) was utilized. From (4.7), the expression in (4.8) can be rewritten as follows

$$M(q)\dot{\eta}_1 = -V_m(q,\dot{q})\eta_1 - Y_1(q,\dot{q})\theta + \tau \qquad (4.9)$$

where θ was introduced in (4.6) and $Y_1(q,\dot{q}) \in \mathbb{R}^{n\times p}$ denotes a measurable regression matrix that is defined as follows

$$Y_1(q,\dot{q})\theta \triangleq M(q)\frac{\partial\vartheta(q)}{\partial q}\dot{q} + V_m(q,\dot{q})\vartheta(q) + G(q). \qquad (4.10)$$

Based on the open-loop error system in (4.9), a number of control designs could be utilized to ensure velocity field tracking (i.e., $\|\eta_1(t)\| \to 0$) given the assumption in (4.1). Motivated by the desire to eliminate the assumption in (4.1), a norm squared gradient term is incorporated in an adaptive controller introduced in [67] as follows

$$\tau(t) \triangleq -\left(K + \left\|\frac{\partial V(q)}{\partial q}\right\|^2 I_n\right)\eta_1 + Y_1(q,\dot{q})\hat{\theta}_1 \qquad (4.11)$$

where $K \in \mathbb{R}^{n\times n}$ is a constant, positive definite diagonal matrix, $I_n \in \mathbb{R}^{n\times n}$ is the standard $n \times n$ identity matrix, and $\frac{\partial V(q)}{\partial q}$ was introduced in (4.2). In (4.11), $\hat{\theta}_1(t) \in \mathbb{R}^p$ denotes a parameter estimate that is generated by the following gradient update law

$$\dot{\hat{\theta}}_1(t) = -\Gamma_1 Y_1^T(q,\dot{q})\eta_1 \qquad (4.12)$$

where $\Gamma_1 \in \mathbb{R}^{p\times p}$ is a constant, positive definite diagonal matrix. After substituting (4.11) into (4.9), the following closed-loop error system can be obtained

$$M(q)\dot{\eta}_1 = -V_m(q,\dot{q})\eta_1 - Y_1(q,\dot{q})\tilde{\theta}_1 - \left(K + \left\|\frac{\partial V(q)}{\partial q}\right\|^2 I_n\right)\eta_1 \qquad (4.13)$$

where the parameter estimation error signal $\tilde{\theta}_1(t) \in \mathbb{R}^p$ is defined as follows

$$\tilde{\theta}_1(t) \triangleq \theta - \hat{\theta}_1. \qquad (4.14)$$

Remark 4.1 *It is required for the selection of a particular $\vartheta(q)$ and $V(q)$ that the inequality as defined in (4.2) must hold. In the event that this condition does not hold, the tracking objective is not guaranteed as described by the subsequent stability analysis.*

Remark 4.2 *While the control development is based on a modification of the adaptive controller introduced in [67], the norm squared gradient term could also be incorporated in other benchmark controllers to yield similar results (e.g., sliding mode controllers).*

Stability Analysis

To facilitate the subsequent stability analysis, the following preliminary theorem is utilized.

Theorem 4.1 *Let $\bar{V}(t) \in \mathbb{R}$ denote the following nonnegative, continuous differentiable function*

$$\bar{V}(t) \triangleq V(q) + P(t)$$

where $V(q) \in \mathbb{R}$ denotes a nonnegative, continuous differentiable function that satisfies (4.2) and the following inequalities

$$0 \leq \gamma_1(\|q\|) \leq V(q) \leq \gamma_2(\|q\|)$$

where $\gamma_1(\cdot)$, $\gamma_2(\cdot)$ are class $\mathcal{K}$ functions, and $P(t) \in \mathbb{R}$ denotes the following nonnegative, continuous differentiable function

$$P(t) \triangleq \gamma - \int_{t_0}^{t} \varepsilon^2(\sigma)d\sigma \qquad (4.15)$$

where $\gamma \in \mathbb{R}$ is a positive constant, and $\varepsilon(t) \in \mathbb{R}$ is defined as follows

$$\varepsilon \triangleq \left\| \frac{\partial V(q)}{\partial q} \right\| \|\eta_1\|. \qquad (4.16)$$

If $\varepsilon(t)$ is a square integrable function, where

$$\int_{t_0}^{t} \varepsilon^2(\sigma)d\sigma \leq \gamma,$$

and if after utilizing (4.7), the time derivative of $\bar{V}(t)$ satisfies the following inequality

$$\dot{\bar{V}}(t) \leq -\gamma_3(\|q\|) + \xi_0 \qquad (4.17)$$

where $\gamma_3(q)$ is the class $\mathcal{K}$ function introduced in (4.2), and $\xi_0 \in \mathbb{R}$ denotes a positive constant, then $q(t)$ is globally uniformly bounded (GUB).

Proof: The time derivative of $\bar{V}(t)$ can be expressed as follows

$$\dot{\bar{V}}(t) = \frac{\partial V(q)}{\partial q} \vartheta(q) + \frac{\partial V(q)}{\partial q} \eta_1 - \varepsilon^2(t)$$

where (4.7) and (4.15) were utilized. By exploiting the inequality introduced in (4.2) and the definition for $\varepsilon(t)$ provided in (4.16), the following inequality can be obtained

$$\dot{V}(t) \le -\gamma_3(\|q\|) + \zeta_0 + \left[\varepsilon(t) - \varepsilon^2(t)\right]. \tag{4.18}$$

After completing the squares on the bracketed terms in (4.18), the inequality introduced in (4.17) is obtained where $\xi_0 \triangleq \zeta_0 + \frac{1}{4}$. Hence, if $\varepsilon(t) \in \mathcal{L}_2$, Section B.2.2 in Appendix B can be used to prove that $q(t)$ is GUUB. ∎

In the following analysis, we first prove that $\varepsilon(t) \in \mathcal{L}_2$. Based on the conclusion that $\varepsilon(t) \in \mathcal{L}_2$, the result from Theorem 4.1 is utilized to ensure that $q(t)$ is bounded under the proposed adaptive controller given in (4.11) and (4.12).

Theorem 4.2 *The adaptive VFC given in (4.11) and (4.12) yields global velocity field tracking in the sense that*

$$\|\eta_1(t)\| \to 0 \quad as \quad t \to \infty. \tag{4.19}$$

Proof: Let $V_1(t) \in \mathbb{R}$ denote the following nonnegative function

$$V_1 \triangleq \frac{1}{2}\eta_1^T M \eta_1 + \frac{1}{2}\tilde{\theta}_1^T \Gamma_1^{-1} \tilde{\theta}_1. \tag{4.20}$$

After taking the time derivative of (4.20) the following expression can be obtained

$$\dot{V}_1 = -\eta_1^T \left(Y_1(q,\dot{q})\tilde{\theta}_1 + \left(K + \left\|\frac{\partial V(q)}{\partial q}\right\|^2 I_n \right) \eta_1 \right) - \tilde{\theta}_1^T \Gamma_1^{-1} \dot{\tilde{\theta}}_1 \tag{4.21}$$

where (4.5) and (4.13) were utilized. After utilizing the parameter update law given in (4.12), the expression given in (4.21) can be rewritten as follows

$$\dot{V}_1 = -\eta_1^T \left(K + \left\|\frac{\partial V(q)}{\partial q}\right\|^2 I_n \right) \eta_1. \tag{4.22}$$

The expressions given in (4.16), (4.20), and (4.22) can be used to conclude that $\eta_1(t), \tilde{\theta}_1(t) \in \mathcal{L}_\infty$ and $\eta_1(t), \varepsilon(t) \in \mathcal{L}_2$. Based on the fact that $\varepsilon(t) \in \mathcal{L}_2$, the results from Theorem 4.1 can be used to prove that $q(t) \in \mathcal{L}_\infty$. Since $q(t) \in \mathcal{L}_\infty$, the assumption that $\vartheta(q)$ and $\frac{\partial \vartheta(q)}{\partial q} \in \mathcal{L}_\infty$ can be used to conclude that $\dot{q}(t) \in \mathcal{L}_\infty$, where the expression in (4.7) was utilized. Based on the fact that $\tilde{\theta}_1(t) \in \mathcal{L}_\infty$, the expression in (4.14) can be used to prove that $\hat{\theta}_1(t) \in \mathcal{L}_\infty$. Based on the fact that $q(t), \dot{q}(t) \in \mathcal{L}_\infty$ as well as the fact that $\vartheta(q)$, $\dfrac{\partial \vartheta(q)}{\partial q}$, $M(q)$, $V_m(q, \dot q)$, $G(q) \in \mathcal{L}_\infty$ when their arguments are bounded, (4.10) can be used to prove that $Y_1(q, \dot q) \in \mathcal{L}_\infty$. Since $Y_1(q, \dot q)$, $\hat\theta_1(t)$, $\eta_1(t)$, $\dfrac{\partial V(q)}{\partial q} \in \mathcal{L}_\infty$, (4.11) can be used to prove that $\tau(t) \in \mathcal{L}_\infty$. Based on the previous bounding statements, the expression given in (4.13) can be used to prove that $\dot\eta_1(t) \in \mathcal{L}_\infty$. Given that $\eta_1(t)$, $\dot\eta_1(t) \in \mathcal{L}_\infty$ and $\eta_1(t) \in \mathcal{L}_2$, Barbalat's Lemma [67] can be utilized to prove (4.19). ∎

4.2.3   Navigation Function Control Extension

Control Objective

The objective in this extension is to navigate a robot's end-effector along a collision-free path to a constant goal point, denoted by $q^* \in \mathcal{D}$, where the set $\mathcal{D}$ denotes a free configuration space that is a subset of the whole configuration space with all configurations removed that involve a collision with an obstacle, and $q^* \in \mathbb{R}^n$ denotes the constant goal point in the interior of $\mathcal{D}$. Mathematically, the primary control objective can be stated as the desire to ensure that

$$q(t) \to q^* \quad \text{as } t \to \infty \tag{4.23}$$

where the secondary control is to ensure that $q(t) \in \mathcal{D}$. To achieve these two control objectives, we define $\varphi(q) \in \mathbb{R}$ as a function $\varphi(q) : \mathcal{D} \to [0,\,1]$ that is assumed to satisfy the following properties: [39]

Property 4.2.4: The function $\varphi(q)$ is first order and second order differentiable Morse function (i.e., $\dfrac{\partial}{\partial q}\varphi(q)$ and $\dfrac{\partial}{\partial q}\left(\dfrac{\partial}{\partial q}\varphi(q)\right)$ exist on $\mathcal{D}$).

Property 4.2.5: The function $\varphi(q)$ obtains its maximum value on the boundary of $\mathcal{D}$.

Property 4.2.6: The function $\varphi(q)$ has a unique global minimum at $q(t) = q^*$.

Property 4.2.7: If $\dfrac{\partial}{\partial q}\varphi(q) = 0$, then $q(t) = q^*$.

Based on (4.23) and the above definition, an auxiliary tracking error signal, denoted by $\eta_2(t) \in \mathbb{R}^n$, can be defined as follows to quantify the control objective

$$\eta_2(t) \triangleq \dot q(t) + \nabla\varphi(q) \tag{4.24}$$

where $\nabla \varphi(q) = \dfrac{\partial}{\partial q} \varphi(q)$ denotes the gradient vector of $\varphi(q)$ defined as follows

$$\nabla \varphi(q) \triangleq \left[\begin{array}{cccc} \dfrac{\partial \varphi}{\partial q_1} & \dfrac{\partial \varphi}{\partial q_2} & \cdots & \dfrac{\partial \varphi}{\partial q_n} \end{array} \right]^T. \qquad (4.25)$$

Remark 4.3 *As discussed in [64], the construction of the navigation function $\varphi(q)$ that satisfies all of the above properties for a general obstacle avoidance problem is nontrivial. Indeed, for a typical obstacle avoidance, it does not seem possible to construct $\varphi(q)$ such that $\dfrac{\partial}{\partial q}\varphi(q) = 0$ only at $q(t) = q^*$. That is, as discussed in [64], the appearance of interior saddle points seems to be unavoidable; however, since the set of saddle points has measure zero, the practical impact is minimal and consequently, $\varphi(q)$ can be constructed as shown in [64] such that only a minuscule set of initial conditions will result in convergence to the unstable equilibria.*

Remark 4.4 *The two control developments presented in the preceding sections appear to be mathematically similar (i.e., (4.7) and (4.24)), but the control objectives are very different. The VFC objective is to achieve robot end-effector velocity tracking with a desired trajectory generated by a velocity field, $\vartheta(q)$, hence, there is no explicit goal point. The navigation function control development utilizes a special function $\varphi(q)$, that has specific properties such that the robot's end-effector finds a collision free path to a known goal point, q^* and stops. Each signal, $\vartheta(q)$ for the VFC development, and $\varphi(q)$ for the navigation function control development must meet a set of qualifying conditions (i.e. the inequality of (4.2) for the VFC and Properties 4.2.4 – 4.2.7 for the NF control), but these conditions are not the same, therefore the two objectives are very different.*

Benchmark Control Modification

To develop the open-loop error dynamics for $\eta_2(t)$, we take the time derivative of (4.24) and premultiply the resulting expression by the inertia matrix as follows

$$M\dot{\eta}_2 = -V_m(q, \dot{q})\eta_2 + Y_2(q, \dot{q})\theta + \tau. \qquad (4.26)$$

where (4.3) and (4.24) were utilized. In (4.26), the linear parameterization $Y_2(q, \dot{q})\theta$ is defined as follows

$$Y_2(q, \dot{q})\theta \triangleq M(q)f(q, \dot{q}) + V_m(q, \dot{q}) \nabla \varphi(q) - G(q) \qquad (4.27)$$

where $Y_2(q, \dot{q}) \in \mathbb{R}^{n \times m}$ denotes a measurable regression matrix, $\theta \in \mathbb{R}^m$ was introduced in (4.6), and the auxiliary signal $f(q, \dot{q}) \in \mathbb{R}^n$ is defined as

$$f(q, \dot{q}) \triangleq \frac{d}{dt}\left(\nabla \varphi(q)\right)$$
$$= H(q)\dot{q} \tag{4.28}$$

where the Hessian matrix $H(q) \in \mathbb{R}^{n \times n}$ is defined as follows

$$H(q) \triangleq \begin{bmatrix} \dfrac{\partial^2 \varphi}{\partial q_1^2} & \dfrac{\partial^2 \varphi}{\partial q_1 \partial q_2} & \cdots & \dfrac{\partial^2 \varphi}{\partial q_1 \partial q_n} \\ \dfrac{\partial^2 \varphi}{\partial q_2 \partial q_1} & \dfrac{\partial^2 \varphi}{\partial q_2^2} & \cdots & \dfrac{\partial^2 \varphi}{\partial q_2 \partial q_n} \\ \cdots & \cdots & \cdots & \cdots \\ \dfrac{\partial^2 \varphi}{\partial q_n \partial q_1} & \cdots & \cdots & \dfrac{\partial^2 \varphi}{\partial q_n^2} \end{bmatrix}.$$

Based on (4.26) and the subsequent stability analysis, the following adaptive controller introduced in [67] can be utilized

$$\tau \triangleq -k\eta_2 - Y_2(q, \dot{q})\hat{\theta}_2 \tag{4.29}$$

where $k \in \mathbb{R}$ is a positive constant gain, and $\hat{\theta}_2(t) \in \mathbb{R}^m$ denotes a parameter update law that is generated from the following expression

$$\dot{\hat{\theta}}_2(t) \triangleq \Gamma_2 Y_2^T(q, \dot{q})\eta_2 \tag{4.30}$$

where $\Gamma_2 \in \mathbb{R}^{m \times m}$ is a positive definite, diagonal gain matrix. Note that the trajectory planning is incorporated in the controller through the gradient terms included in (4.27) and (4.28). After substituting (4.29) into (4.26), the following closed loop error systems can be obtained

$$M\dot{\eta}_2 = -V_m(q, \dot{q})\eta_2 - k\eta_2 + Y_2(q, \dot{q})\tilde{\theta}_2 \tag{4.31}$$

where $\tilde{\theta}_2(t) \in \mathbb{R}^p$ is defined as follows

$$\tilde{\theta}_2(t) \triangleq \theta - \hat{\theta}_2. \tag{4.32}$$

Stability Analysis

Theorem 4.3 *The adaptive controller given in (4.29) and (4.30) ensures that the robot manipulator tracks an obstacle free path to the unique goal configuration in sense that*

$$q(t) \to q^* \ as \ t \to \infty$$

provided the control gain k introduced in (4.29) is selected sufficiently large.

Proof: Let $V_2(q, \eta_2, \tilde{\theta}_2) \in \mathbb{R}$ denote the following nonnegative function

$$V_2 \triangleq \varphi(q) + \gamma \left[\frac{1}{2} \eta_2^T M \eta_2 + \tilde{\theta}_2^T \Gamma_2^{-1} \tilde{\theta}_2 \right]. \tag{4.33}$$

where $\gamma \in \mathbb{R}$ is an adjustable, positive constant. After taking the time derivative of (4.33) the following expression can be obtained

$$\dot{V}_2 = [\nabla\varphi(q)]^T \dot{q} + \gamma \eta_2^T \left(-k\eta_2 + Y_2(q, \dot{q})\tilde{\theta}_2 \right) - \gamma \tilde{\theta}_2^T \Gamma_2^{-1} \dot{\hat{\theta}}_2$$

where (4.5), (4.25), (4.31), and (4.32) were utilized. By utilizing (4.24), (4.30), the following expression can be obtained

$$\dot{V}_2 = -\|\nabla\varphi(q)\|^2 - \gamma k \|\eta_2\|^2 + [\nabla\varphi(q)]^T \eta_2.$$

The expression above can be further simplified as follows

$$\dot{V}_2 \leq -\frac{1}{2} \|\nabla\varphi(q)\|^2 - (\gamma k - 2) \|\eta_2\|^2 \tag{4.34}$$

where the following upper bound was utilized

$$[\nabla\varphi(q)]^T \eta_2 \leq \frac{1}{2} \|\nabla\varphi(q)\|^2 + 2 \|\eta_2\|^2.$$

Provided k is selected sufficiently large to satisfy

$$k > \frac{2}{\gamma}, \tag{4.35}$$

it is clear from (4.4), (4.33), and (4.34) that

$$0 \leq \varphi(q(t)) + \gamma\zeta(q,t) \leq \varphi(q(0)) + \gamma\zeta(q(0),0) \tag{4.36}$$

where $\zeta(q,t) \in \mathbb{R}$ is defined as

$$\zeta(q,t) \triangleq \left[\frac{m_2(q)}{2} \|\eta_2(t)\|^2 + \lambda_{\max}\{\Gamma_2^{-1}\} \left\| \tilde{\theta}_2(t) \right\|^2 \right]. \tag{4.37}$$

From (4.32), (4.36), and (4.37) it is clear that $\eta_2(t), \varphi(q), \tilde{\theta}_2(t), \hat{\theta}_2(t) \in \mathcal{L}_\infty$. Let the region $\mathcal{D}_0$ be defined as follows

$$\mathcal{D}_0 \triangleq \{ q(t) | 0 \leq \varphi(q(t)) \leq \varphi(q(0)) + \gamma\zeta(q(0),0) \}. \tag{4.38}$$

Hence, (4.33), (4.34), and (4.36) can be utilized to show that $q(t) \in \mathcal{D}_0$ provided $q(0) \in \mathcal{D}_0$ (i.e., $q(t) \in \mathcal{D}_0 \; \forall q(0) \in \mathcal{D}_0$). Hereafter, we restrict the remainder of the analysis to be valid in the region $\mathcal{D}_0$. Based on Property 4.2.4 given above, we know that $\nabla\varphi(q) \in \mathcal{L}_\infty$. Since $\eta_2(t), \nabla\varphi(q) \in \mathcal{L}_\infty$,

(4.24) can be used to conclude that $\dot{q}(t) \in \mathcal{L}_\infty$; hence, Property 4.2.4 and (4.28) can be used to conclude that $f(q, \dot{q}) \in \mathcal{L}_\infty$. Based on the fact that $M(q)$, $V_m(q, \dot{q})$, $G(q)$, $\nabla\varphi(q)$, $f(q, \dot{q}) \in \mathcal{L}_\infty$, (4.27) can be used to prove that $Y_2(q, \dot{q}) \in \mathcal{L}_\infty$. Since $\eta_2(t)$, $Y_2(q, \dot{q})$, $\hat{\theta}_2(t) \in \mathcal{L}_\infty$, (4.29) can be used to prove that $\tau(t) \in \mathcal{L}_\infty$. Based on the previous boundedness statements, (4.31) can be used to show that $\dot{\eta}_2(t) \in \mathcal{L}_\infty$; hence, $\nabla\varphi(q)$, $\eta_2(t)$ are uniformly continuous. From (4.34), it can also be determined that $\nabla\varphi(q), \eta_2(t) \in \mathcal{L}_2$. From these facts, Barbalat's Lemma [67] can be used to show that $\nabla\varphi(q), \eta_2(t) \to 0$ as $t \to \infty$. Since $\nabla\varphi(q) \to 0$, Property 4.2.7 can be used to prove that $q(t) \to q^*$ as $t \to \infty$. To ensure that $q(t)$ will remain in a collision-free region, we must account for the effects of the $\gamma\zeta(q(0), 0)$ term introduced in the definition of the region $\mathcal{D}_0$ given in (4.38). To this end, we first define the region $\mathcal{D}_1$ as follows

$$\mathcal{D}_1 \triangleq \{q(t)|\, 0 \le \varphi(q(t)) < 1\} \tag{4.39}$$

where $\mathcal{D}_1$ denotes the largest collision-free region, which is based on the definition of the function $\varphi(q) : \mathcal{D} \to [0, 1]$. It is now clear from (4.38) and (4.39) that if the weighting constant γ is selected sufficiently small to satisfy

$$\varphi(q(0)) + \gamma\zeta(q(0), 0) < 1, \tag{4.40}$$

this would make the upper bound of $\mathcal{D}_1$ greater than the upper bound of $\mathcal{D}_0$; then $\mathcal{D}_0 \subset \mathcal{D}_1$, and therefore, the robot manipulator tracks an obstacle-free path. ∎

4.2.4 Experimental Verification

Experimental results were obtained by implementing the adaptive VFC and the navigation function controller on a Barrett Whole Arm Manipulator (WAM). The experimental testbed and results from implementing the controllers are provided in the following sections.

Experimental Setup

The WAM testbed depicted in Figure 4.1 was utilized to implement the VFC and the navigation function controller. For simplicity, 5 links of the robot were locked at a fixed, specified angle during the experiment, and the remaining links of the manipulator were used to enable the manipulator to move along a planar trajectory. Specifically, a joint-space proportional derivative (PD) controller was utilized to servo the WMR to the following initial joint configuration for the adaptive VFC experiment (in [deg])

$$q(0) = \begin{bmatrix} 0 & 90 & -90 & 60 & 90 & 20 & 0 \end{bmatrix}^T$$

and to the following joint configuration for the navigation function experiment (in [deg])

$$q(0) = \begin{bmatrix} -58.84 & 90 & 90 & 140.72 & 11.5 & 84.5 & 0 \end{bmatrix}^T.$$

Once the WAM was servoed to the initial joint configuration, links 2, 3, 5, 6 and 7 were locked, resulting in a planar configuration with links 1 and 4 (see Figure 4.1). The resulting forward kinematics and manipulator Jacobian for the planar-WAM are given as follows

$$\begin{bmatrix} x_1 \\ x_2 \end{bmatrix} = \begin{bmatrix} \ell_1 \cos(q_1) + \ell_4 \cos(q_1 + q_4) \\ \ell_1 \sin(q_1) + \ell_4 \sin(q_1 + q_4) \end{bmatrix}$$

$$J(q) = \begin{bmatrix} -\ell_1 \sin(q_1) - \ell_4 \sin(q_1 + q_4) & -\ell_4 \sin(q_1 + q_4) \\ \ell_1 \cos(q_1) + \ell_4 \cos(q_1 + q_4) & \ell_4 \cos(q_1 + q_4) \end{bmatrix} \quad (4.41)$$

where $\ell_1 = 0.558$ [m] and $\ell_4 = 0.291$ [m]. The dynamics of the planar-WAM can be expressed in the following form [68]

$$\tau = \begin{bmatrix} M_{11} & M_{12} \\ M_{21} & M_{22} \end{bmatrix} \begin{bmatrix} \ddot{q}_1 \\ \ddot{q}_4 \end{bmatrix} + \begin{bmatrix} V_{m_{11}} & V_{m_{12}} \\ V_{m_{21}} & V_{m_{22}} \end{bmatrix} \begin{bmatrix} \dot{q}_1 \\ \dot{q}_4 \end{bmatrix}$$
$$+ \begin{bmatrix} f_{d_1} & 0 \\ 0 & f_{d_4} \end{bmatrix} \begin{bmatrix} \dot{q}_1 \\ \dot{q}_4 \end{bmatrix}. \quad (4.42)$$

In (4.42), the elements of the inertia and centripetal-Coriolis matrices are given as

$$M_{11} = p_1 + 2p_2 cos(q_4)$$
$$M_{12} = p_3 + p_2 cos(q_4)$$
$$M_{21} = p_3 + p_2 cos(q_4)$$
$$M_{22} = p_3$$

$$V_{m_{11}} = -p_2 sin(q_4)\dot{q}_4$$
$$V_{m_{12}} = -p_2 sin(q_4)\dot{q}_1 - p_2 sin(q_4)\dot{q}_4$$
$$V_{m_{21}} = p_2 sin(q_4)\dot{q}_1$$
$$V_{m_{22}} = 0$$

where p_1, p_2, p_3 denote unknown constant inertial parameters, and $f_{d_1} = 6.8$ [Nm·s] and $f_{d_4} = 3.8$ [Nm·s]. The gravitational effects are not included in (4.42) due to the plane of motion of the manipulator.

The links of the WAM are driven by brushless motors supplied with sinusoidal electronic commutation. Each axis has encoders located at the motor shaft for link position measurements. Since no tachometers are present for velocity measurements, link velocity signals are calculated via a filtered backwards difference algorithm. An AMD Athlon 1.2GHz PC operating

QNX 6.2.1 RTP (Real Time Platform), a real-time micro-kernel based operating system, hosts the control, detection, and identification algorithms which were written in C++. Qmotor 3.0 [51] was used to facilitate real time graphing, data logging and on-line gain adjustment. Data acquisition and control implementation were performed at a frequency of 1.0 [kHz] using the ServoToGo I/O board.

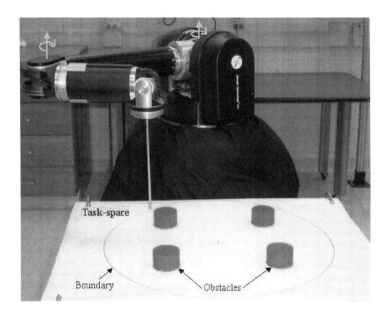

FIGURE 4.1. Front View of the Experimental Setup

Adaptive VFC Experiment

The following task-space velocity field for a planar, circular contour was utilized for the experiment [4]

$$\vartheta(x) = -2K(x)f(x)\begin{bmatrix}(x_1 - x_{c1})\\(x_2 - x_{c2})\end{bmatrix} + 2c(x)\begin{bmatrix}-(x_2 - x_{c2})\\(x_1 - x_{c1})\end{bmatrix} \qquad (4.43)$$

where $x_{c1} = 0.54$ [m] and $x_{c2} = 0.04$ [m] denote the circle center, and the functions $f(x)$, $K(x)$, and $c(x) \in \mathbb{R}$ are defined as follows [4]

$$f(x) = (x_1 - x_{c1})^2 + (x_2 - x_{c2})^2 - r_o^2 \qquad (4.44)$$

$$K(x) = \frac{k_0^*}{\sqrt{f^2(x)} \left\| \frac{\partial f(x)}{\partial x} \right\| + \epsilon}$$

$$c(x) = \frac{c_0 \exp\left(-\mu \sqrt{f^2(x)}\right)}{\left\| \frac{\partial f(x)}{\partial x} \right\|}.$$

In (4.44), $r_o = 0.2$ [m] denotes the circle diameter, and the parameters $\epsilon = 0.005$ [m^3], $\mu = 20$ [m^{-1}], $k_0^* = 0.25$ [ms^{-1}], and $c_0 = 0.25$ [ms^{-1}] were selected according to [4]. The task-space velocity field is depicted in Figure 4.2. The development in Section B.2.1 of Appendix B indicates that the velocity field in (4.43) satisfies the condition given in (4.2). To implement the adaptive VFC given in (4.11) and (4.12), the task-space velocity field is transformed into a joint-space velocity field as follows $\vartheta(q) = J^{-1}(q)\vartheta(x)$. The controller parameters were recorded as follows

$$K = diag(25, 15) \qquad \Gamma_1 = diag(3, 1, 5)$$

where $diag(\cdot)$ denotes a diagonal matrix. The resulting velocity field tracking errors are given in Figure 4.3. Figure 4.4 depicts the parameter estimates, and Figure 4.5 depicts the control torque inputs.

Adaptive Navigation Function Control Experiment

For the navigation function control experiment, four circular obstacles with known dimensions were placed in known locations in the task-space (see Figure 4.1). The actual size of the obstacles and task-space was then modified in the algorithm to accommodate for the term $\gamma \zeta(q(0), 0)$ given in (4.36) and (4.37) (i.e., the configuration-space was reduced to ensure obstacle avoidance). To modify the configuration-space according to (4.36) and (4.37), exact knowledge of the inertial parameters is required. Since these parameters are unknown, an upper bound for $\zeta(q(0), 0)$ was utilized based on known upper bounds for the inertial parameters. The modifications to the configuration-space are depicted in Figure 4.6. A task-space navigation function was developed to encapsulate the obstacles and the task-space boundary as follows

$$\varphi(x) = \frac{\|x - x^*\|^2}{\left(\|x - x^*\|^{2\kappa} + \beta_0 \beta_1 \beta_2 \beta_3 \beta_4\right)^{1/\kappa}} \qquad (4.45)$$

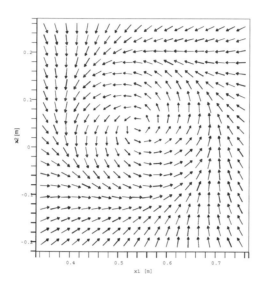

FIGURE 4.2. Desired Trajectory

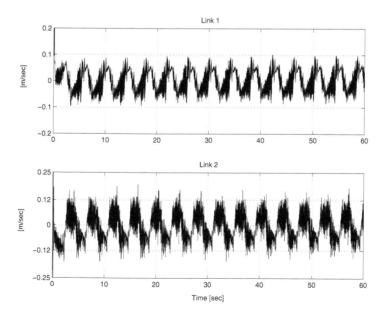

FIGURE 4.3. Velocity Field Tracking Errors

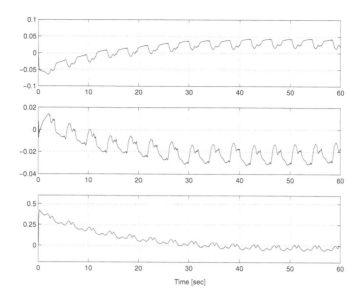

FIGURE 4.4. Parameter Estimates: (Top) $\hat{\theta}_1$, (Middle) $\hat{\theta}_2$, (Bottom) $\hat{\theta}_3$

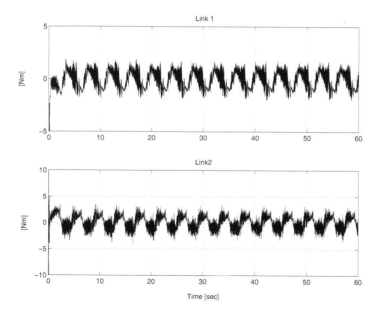

FIGURE 4.5. Control Torque Inputs

where $x(t) \triangleq [x_1(t), x_2(t)]^T \in \mathbb{R}^2$ denote the actual task-space position of the WAM end-effector, $x^* \triangleq [x_1^*, x_2^*]^T \in \mathbb{R}^2$ denotes the task-space goal position, and κ is a parameter. In [39], Koditschek proved that $\varphi(x)$ in (4.45) is a navigation function satisfying Properties 4.2.4 – 4.2.7, provided that κ is big enough; for this simulation, κ is chosen to be 14. In (4.45), the boundary function $\beta_0(x) \in \mathbb{R}$ and the obstacle functions $\beta_1(x)$, $\beta_2(x)$, $\beta_3(x)$, $\beta_4(x) \in \mathbb{R}$ are defined as follows

$$\begin{aligned}
\beta_0 &= r_0^2 - (x_1 - x_{1r_0})^2 - (x_2 - x_{2r_0})^2 & (4.46) \\
\beta_1 &= (x_1 - x_{r_1})^2 + (x_2 - x_{2r_1})^2 - r_1^2 \\
\beta_2 &= (x_1 - x_{r_2})^2 + (x_2 - x_{2r_2})^2 - r_2^2 \\
\beta_3 &= (x_1 - x_{r_3})^2 + (x_2 - x_{2r_3})^2 - r_3^2 \\
\beta_4 &= (x_1 - x_{r_4})^2 + (x_2 - x_{2r_4})^2 - r_4^2.
\end{aligned}$$

In (4.46), $(x_1 - x_{1r_i})$ and $(x_2 - x_{2r_i})$ where $i = 0, 1, 2, 3, 4$ are the respective centers of the boundary and obstacles, and r_0, r_1, r_2, r_3, $r_4 \in \mathbb{R}$ are the respective radii of the boundary and obstacles. From (4.45) and (4.46), it is clear that the model-space is a circle that excludes four smaller circles described by the obstacle functions $\beta_1(x)$, $\beta_2(x)$, $\beta_3(x)$, $\beta_4(x)$. If more obstacles are present, the corresponding obstacle functions can be easily incorporated into the navigation function [39]. Based on the known location and size of the obstacles and task-space boundary, the model-space configuration parameters were selected as follows (in [m])

$$\begin{aligned}
x_{1r_0} &= 0.5064 & x_{2r_0} &= -0.0275 & r_0 &= 0.28 \\
x_{1r_1} &= 0.63703 & x_{2r_1} &= 0.11342 & r_1 &= 0.03 \\
x_{1r_2} &= 0.4011 & x_{2r_2} &= 0.0735 & r_2 &= 0.03 \\
x_{1r_3} &= 0.3788 & x_{2r_3} &= -0.1529 & r_3 &= 0.03 \\
x_{1r_4} &= 0.6336 & x_{2r_4} &= -0.12689 & r_4 &= 0.03.
\end{aligned}$$

To implement the navigation function based controller given in (4.29) and (4.30) the joint-space dynamic model given in (4.42) was transformed to the task-space as follows [19]

$$\tau^* = M^*(x)\ddot{x} + V_m^*(x, \dot{x})\dot{x} + F_d^*\dot{x}$$

where

$$\begin{aligned}
\tau^* &= J^{-T}\tau, \quad M^* = J^{-T} \begin{bmatrix} M_{11} & M_{12} \\ M_{21} & M_{22} \end{bmatrix} J^{-1} \\
V_m^* &= J^{-T} \left(\begin{bmatrix} V_{m11} & V_{m12} \\ V_{m21} & V_{m22} \end{bmatrix} - \begin{bmatrix} M_{11} & M_{12} \\ M_{21} & M_{22} \end{bmatrix} J^{-1}\dot{j} \right) J^{-1} \\
F_d^* &= J^{-T} \begin{bmatrix} f_{d_1} & 0 \\ 0 & f_{d_4} \end{bmatrix} J^{-1}
\end{aligned}$$

where $J^{-1}(q)$ can be determined from (4.41) as follows

$$J^{-1}(q) = \begin{bmatrix} \dfrac{\cos(q_1 + q_4)}{\ell_1 \sin q_4} & \dfrac{\sin(q_1 + q_4)}{\ell_1 \sin q_4} \\ -\dfrac{\ell_1 \cos q_1 + \ell_4 \cos(q_1 + q_4)}{\ell_1 \ell_4 \sin q_4} & -\dfrac{\ell_1 \sin q_1 + \ell_4 \sin(q_1 + q_4)}{\ell_1 \ell_4 \sin q_4} \end{bmatrix}.$$

After adjusting the control gains to ensure the gain conditions (4.35) and (4.40) are satisfied, the following values were recorded

$$k = 45 \qquad \Gamma_2 = diag\,(0.02, 0.01, 0.01)\,;$$

the resulting actual trajectory of the WAM end-effector is provided in Figure 4.6. Figure 4.6 illustrates that the WAM end-effector avoids the actual obstacles as it moves to the goal point. The parameter estimates and control torque inputs are provided in Figures 4.7 and 4.8, respectively.

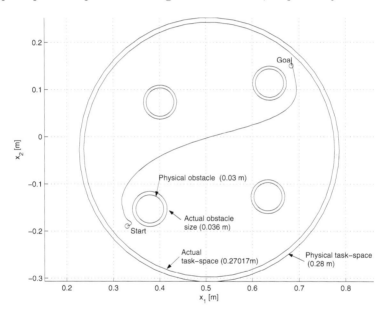

FIGURE 4.6. Actual Trajectory of the WAM Robot

Remark 4.5 *The adaptive control results achieved in Sections 4.2.2 and 4.2.3 prove only that $\dot{\hat{\theta}}_1(t)$ and $\dot{\hat{\theta}}_2(t) \to 0$ as $t \to \infty$; therefore, the unknown parameters are not guaranteed to be identified. The values of the estimates that are reached could be different from one experimental run to another. In practice, the parameter estimates may not become constant due to steady-state tracking errors.*

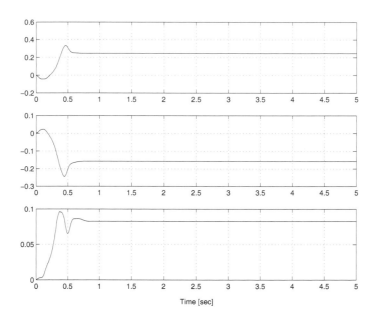

FIGURE 4.7. Parameter Estimates: (Top) $\hat{\theta}_{21}$, (Middle) $\hat{\theta}_{22}$, (Bottom) $\hat{\theta}_{23}$

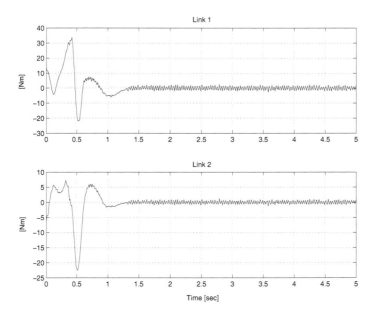

FIGURE 4.8. Control Torque Inputs

4.3 Velocity Field and Navigation Function Control for WMRs

This part describes in detail the integration of control techniques for WMRs with the path planning algorithms presented in the preceding section. Velocity Field Control is tackled first. Specifically, a velocity field is developed for the constrained WMR trajectory, and a differentiable controller is formulated to prove global asymptotic velocity field tracking. Motivated by the desire to improve the robustness of the system, the developed differentiable kinematic controller is embedded inside of an adaptive controller that fosters global asymptotic tracking despite parametric uncertainty associated with the dynamic model. In the latter half of this section, several approaches for incorporating navigation functions into different controllers are developed for task execution by a WMR in the presence of known obstacles. The first approach is based on the use of a 3-dimensional (position and orientation) navigation function that is based on desired trajectory information. The navigation function yields a path from an initial condition inside the free configuration space of the mobile robot to a stable equilibrium point. A differentiable, oscillator-based controller is then used to enable the mobile robot to follow the path and stop at the goal position. A second approach is developed for a 2-dimensional (position-based) navigation function that is constructed using sensor feedback. A differentiable controller is proposed based on this navigation function that yields asymptotic convergence. Simulation results are provided to illustrate the performance of the controller.

4.3.1 Kinematic Model

The kinematic model for a WMR subject to the nonholonomic constraints of pure rolling and nonslipping can be expressed as follows (See Figure 4.9)

$$\dot{q} = S(q)v \tag{4.47}$$

where $q(t)$, $\dot{q}(t) \in \mathbb{R}^3$ are defined as

$$q = \begin{bmatrix} x_c & y_c & \theta \end{bmatrix}^T \qquad \dot{q} = \begin{bmatrix} \dot{x}_c & \dot{y}_c & \dot{\theta} \end{bmatrix}^T. \tag{4.48}$$

In (4.47) and (4.48), $x_c(t), y_c(t)$, and $\theta(t) \in \mathbb{R}$ denote the Cartesian position and orientation of the WMR, $\dot{x}_c(t), \dot{y}_c(t)$ denote the Cartesian components of the linear velocity of the COM, $\dot{\theta}(t) \in \mathbb{R}$ denotes the angular velocity, the velocity vector is defined as $v(t) = \begin{bmatrix} v_c & \omega_c \end{bmatrix}^T \in \mathbb{R}^2$ with $v_c(t), \omega_c(t) \in \mathbb{R}$

denoting the linear and angular velocity of the system, while the matrix $S(q) \in \mathbb{R}^{3 \times 2}$ is defined as follows

$$S(q) = \begin{bmatrix} \cos\theta & 0 \\ \sin\theta & 0 \\ 0 & 1 \end{bmatrix}. \tag{4.49}$$

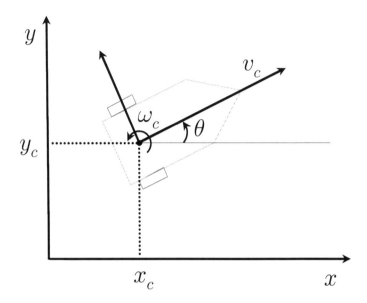

FIGURE 4.9. Mobile Robot Coordinate Frames

4.3.2   WMR Velocity Field Control

As described previously, many robotic tasks can be effectively encapsulated as a velocity field where the control objective can be described as the desire for the trajectory of a system to track a state-dependent desired trajectory. To quantify this objective for the WMR, a velocity field tracking error, denoted by $\eta_v(t) \in \mathbb{R}^3$, is defined as follows

$$\eta_v(t) \triangleq \dot{q}(t) - \vartheta(q) \tag{4.50}$$

where $\vartheta(\cdot) \in \mathbb{R}^3$ denotes the desired WMR velocity field that is defined as follows

$$\vartheta(q) = \begin{bmatrix} \dot{x}_{cd} \\ \dot{y}_{cd} \\ \dot{\theta}_d \end{bmatrix} = \begin{bmatrix} \cos\theta_d(q) & 0 \\ \sin\theta_d(q) & 0 \\ 0 & 1 \end{bmatrix} v_d(q). \tag{4.51}$$

In (4.51), $v_d(q) = [v_{cd} \quad \omega_{cd}]^T \in \mathbb{R}^2$ denotes the desired heading and rotational velocity of the center of mass (COM) of the WMR, while $x_{cd}(t)$, $y_{cd}(t)$, and $\theta_d(t) \in \mathbb{R}$ denote the desired Cartesian position and orientation of the WMR, respectively. The velocity field is assumed to be designed so that $\vartheta(q)$, $\dfrac{\partial \vartheta(q)}{\partial q}$, $x_{cd}(q)$, and $y_{cd}(q)$ remain bounded provided q remains bounded. The velocity field is also assumed to be designed such that $\lim\limits_{t \to \infty} \|v_d(t)\| \neq 0$. As an example, a velocity field that leads to a desired WMR circular contour centered about the origin with radius R can be chosen as follows

$$v_{cd}(q) = \sqrt{\rho_1^2(q) + \rho_2^2(q)} \tag{4.52}$$

$$\omega_{cd}(q) = \dot{\theta}_d(q) = \begin{cases} \dfrac{d}{dt} \arctan 2\left(\rho_2(q), \rho_1(q)\right) & \forall -\pi < \theta_d < \pi \\ 1 & \forall \, \theta_d = \pi \end{cases}$$

where $\arctan 2\,(\cdot)$ is the four quadrant inverse tangent function that is confined to the region $-\pi < \theta_d \leqslant \pi$ while the auxiliary functions $\rho_1(q)$, $\rho_2(q) \in \mathbb{R}$ are defined as follows

$$\rho_1(q) = (R^2 - x_c^2 - y_c^2)x_c + y_c \tag{4.53}$$

$$\rho_2(q) = (R^2 - x_c^2 - y_c^2)y_c - x_c. \tag{4.54}$$

Although the $\arctan 2\,(\cdot)$ function exhibits a discontinuity between periods, *ad hoc* adjustments can be used to implement the function. For $R = 1$, the velocity field depicted in Figure 4.10 is produced.

Kinematic Transformation

To express the WMR model in a form that is more amenable to the subsequent control design and stability analysis, the new state variables $z(t) = [\; z_1(t) \quad z_2(t) \;]^T \in \mathbb{R}^2$ and $w(t) \in \mathbb{R}$ are defined in terms of the original states through the following global diffeomorphism [19]

$$\begin{bmatrix} z_1 \\ z_2 \\ w \end{bmatrix} = G \begin{bmatrix} \tilde{x} \\ \tilde{y} \\ \tilde{\theta} \end{bmatrix} \tag{4.55}$$

where $G(q) \in \mathbb{R}^{3 \times 3}$ is defined as follows

$$G(q) = \begin{bmatrix} 0 & 0 & 1 \\ \cos\theta & \sin\theta & 0 \\ -\tilde{\theta}\cos\theta + 2\sin\theta & -\tilde{\theta}\sin\theta - 2\cos\theta & 0 \end{bmatrix}. \tag{4.56}$$

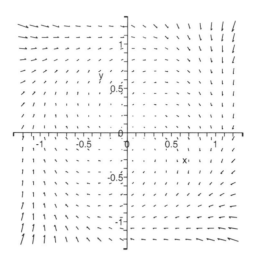

FIGURE 4.10. Example Circular Velocity Field for a WMR

In (4.55), $\tilde{x}(t), \tilde{y}(t), \tilde{\theta}(t) \in \mathbb{R}$ denote the difference between the actual and desired Cartesian position and orientation of the WMR as follows

$$\tilde{x} = x_c - x_{cd} \qquad \tilde{y} = y_c - y_{cd} \qquad \tilde{\theta} = \theta - \theta_d. \qquad (4.57)$$

By using (4.47)–(4.57), the time derivative of (4.55) can be expressed as follows

$$\dot{w} = u^T J^T z + Az \qquad (4.58)$$
$$\dot{z} = u$$

where $J \in \mathbb{R}^{2 \times 2}$ is an auxiliary skew-symmetric matrix defined as

$$J = \begin{bmatrix} 0 & -1 \\ 1 & 0 \end{bmatrix}, \qquad (4.59)$$

and the auxiliary row vector $A(q) \in \mathbb{R}^{1 \times 2}$ is defined as

$$A = \begin{bmatrix} -2v_{cd}\dfrac{\sin(z_1)}{z_1} & 2\omega_{cd} \end{bmatrix}. \qquad (4.60)$$

The auxiliary control variable $u(t) = \begin{bmatrix} u_1(t) & u_2(t) \end{bmatrix}^T \in \mathbb{R}^2$ introduced in (4.58) is used to simplify the transformed dynamics and is explicitly defined in terms of the WMR position and orientation, the WMR linear velocities, and the velocity field as follows

$$u = T^{-1}v - \begin{bmatrix} \omega_{cd} \\ v_{cd}\cos\tilde\theta \end{bmatrix} \qquad v = Tu + \Pi \tag{4.61}$$

where the auxiliary variables $T(q) \in \mathbb{R}^{2\times2}$ and $\Pi(q) \in \mathbb{R}^2$ are defined as follows

$$T = \begin{bmatrix} (\tilde x\sin\theta - \tilde y\cos\theta) & 1 \\ 1 & 0 \end{bmatrix} \tag{4.62}$$

and

$$\Pi = \begin{bmatrix} v_{cd}\cos\tilde\theta + \omega_{cd}(\tilde x\sin\theta - \tilde y\cos\theta) \\ \omega_{cd} \end{bmatrix}. \tag{4.63}$$

Dynamic Model

The dynamic model for the kinematic wheel can be expressed in the following form

$$M\dot v + F(v) = B\tau \tag{4.64}$$

where $M \in \mathbb{R}^{2\times2}$ represents the constant inertia matrix, $F(v) \in \mathbb{R}^2$ represents the friction effects, $\tau(t) \in \mathbb{R}^2$ represents the torque input vector, and $B \in \mathbb{R}^{2\times2}$ represents an input matrix that governs torque transmission. After premultiplying (4.64) by $T^T(q)$ defined in (4.62), and substituting (4.61) for $v(t)$, the following convenient dynamic model can be obtained [19]

$$\bar M\dot u + \bar V_m u + \bar N = \bar B\tau \tag{4.65}$$

where

$$\begin{aligned} \bar M &= T^T MT, \quad \bar V_m = T^T M\dot T, \\ \bar B &= T^T B, \quad \bar N = T^T\left(F(Tu + \Pi) + M\dot\Pi\right). \end{aligned} \tag{4.66}$$

The dynamic model in (4.65) exhibits the following properties which will be employed during the subsequent control development and stability analysis.

Property 4.3.1: The transformed inertia matrix $\bar M(q)$ is symmetric, positive definite, and satisfies the following inequalities

$$m_1\|\xi\|^2 \le \xi^T\bar M\xi \le m_2(z,w)\|\xi\|^2 \quad \forall\xi \in \Re^2 \tag{4.67}$$

where m_1 is a known positive constant, $m_2(z,w) \in \mathbb{R}$ is a known, positive bounding function which is assumed to be bounded provided its arguments are bounded, and $\|\cdot\|$ is the standard Euclidean norm.

Property 4.3.2: A skew-symmetric relationship exists between the transformed inertia matrix and the auxiliary matrix $\bar{V}_m(q, \dot{q})$ as follows

$$\xi^T \left(\frac{1}{2} \dot{\bar{M}} - \bar{V}_m \right) \xi = 0 \qquad \forall \xi \in \mathbb{R}^2 \tag{4.68}$$

where $\dot{\bar{M}}(q, \dot{q})$ represents the time derivative of the transformed inertia matrix.

Property 4.3.3: The robot dynamics given in (4.65) can be linearly parameterized as follows

$$Y_o \psi = \bar{M}\dot{u} + \bar{V}_m u + \bar{N} \tag{4.69}$$

where $\psi \in \mathbb{R}^p$ contains the unknown constant mechanical parameters (i.e., inertia, mass, and friction effects), and $Y_o(\dot{u}, u) \in \mathbb{R}^{2 \times p}$ denotes a regression matrix.

Property 4.3.4: The transformed torque transmission matrix $\bar{B}(q)$ is globally invertible.

Control Development

Control Objective

The velocity field tracking control objective is to ensure that

$$\eta_v(t) \to 0 \text{ as } t \to \infty \tag{4.70}$$

despite parametric uncertainty in the WMR dynamic model. To achieve this objective, the subsequent control development will focus on proving that the auxiliary states $w(t)$ and $z(t)$ and the respective time derivatives asymptotically approach zero. The auxiliary error signal $\tilde{z}(t) \in \mathbb{R}^2$ is introduced to facilitate the control development, where $\tilde{z}(t)$ is defined as the difference between the subsequently designed auxiliary signal $z_d(t) \in \mathbb{R}^2$ and the transformed variable $z(t)$, defined in (4.55), as follows

$$\tilde{z} = z_d - z. \tag{4.71}$$

Given the nonholonomic motion constraints, the control development for a WMR is facilitated by designing a velocity control input (e.g., the control input to the kinematic system in (4.58) is given by $u(t)$). Since the actual control input to the WMR can be expressed as a force/torque, a backstepping error, denoted by $\eta(t) \in \mathbb{R}^2$, is introduced as follows

$$\eta = u_d - u \tag{4.72}$$

where $u_d(t) \in \mathbb{R}^2$ denotes the desired kinematic control signal. Given that uncertainty exists in the WMR dynamic model, the subsequent control development will include the design of an adaptive estimate, denoted by $\hat{\psi}(t) \in \mathbb{R}^p$. To quantify the performance of the adaptive estimate, a parameter estimation error signal, denoted by $\tilde{\psi}(t) \in \mathbb{R}^p$, is defined as follows

$$\tilde{\psi} = \psi - \hat{\psi}. \tag{4.73}$$

Control Formulation

Given the transformed dynamic model of (4.65), the subsequent stability analysis, and the assumption that $q(t)$ and $\dot{q}(t)$ are measurable, the control torque input $\tau(t)$ is designed as follows

$$\tau = \bar{B}^{-1}\left(Y\hat{\psi} + K_a\eta + Jzw + \tilde{z}\right). \tag{4.74}$$

Here, $K_a \in \mathbb{R}^{2\times2}$ is a positive definite, diagonal control gain matrix, $\hat{\psi}(t)$ is dynamically generated by the following gradient update law

$$\dot{\hat{\psi}} = \Gamma Y^T \eta \tag{4.75}$$

and the regression matrix $Y(\dot{u}_d, u_d, u) \in \mathbb{R}^{n\times p}$ is defined as follows

$$Y\psi = \bar{M}\dot{u}_d + \bar{V}_m u_d + \bar{N} \tag{4.76}$$

where $u_d(t)$ is introduced in (4.72). From the kinematic equations given in (4.58) and the subsequent stability analysis, the desired kinematic control signal $u_d(t)$ given in (4.76) is designed as follows

$$u_d = u_a - k_3 z + u_c. \tag{4.77}$$

In (4.77), the auxiliary control terms $u_a(t) \in \mathbb{R}^2$ and $u_c(t) \in \mathbb{R}^2$ are defined as

$$u_a = k_1 w J z_d + \Omega_1 z_d \tag{4.78}$$

and

$$u_c = -(I_2 + 2wJ)^{-1}(2wA^T) \tag{4.79}$$

respectively, the auxiliary signal $z_d(t)$ is defined by the following dynamic oscillator-like relationship

$$\dot{z}_d = \left(k_1\left(w^2 - z_d^T z_d\right) - k_2\right)z_d + J\Omega_2 z_d + \frac{1}{2}u_c \tag{4.80}$$

$$\beta = z_d^T(0)z_d(0),$$

and the auxiliary terms $\Omega_1(t) \in \mathbb{R}$ and $\Omega_2(t) \in \mathbb{R}$ are defined as

$$\Omega_1 = k_1 w^2 + k_1 \left(w^2 - z_d^T z_d \right) - k_2 + k_3 \qquad (4.81)$$

and

$$\Omega_2 = k_1 w + w \Omega_1 \qquad (4.82)$$

respectively. In (4.77)–(4.82), k_1, k_2, $k_3 \in \mathbb{R}$ are positive, constant control gains, I_2 represents the standard 2×2 identity matrix, $\beta \in \mathbb{R}$ is a positive constant, and $A(q)$ was defined in (4.60). Note that it is straightforward to show that the matrix $I_2 + 2w J^T$ used in (4.79) is always invertible provided $w(t)$ remains bounded.

Closed-Loop Error System

The closed-loop error system for $\eta(t)$ is obtained by taking the time derivative of (4.72), premultiplying the resulting expression by the transformed inertia matrix, substituting (4.65) for $\bar{M}(q)\dot{u}(t)$, and utilizing (4.76) to obtain the following expression

$$\bar{M}\dot{\eta} = -\bar{V}_m \eta + Y\psi - \bar{B}\tau \qquad (4.83)$$

After substituting (4.74) into (4.83) and utilizing (4.73), the following expression for the closed-loop error system can be obtained

$$\bar{M}\dot{\eta} = -K_a \eta - \bar{V}_m \eta + Y\tilde{\psi} - Jzw - \tilde{z}. \qquad (4.84)$$

The first term in (4.84) is a stabilizing feedback term, and the second term in (4.84) can be canceled through a Lyapunov-analysis by virtue of Property 2. The design for the adaptive update law given in (4.75) is motivated by the desire to cancel the third term in (4.84) through a Lyapunov-based analysis. The remaining two terms represent coupling terms that must be canceled through the closed-loop error systems developed for $w(t)$ and $\tilde{z}(t)$; hence, the control development given in (4.77)–(4.82) is motivated by the need to cancel these terms.

  To facilitate the closed-loop error system development for $w(t)$, the desired kinematic control input $u_d(t)$ is injected into the open-loop dynamics of $w(t)$ given by (4.58) by adding and subtracting the term $u_d^T Jz$ to the right side of (4.58) and utilizing (4.72) to obtain the following expression

$$\dot{w} = \eta^T Jz - u_d^T Jz + Az. \qquad (4.85)$$

After substituting (4.77) for $u_d(t)$, adding and subtracting $u_a^T Jz_d$ to the resulting expression, utilizing (4.71), and exploiting the skew symmetry of

J defined in (4.59), we can rewrite the closed-loop dynamics for $w(t)$ as follows

$$\dot{w} = \eta^T J z + u_a^T J \tilde{z} - u_a^T J z_d + A z - u_c^T J z. \tag{4.86}$$

After substituting (4.78) for only the second occurrence of $u_a(t)$ in (4.86), substituting (4.79) for $u_c(t)$, utilizing the skew symmetry of J defined in (4.59), and the fact that $J^T J = I_2$, we can obtain the final expression for the closed-loop error system as follows

$$
\begin{aligned}
\dot{w} &= \eta^T J z + u_a^T J \tilde{z} - k_1 w z_d^T z_d + A z \\
&\quad + 2 w A (I_2 + 2 w J^T)^{-1} J z.
\end{aligned}
\tag{4.87}
$$

The closed-loop error system for $\tilde{z}(t)$ can be obtained by taking the time derivative of (4.71), substituting (4.80) for $\dot{z}_d(t)$, and then substituting (4.58) for $\dot{z}(t)$ to obtain

$$
\begin{aligned}
\dot{\tilde{z}} &= \left(k_1 \left(w^2 - z_d^T z_d \right) - k_2 \right) z_d \\
&\quad + J \Omega_2 z_d + \frac{1}{2} u_c + \eta - u_d
\end{aligned}
\tag{4.88}
$$

where (4.72) was utilized. The following expression can be obtained by substituting (4.77) for $u_d(t)$ and then substituting (4.78) in the resulting expression

$$
\begin{aligned}
\dot{\tilde{z}} &= \left(k_1 \left(w^2 - z_d^T z_d \right) - k_2 \right) z_d + J \Omega_2 z_d \\
&\quad - \frac{1}{2} u_c - k_1 w J z_d - \Omega_1 z_d + k_3 z + \eta.
\end{aligned}
\tag{4.89}
$$

After substituting (4.81) and (4.82) for $\Omega_1(t)$ and $\Omega_2(t)$ into (4.89), respectively, and then using the fact that $JJ = -I_2$, the following expression is obtained

$$\dot{\tilde{z}} = -k_3 \tilde{z} + w J \left[\Omega_1 z_d + k_1 w J z_d \right] - \frac{1}{2} u_c + \eta \tag{4.90}$$

where (4.71) has been utilized. Substituting (4.79) for $u_c(t)$ yields the final expression for the closed-loop error system as follows

$$\dot{\tilde{z}} = -k_3 \tilde{z} + (I_2 + 2 w J)^{-1} w A^T + w J u_a + \eta \tag{4.91}$$

since the bracketed term in (4.90) is equal to $u_a(t)$ defined in (4.78).

Stability Analysis

Theorem 4.4 *The controller introduced in (4.74)–(4.82) ensures global asymptotic tracking of the velocity field in the sense that*

$$\eta_v(t) \to 0 \text{ as } t \to \infty \tag{4.92}$$

provided the reference trajectory is selected such that

$$\lim_{t\to\infty} \|v_d(t)\| \neq 0. \tag{4.93}$$

Proof: Consider the following non-negative function denoted by $V(w, z_d, \tilde{z}, \eta, \tilde{\psi}) \in \mathbb{R}$ as follows

$$V(t) = \frac{1}{2}w^2 + \frac{1}{2}z_d^T z_d + \frac{1}{2}\tilde{z}^T \tilde{z} + \frac{1}{2}\eta^T \bar{M}\eta + \frac{1}{2}\tilde{\psi}^T \Gamma^{-1}\tilde{\psi}. \tag{4.94}$$

The following expression can be obtained after taking the time derivative of (4.94) and making the appropriate substitutions from (4.75), (4.79), (4.80), (4.84), (4.87), (4.91), and utilizing the fact that $\dot{\tilde{\psi}}(t) = -\dot{\hat{\psi}}(t)$

$$\begin{aligned}
\dot{V} =\ & w\left[\eta^T J z + u_a^T J\tilde{z} - k_1 w z_d^T z_d\right] \\
& + w\left[Az + 2wA(I_2 + 2wJ^T)^{-1}Jz\right] \\
& + z_d^T\left[\left(k_1\left(w^2 - z_d^T z_d\right) - k_2\right)z_d\right] \\
& + z_d^T\left[J\Omega_2 z_d - (I_2 + 2wJ)^{-1}wA^T\right] \\
& + \tilde{z}^T\left[-k_3\tilde{z} + (I_2 + 2wJ)^{-1}wA^T\right] \\
& + \tilde{z}^T\left[wJu_a + \eta\right] + \frac{1}{2}\eta^T \dot{\bar{M}}\eta - \tilde{\psi}^T\left[Y^T\eta\right] \\
& + \eta^T\left[Y\tilde{\psi} - Jzw - \tilde{z} - K_a\eta - \bar{V}_m\eta\right].
\end{aligned} \tag{4.95}$$

After utilizing (4.68), the skew symmetry property of J, and cancelling common terms, (4.95) can be rewritten as

$$\begin{aligned}
\dot{V} =\ & -k_1\|z_d\|^4 - k_2 z_d^T z_d - k_3\tilde{z}^T\tilde{z} - \eta^T K_a\eta \\
& + \left[wAz + \left(wA(I_2 + 2wJ^T)^{-1}(2wJ)\right)z\right. \\
& \left. - \left(wA(I_2 + 2wJ^T)^{-1}\right)z\right]
\end{aligned} \tag{4.96}$$

where (4.71) has been utilized. Furthermore, after combining the bracketed terms in (4.96) and making algebraic simplifications, one can obtain the following simplified expression for $\dot{V}(t)$

$$\dot{V} = -k_1\|z_d\|^4 - k_2 z_d^T z_d - k_3\tilde{z}^T\tilde{z} - \eta^T K_a\eta. \tag{4.97}$$

Based on (4.94) and (4.97), it can be concluded that $V(t) \in \mathcal{L}_\infty$; thus, $w(t)$, $z_d(t)$, $\tilde{z}(t)$, $\eta(t)$, $\tilde{\psi}(t) \in \mathcal{L}_\infty$. Since $w(t)$, $z_d(t)$, $\tilde{z}(t)$, $\eta(t) \in \mathcal{L}_\infty$, one can utilize (4.60), (4.71), (4.77)–(4.82), (4.87), and (4.91) to conclude that $A(q)$, $z(t)$, $u_d(t)$, $u(t)$, $u_a(t)$, $u_c(t)$, $\Omega_1(t)$, $\Omega_2(t)$, $\dot{z}_d(t)$, $\dot{w}(t)$, $\dot{\tilde{z}}(t) \in \mathcal{L}_\infty$. Since $\dot{z}_d(t)$, $\dot{\tilde{z}}(t) \in \mathcal{L}_\infty$, (4.71) can be utilized to show that $\dot{z}(t) \in \mathcal{L}_\infty$ (since $\dot{w}(t)$, $\dot{z}_d(t)$, $\dot{\tilde{z}}(t)$, $\dot{z}(t) \in \mathcal{L}_\infty$, it follows that $w(t)$, $z_d(t)$, $\tilde{z}(t)$, and $z(t)$

are uniformly continuous). To illustrate that the Cartesian position and orientation tracking signals are bounded, we calculate the inverse transformation of (4.55) as follows

$$
\begin{bmatrix} \tilde{x} \\ \tilde{y} \\ \tilde{\theta} \end{bmatrix} = G^{-1} \begin{bmatrix} z_1 \\ z_2 \\ w \end{bmatrix}.
\tag{4.98}
$$

Since $z(t) \in \mathcal{L}_\infty$, it is clear from (4.57) and (4.98) that $\tilde{\theta}(t) \in \mathcal{L}_\infty$. Furthermore, from (4.57), (4.98), and the fact that $w(t)$, $z(t)$, $\tilde{\theta}(t) \in \mathcal{L}_\infty$, we can conclude that $\tilde{x}(t)$, $\tilde{y}(t) \in \mathcal{L}_\infty$. Since $x_{cd}(q)$ and $y_{cd}(q)$ are assumed to be bounded, we can conclude that $x_c(t)$ and $y_c(t) \in \mathcal{L}_\infty$. We can utilize (4.61), the assumption that $v_d(q)$ is bounded, and the fact that $u(t)$, $\tilde{x}(t)$, $\tilde{y}(t) \in \mathcal{L}_\infty$, to show that $v(t) \in \mathcal{L}_\infty$; therefore, it follows from (4.47)–(4.49) that $\dot{\theta}(t)$, $\dot{x}_c(t)$, $\dot{y}_c(t) \in \mathcal{L}_\infty$. Based on the boundedness of the aforementioned signals, we can take the time derivative of (4.80) and show that $\ddot{z}_d(t) \in \mathcal{L}_\infty$. The expressions in (4.62)–(4.66) can then be used to prove that $\bar{M}(q), \bar{V}_m(q,\dot{q}), \bar{N}(q,\dot{q}) \in \mathcal{L}_\infty$. The time derivative of (4.77)–(4.82) can be used to prove that $\dot{u}_d(t) \in \mathcal{L}_\infty$; hence, (4.74)–(4.76) and (4.84) can be used to prove that $Y(t)$, $\dot{\eta}(t)$, $\dot{\psi}(t)$, $\tau(t) \in \mathcal{L}_\infty$. Based on the boundedness of the closed-loop signals, standard signal chasing arguments can now be used to show that all remaining signals remain bounded during closed-loop operation.

From (4.94) and (4.97), it is easy to show that $z_d(t)$, $\tilde{z}(t)$, $\eta(t) \in \mathcal{L}_2$; hence, since $z_d(t)$, $\tilde{z}(t)$, $\eta(t)$ are uniformly continuous, a corollary to Barbalat's Lemma [66] can be used to show that $\lim_{t\to\infty} z_d(t)$, $\tilde{z}(t)$, $z(t)$, $\eta(t) = 0$. Next, since $\ddot{z}_d(t) \in \mathcal{L}_\infty$, we know that $\dot{z}_d(t)$ is uniformly continuous. Since we know that $\lim_{t\to\infty} z_d(t) = 0$ and $\dot{z}_d(t)$ is uniformly continuous, we can use the following equality

$$
\lim_{t\to\infty} \int_0^t \frac{d}{d\tau}(z_d(\tau))\, d\tau = \lim_{t\to\infty} z_d(t) + \text{constant}
\tag{4.99}
$$

and Barbalat's Lemma [66] to conclude that $\lim_{t\to\infty} \dot{z}_d(t) = 0$. Based on the fact that $\lim_{t\to\infty} z_d(t)$, $\dot{z}_d(t) = 0$, it is straightforward from (4.79) and (4.80) to see that $\lim_{t\to\infty} wA^T = 0$. Finally, based on (4.60) and (4.93) we can conclude that $\lim_{t\to\infty} w(t) = 0$. From (4.98), we can now conclude that $\lim_{t\to\infty} \tilde{x}(t)$, $\tilde{y}(t)$, $\tilde{\theta}(t) = 0$. Given that $\lim_{t\to\infty} z(t)$, $\eta(t)$, $w(t) = 0$, (4.72) and (4.77)–(4.79) can be used to prove that $\lim_{t\to\infty} u(t) = 0$. Hence, from (4.58) and (4.85), $\lim_{t\to\infty} \dot{z}(t)$, $\dot{w}(t) = 0$. To prove that $\eta_v(t) \to 0$, one can take the time derivative of (4.55) as follows

$$\frac{d}{dt}\begin{bmatrix} z_1 \\ z_2 \\ w \end{bmatrix} = \dot{G}\begin{bmatrix} \tilde{x} \\ \tilde{y} \\ \tilde{\theta} \end{bmatrix} + G\frac{d}{dt}\begin{bmatrix} \tilde{x} \\ \tilde{y} \\ \tilde{\theta} \end{bmatrix}. \tag{4.100}$$

Since $\lim_{t\to\infty} \dot{z}(t), \dot{w}(t), \tilde{x}(t), \tilde{y}(t), \tilde{\theta}(t) = 0$, the fact that the $G(q)$ is a global diffeomorphism can be used along with (4.100) to conclude that $\lim_{t\to\infty} \dot{\tilde{x}}(t)$, $\dot{\tilde{y}}(t), \dot{\tilde{\theta}}(t) = 0$. From (4.50), (4.51), and (4.57), the velocity field tracking result given in (4.92) is now obtained. ∎

4.3.3 WMR Navigation Function Control Objective

Given the kinematic model described in Section 4.3.1, the objective here is to navigate the wheeled mobile robot along a collision-free path to a constant goal position and orientation, denoted by $q^* \triangleq [x_c^* \quad y_c^* \quad \theta^*]^T \in \mathbb{R}^3$, in a cluttered environment with known obstacles. Specifically, the objective is to control the nonholonomic system along a path from the initial position and orientation to $q^* \in \mathcal{D}$, where $\mathcal{D}$ denotes a free configuration space. The free configuration space $\mathcal{D}$ is a subset of the whole configuration space with all configurations removed that involve a collision with an obstacle. The navigation control objective is to drive the errors $\tilde{x}(t), \tilde{y}(t), \tilde{\theta}(t)$ (as defined in 4.57) to zero in the limit. Here, the desired position and orientation of the WMR, denoted by $q_d(t) \triangleq [x_{cd}(t) \quad y_{cd}(t) \quad \theta_d(t)]^T$, is designed such that $q_d(t) \to q^*$. As previously elaborated in Section 4.2.3, the navigation function used here to generate $q_d(t)$ is assumed to satisfy Properties 4.2.4–4.2.7.

Off-line 3D Navigation

Trajectory Planning

The off-line 3D desired trajectory can be generated as follows

$$\dot{q}_d = -\nabla\varphi(q_d) \triangleq -\begin{bmatrix} \dfrac{\partial\varphi}{\partial x_{cd}} & \dfrac{\partial\varphi}{\partial y_{cd}} & \dfrac{\partial\varphi}{\partial\theta_d} \end{bmatrix}^T \tag{4.101}$$

where $\varphi(q_d) \in \mathbb{R}$ denotes a navigation function defined in $\mathcal{D}$ with a minimum at $q_d = q^*$ (see Properties 4.2.4–4.2.7 for details), and $x_{cd}(t)$, $y_{cd}(t)$, $\theta_d(t)$ were introduced in (4.51).

Lemma 4.5 *Provided $q_d(0) \in \mathcal{D}$, the desired trajectory generated by (4.101) ensures that $q_d(t) \in \mathcal{D}$ and (4.101) has the asymptotically stable equilibrium point q^*.*

Proof: Let $V_1(q_d) : \mathcal{D} \to \mathbb{R}$ denote a non-negative function defined as follows

$$V_1(q_d) \triangleq \varphi(q_d). \tag{4.102}$$

After taking the time derivative of (4.102) and utilizing (4.101), the following expression can be obtained

$$\dot{V}_1(q_d) = (\nabla\varphi(q_d))^T \dot{q}_d = -\|\nabla\varphi(q_d)\|^2. \tag{4.103}$$

From (4.103), it is clear that $V_1(q_d(t))$ is a non-increasing function in the sense that

$$V_1(q_d(t)) \le V_1(q_d(0)). \tag{4.104}$$

From (4.104), it is clear that for any initial condition $q_d(0) \in \mathcal{D}$, that $q_d(t) \in \mathcal{D} \; \forall t > 0$; therefore, $\mathcal{D}$ is a positively invariant set. Let $E_1 \subset \mathcal{D}$ denote a set defined as follows $E_1 \triangleq \{q_d(t)| \; \dot{V}_1(q_d) = 0\}$. Based on (4.103), it is clear that $\nabla\varphi(q_d) = 0$ in E_1; hence, from (4.101) it can be determined that $\dot{q}_d(t) = 0$ in E_1, and that E_1 is the largest invariant set. By invoking LaSalle's Theorem [35], it can be determined that every solution $q_d(t) \in \mathcal{D}$ approaches E_1 as $t \to \infty$, and hence, $\nabla\varphi(q_d) \to 0$. Based on Property 4.2.7, it can be determined that if $\nabla\varphi(q_d) \to 0$ then $q_d(t) \to q^*$. ■

Model Transformation

To achieve the navigation control objective, a controller must be designed to track the desired trajectory developed in (4.101) and stop at the goal position q^*. To this end, the unified tracking and regulation controller presented in [18] can be used. To develop the controller in [18], the open-loop error system defined in (4.57) must be transformed into a suitable form. Specifically, the position and orientation tracking error signals defined in (4.57) are related to the auxiliary tracking error variables $w(t) \in \mathbb{R}$ and $z(t) \triangleq \begin{bmatrix} z_1(t) & z_2(t) \end{bmatrix}^T \in \mathbb{R}^2$ through the global invertible transformation of (4.55). As similarly done in Section 4.3.2, one can take the time derivative of (4.55) to obtain the following transformed kinematic system

$$\begin{aligned} \dot{w} &= u^T J^T z + f \\ \dot{z} &= u \end{aligned} \tag{4.105}$$

where we have utilized (4.101) as well as the robot kinematics of (4.47), $J \in \mathbb{R}^{2 \times 2}$ has been previously defined in (4.59), and $f(\theta, z_2, \dot{q}_d) \in \mathbb{R}$ is defined as

$$f \triangleq 2 \begin{bmatrix} -\sin\theta & \cos\theta & z_2 \end{bmatrix} \dot{q}_d. \tag{4.106}$$

The auxiliary control input $u(t) \triangleq \begin{bmatrix} u_1(t) & u_2(t) \end{bmatrix}^T \in \mathbb{R}^2$ introduced in (4.105) is defined in terms of the position and orientation, the linear and angular velocities, and the gradient of the navigation function as follows

$$
u = \begin{bmatrix} 0 & 1 \\ 1 & -\tilde{x}\sin\theta + \tilde{y}\cos\theta \end{bmatrix} v
$$
$$
+ \begin{bmatrix} \dfrac{\partial\varphi}{\partial\theta_d} \\ \dfrac{\partial\varphi}{\partial x_{cd}}\cos\theta + \dfrac{\partial\varphi}{\partial y_{cd}}\sin\theta \end{bmatrix} . \tag{4.107}
$$

Control Development

Based on the form of the open-loop error system in (4.105)–(4.107), $u(t)$ can be designed as follows [18]

$$
u \triangleq u_a - k_2 z \tag{4.108}
$$

where $k_2 \in \mathbb{R}$ is a positive, constant control gain. The auxiliary control term $u_a(t) \in \mathbb{R}^2$ introduced in (4.108) is defined as

$$
u_a \triangleq \left(\frac{k_1 w + f}{\delta_d^2} \right) J z_d + \Omega_1 z_d . \tag{4.109}
$$

In (4.109), $z_d(t)$ is defined by the following differential equation and initial condition

$$
\dot{z}_d = \frac{\dot{\delta}_d}{\delta_d} z_d + \left(\frac{k_1 w + f}{\delta_d^2} + w\Omega_1 \right) J z_d \tag{4.110}
$$
$$
z_d^T(0) z_d(0) = \delta_d^2(0) ,
$$

the auxiliary terms $\Omega_1(w, f, t) \in \mathbb{R}$ and $\delta_d(t) \in \mathbb{R}$ are defined as

$$
\Omega_1 \triangleq k_2 + \frac{\dot{\delta}_d}{\delta_d} + w \left(\frac{k_1 w + f}{\delta_d^2} \right) , \tag{4.111}
$$

$$
\delta_d \triangleq \alpha_0 \exp(-\alpha_1 t) + \varepsilon_1, \tag{4.112}
$$

while $k_1, \alpha_0, \alpha_1, \varepsilon_1 \in \mathbb{R}$ are positive, constant control gains, and $f(\theta, z_2, \dot{q}_d)$ was defined in (4.106).

Based on the control design given in (4.108)–(4.112), the following stability result can be obtained.

Theorem 4.6 *The kinematic control law given in (4.108)–(4.112) ensures uniformly ultimately bounded (UUB) position and orientation tracking in the sense that*

$$
|\tilde{x}(t)|, |\tilde{y}(t)|, \left| \tilde{\theta}(t) \right| \leqslant \beta_0 \exp(-\gamma_0 t) + \beta_1 \varepsilon_1 \tag{4.113}
$$

where ε_1 was given in (4.112), $\beta_0 \triangleq \sqrt{w^2(0) + z_1^2(0) + z_2^2(0)}$, and β_1 and γ_0 are known positive constants.

Proof: See [18].

Remark 4.6 *Although $q_d(t)$ is a collision-free path, the stability result in Theorem 4.6 only ensures practical tracking of the path in the sense that the actual WMR trajectory is only guaranteed to remain in a neighborhood of the desired path. From (4.57) and (4.113), the following bound can be developed*

$$\|q\| \leq \|q_d\| + \sqrt{3}\beta_0 \exp(-\gamma_0 t) + \sqrt{3}\beta_1 \varepsilon_1 \tag{4.114}$$

where $q_d(t) \in \mathcal{D}$ based on the proof for Lemma 4.5. To ensure that $q(t) \in \mathcal{D}$, the free configuration space needs to be reduced to incorporate the effects of the second and third terms on the right-hand side of (4.114). To this end, the size of the obstacles needs to be increased by $\sqrt{3}(\beta_0 + \beta_1 \varepsilon_1)$, where $\beta_1 \varepsilon_1$ can be made arbitrarily small by adjusting the control gains. To minimize the effects of β_0, the initial errors $w(0)$ and $z(0)$ need to be chosen sufficiently small to yield a feasible path to the goal.

On-Line 2D Navigation

In the previous approach, the size of the obstacles is required to be increased due to the fact that the navigation function is formulated in terms of the desired trajectory. In the following approach, the navigation function is formulated based on current position feedback, and hence, $q(t)$ can be proven to be a member of $\mathcal{D}$ without placing restrictions on the initial conditions. However, it is important to note that the orientation control for this approach requires additional development in order to align the WMR with a desired orientation. The reader is referred to the simulation results (Section 4.3.3) as well as the notes at the end of the chapter for more details on this aspect.

Trajectory Planning

Let $\varphi(x_c, y_c) \in \mathbb{R}$ denote a 2D position-based navigation function defined in $\mathcal{D}$ that is generated on-line, where the gradient vector of $\varphi(x_c, y_c)$ is defined as follows

$$\nabla\varphi(x_c, y_c) \triangleq \begin{bmatrix} \dfrac{\partial\varphi}{\partial x_c} & \dfrac{\partial\varphi}{\partial y_c} \end{bmatrix}^T. \tag{4.115}$$

Let $\theta_d(x_c, y_c) \in \mathbb{R}$ denote a desired orientation that is defined as a function of the negated gradient of the 2D navigation function as follows

$$\theta_d \triangleq \arctan 2\left(-\dfrac{\partial\varphi}{\partial y_c}, -\dfrac{\partial\varphi}{\partial x_c} \right) \tag{4.116}$$

where $\arctan 2 \left(\cdot \right) : \mathbb{R}^2 \rightarrow \mathbb{R}$ denotes the four quadrant inverse tangent function [72], where $\theta_d \left(t \right)$ is confined to the following region

$$-\pi < \theta_d \leqslant \pi.$$

By defining $\theta_d|_{(x_c^*, y_c^*)} = \arctan 2 \left(0, 0 \right) = \theta|_{(x_c^*, y_c^*)}$, $\theta_d(t)$ remains continuous along any approaching direction to the goal position [16].

Control Development

Based on the open-loop system introduced in (4.47)–(4.49) and the subsequent stability analysis, the linear velocity control input $v_c \left(t \right)$ is designed as follows

$$v_c \triangleq k_v \left\| \nabla \varphi \right\| \cos \tilde{\theta} \tag{4.117}$$

where $k_v \in \mathbb{R}$ denotes a positive, constant control gain, and $\tilde{\theta}(t)$ was introduced in (4.57). After substituting (4.117) into (4.47), the following closed-loop system can be obtained

$$\begin{bmatrix} \dot{x}_c \\ \dot{y}_c \end{bmatrix} = k_v \begin{bmatrix} \cos \theta \\ \sin \theta \end{bmatrix} \left\| \nabla \varphi \right\| \cos \tilde{\theta}. \tag{4.118}$$

The open-loop orientation tracking error system can be obtained by taking the time derivative of $\tilde{\theta}(t)$ in (4.57) as follows

$$\dot{\tilde{\theta}} = \omega_c - \dot{\theta}_d \tag{4.119}$$

where (4.47) was utilized. Based on (4.119), the angular velocity control input $\omega_c \left(t \right)$ is designed as follows

$$\omega_c \triangleq -k_\omega \tilde{\theta} + \dot{\theta}_d \tag{4.120}$$

where $k_\omega \in \mathbb{R}$ denotes a positive, constant control gain, and $\dot{\theta}_d(t)$ denotes the time derivative of the desired orientation. See Section B.2.3 of Appendix B for an explicit expression for $\dot{\theta}_d \left(t \right)$ based on the previous continuous definition for $\theta_d \left(t \right)$. After substituting (4.120) into (4.119), the closed-loop orientation tracking error system is given by the following differential equation

$$\dot{\tilde{\theta}} = -k_\omega \tilde{\theta}. \tag{4.121}$$

the solution for which can be obtained as

$$\tilde{\theta}(t) = \tilde{\theta}(0) \exp(-k_\omega t). \tag{4.122}$$

After substituting (4.122) into (4.118), the following closed-loop error system can be determined

$$\begin{bmatrix} \dot{x}_c \\ \dot{y}_c \end{bmatrix} = k_v \begin{bmatrix} \cos \theta \\ \sin \theta \end{bmatrix} \left\| \nabla \varphi \right\| \cos \left(\tilde{\theta}(0) \exp(-k_\omega t) \right). \tag{4.123}$$

Stability Analysis

Theorem 4.7 *The control input designed in (4.117) and (4.120) along with the navigation function $\varphi\left(x_c\left(t\right), y_c\left(t\right)\right)$ ensure asymptotic navigation in the sense that*

$$|x\left(t\right) - x^*|, |y\left(t\right) - y^*|, \left|\tilde{\theta}\left(t\right)\right| \to 0. \tag{4.124}$$

Proof: Let $V_2\left(x_c, y_c\right) : \mathcal{D} \to \mathbb{R}$ denote the following non-negative function

$$V_2\left(x_c, y_c\right) \triangleq \varphi\left(x_c, y_c\right). \tag{4.125}$$

After taking the time derivative of (4.125) and utilizing (4.115) and (4.118), the following expression can be obtained

$$\dot{V}_2 = k_v \left(\nabla\varphi\right)^T \begin{bmatrix} \cos\theta \\ \sin\theta \end{bmatrix} \|\nabla\varphi\| \cos\tilde{\theta}. \tag{4.126}$$

Based on the development provided in Section B.2.3 of Appendix B (see (B.42) and (B.45)), the gradient of the navigation function can be expressed as follows

$$\nabla\varphi = -\|\nabla\varphi\| \begin{bmatrix} \cos\theta_d & \sin\theta_d \end{bmatrix}^T. \tag{4.127}$$

After substituting (4.127) into (4.126), the following expression can be obtained

$$\dot{V}_2 = -k_v \|\nabla\varphi\|^2 \left(\cos\theta\cos\theta_d + \sin\theta\sin\theta_d\right)\cos\tilde{\theta}. \tag{4.128}$$

After utilizing a trigonometric identity, (4.128) can be rewritten as follows

$$\dot{V}_2 = -g(t) \triangleq -k_v \|\nabla\varphi\|^2 \cos^2\tilde{\theta} \tag{4.129}$$

Hereafter, the ensuing analysis is valid on the free configuration space defined by the set $\mathcal{D}$. Based on (4.115) and Property 4.2.4, it is clear that $\|\nabla\varphi\left(x_c, y_c\right)\| \in \mathcal{L}_\infty$; hence, (4.117) can be used to conclude that $v_c\left(t\right) \in \mathcal{L}_\infty$. Furthermore, it can be seen from Section B.2.3 of Appendix B (see (B.50)) that $\dot{\theta}_d\left(t\right) \in \mathcal{L}_\infty$. Thus, (4.120) can be used to show that $\omega_c\left(t\right) \in \mathcal{L}_\infty$. Based on the fact that $v_c\left(t\right) \in \mathcal{L}_\infty$, (4.47)–(4.49) can be used to prove that $\dot{x}_c\left(t\right), \dot{y}_c\left(t\right) \in \mathcal{L}_\infty$. After taking the time derivative of (4.115), the following expression can be obtained

$$\frac{d}{dt}\left(\nabla\varphi\left(x_c, y_c\right)\right) = \begin{bmatrix} \dfrac{\partial^2\varphi}{\partial x_c^2} & \dfrac{\partial^2\varphi}{\partial y_c \partial x_c} \\ \dfrac{\partial^2\varphi}{\partial x_c \partial y_c} & \dfrac{\partial^2\varphi}{\partial y_c^2} \end{bmatrix} \begin{bmatrix} \dot{x}_c \\ \dot{y}_c \end{bmatrix}. \tag{4.130}$$

Since $\dot{x}_c(t)$, $\dot{y}_c(t) \in \mathcal{L}_\infty$, and since each element of the Hessian matrix in (4.130) is bounded by virtue of Property 4.2.4, it is clear that $\dot{g}(t) \in \mathcal{L}_\infty$. Based on (4.125), (4.129), and the fact that $\dot{g}(t) \in \mathcal{L}_\infty$, Lemma A.6 of [17] can be invoked to prove that

$$\|\nabla\varphi(x_c, y_c)\|^2 \cos^2\tilde{\theta} \to 0. \tag{4.131}$$

Based on the fact that $\cos^2\tilde{\theta}(t) \to 1$, (4.131) can be used to prove that $\|\nabla\varphi(x_c, y_c)\| \to 0$. Finally, Property 4.2.7 and (4.122) can be used to obtain the result in (4.124). ∎

Remark 4.7 *The control development in this section is based on a 2D position navigation function. To achieve the objective, a desired orientation $\theta_d(t)$ was defined as a function of the negated gradient of the 2D navigation function. The previous development can be used to prove the result in (4.124). If a navigation function $\varphi(x_c, y_c)$ can be found such that $\theta_d|_{(x_c^*, y_c^*)} = \theta^*$, then asymptotic navigation can be achieved by the controller in (4.117) and (4.120); otherwise, a standard regulation controller (e.g., see [19] for several candidates) could be implemented to regulate the orientation of the WMR from $\theta_d|_{(x_c^*, y_c^*)} \to \theta^*$. Alternatively, a dipolar potential field approach [69, 70], or a virtual obstacle [29] could be utilized to align the gradient field of the navigation function to the goal orientation of the WMR.*

Simulation Results

To illustrate the performance of the controller given in (4.117) and (4.120), a numerical simulation was performed to navigate the mobile robot from $q(x_c(0), y_c(0), \theta(0))$ to $q^*(x_c^*, y_c^*, \theta^*)$. Since the properties of a navigation function are invariant under a diffeomorphism, a diffeomorphism is developed to map the WMR free configuration space to a model space [39]. As similarly done in Section 4.2.4, a positive function $\varphi(x_c, y_c)$ was chosen as follows

$$\varphi(x_c, y_c) = \frac{(x_c - x_c^*)^2 + (y_c - y_c^*)^2}{\left(\left((x_c - x_c^*)^2 + (y_c - y_c^*)^2\right)^\kappa + \rho_0\rho_1\right)^{1/\kappa}}. \tag{4.132}$$

where κ is positive integer, and the boundary function $\rho_0(x_c, y_c) \in \mathbb{R}$ and the obstacle function $\rho_1(x_c, y_c) \in \mathbb{R}$ are defined as follows

$$\rho_0 \triangleq r_0^2 - (x_c - x_{r_0})^2 - (y_c - y_{r_0})^2 \tag{4.133}$$
$$\rho_1 \triangleq (x_c - x_{r_1})^2 + (y_c - y_{r_1})^2 - r_1^2.$$

In (4.133), (x_{r_0}, y_{r_0}) and (x_{r_1}, y_{r_1}) are the centers of the boundary and the obstacle respectively, r_0, $r_1 \in \mathbb{R}$ are the radii of the boundary and the obstacle respectively. From (4.132) and (4.133), it is clear that the model space is a unit circle that excludes a circle described by the obstacle function $\rho_1 (x_c, y_c)$. For the simulation, the model space configuration is selected as follows

$$x_{r_0} = 0 \quad y_{r_0} = 0 \quad r_0 = 1$$
$$x_{r_1} = 0 \quad y_{r_1} = 0.1 \quad r_1 = 0.15$$

where the initial and goal configuration were selected as

$$q(0) = \begin{bmatrix} 0.1 & 0.6 & 51.6 \end{bmatrix}^T$$

$$q^* = \begin{bmatrix} -0.2 & -0.4 & -40.1 \end{bmatrix}^T .$$

The control inputs defined in (4.117) and (4.120) were utilized to drive the WMR to the goal point along the negated gradient angle. The control gains k_v and k_ω were chosen to be 0.3 and 17, respectively, in order to yield the best performance. Once the WMR reached the goal position, the regulation controller in [19] was implemented to regulate the WMR from $\theta_d|_{(x_c^*, y_c^*)} \to \theta^*$. The actual Cartesian trajectory of the WMR is shown in Figure 4.11. The outer circle in Figure 4.11 depicts the outer boundary of the obstacle free space and the inner circle represents the boundary around an obstacle. The resulting position and orientation errors for the WMR are depicted in Figure 4.12, where the rotational error shown in Figure 4.12 is the error between the actual orientation and goal orientation. The control input velocities $v_c(t)$ and $\omega_c(t)$ defined in (4.117) and (4.120), respectively, are depicted in Figure 4.13. Note that the angular velocity input was artificially saturated between ± 90 $[\deg \cdot s^{-1}]$.

4.4   Vision Navigation

In the introduction to the chapter, we noted the need for visual servoing controllers for robots negotiating unstructured environments. As previously stated, three approaches, namely, position-based visual servo (PBVS) control, image-based visual servo (IBVS) control, and hybrid control, are commonly employed to perform vision based servoing. A common problem with these approaches is the inability to achieve the control objective while ensuring the target features remain visible. In this section, we present a solution to this problem of a limited field-of-view (FOV) — this solution is motivated by the image space navigation function developed in [13]. To

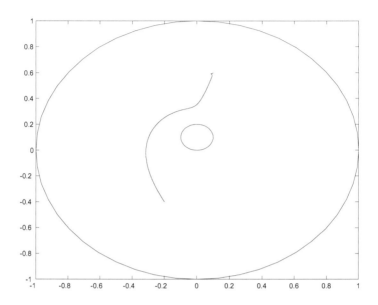

FIGURE 4.11. Actual Cartesian Space Trajectory of the WMR

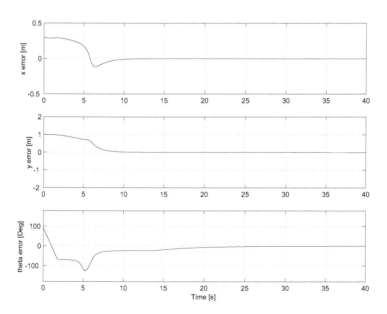

FIGURE 4.12. Position and Orientation Errors

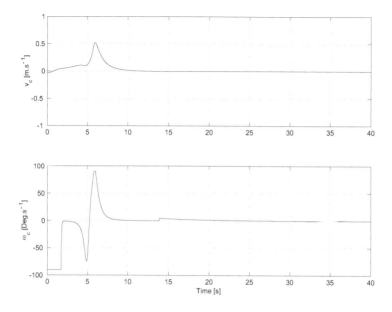

FIGURE 4.13. Linear and Angular Velocity Inputs

begin with, the mapping between the desired camera feature vector and the desired camera pose (i.e., the position and orientation) is investigated to develop a measurable image Jacobian-like matrix. An off-line image-space path planner is then proposed to generate a desired image trajectory based on this measurable image Jacobian-like matrix and an image space navigation function (NF) while satisfying rigid body constraints. An adaptive, homography-based visual servo tracking controller is then developed to navigate the position and orientation of a camera held by the end-effector of a robot manipulator to a goal position and orientation along the desired image-space trajectory, while ensuring the target points remain visible (i.e., the target points avoid self-occlusion and remain in the FOV) under certain technical restrictions. Due to the inherent nonlinear nature of the problem, and the lack of depth information from a monocular system, a Lyapunov-based analysis is used to analyze the path planner and the adaptive controller. Simulation results are provided to illustrate the performance of the proposed approach.

4.4.1   Geometric Modeling

Euclidean Homography

Four feature points, denoted by O_i $\forall i = 1, 2, 3, 4$, are assumed to be located on a reference plane π (see Figure 4.14), and are considered to be coplanar[2] with no three being colinear. The reference plane can be related to the coordinate frames $\mathcal{F}$, $\mathcal{F}_d$, and $\mathcal{F}^*$ depicted in Fig. 4.14 that denote the actual, desired, and goal pose of the camera, respectively. Specifically,

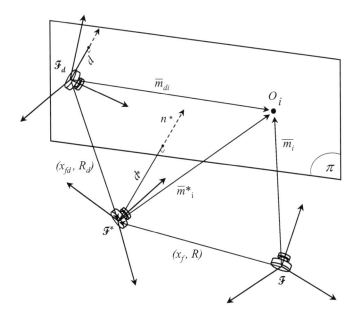

FIGURE 4.14. Coordinate Frame Relationships

the following relationships can be developed from the geometry between the coordinate frames and the feature points located on π

$$\bar{m}_i = x_f + R\bar{m}_i^*$$
$$\bar{m}_{di} = x_{fd} + R_d\bar{m}_i^* \tag{4.134}$$

where $\bar{m}_i(t)$, $\bar{m}_{di}(t)$, and $\bar{m}_i^*$ denote the Euclidean coordinates of O_i expressed in $\mathcal{F}$, $\mathcal{F}_d$, and $\mathcal{F}^*$, respectively. In (4.134), $R(t)$, $R_d(t) \in SO(3)$ denote the rotation between $\mathcal{F}$ and $\mathcal{F}^*$ and between $\mathcal{F}_d$ and $\mathcal{F}^*$, respectively, and $x_f(t)$, $x_{fd}(t) \in \mathbb{R}^3$ denote translation vectors from $\mathcal{F}$ to $\mathcal{F}^*$

---

[2]It should be noted that if four coplanar target points are not available, then the subsequent development can exploit the classic eight-point algorithm [56] with no four of the eight target points being coplanar.

and $\mathcal{F}_d$ to $\mathcal{F}^*$ expressed in the coordinates of $\mathcal{F}$ and $\mathcal{F}_d$, respectively. Since the Euclidean position of $\mathcal{F}$, $\mathcal{F}_d$, and $\mathcal{F}^*$ cannot be directly measured, the expressions in (4.134) need to be related to the measurable image-space coordinates. To this end, $m_i(t)$, $m_{di}(t)$, m_i^* denote the normalized Euclidean coordinates of O_i expressed, respectively, in coordinate frames $\mathcal{F}$, $\mathcal{F}_d$, and $\mathcal{F}^*$ as follows

$$m_i \triangleq \frac{\bar{m}_i}{z_i} \qquad m_{di} \triangleq \frac{\bar{m}_{di}}{z_{di}} \qquad m_i^* \triangleq \frac{\bar{m}_i^*}{z_i^*} \qquad (4.135)$$

under the standard assumption that $z_i(t), z_{di}(t), z_i^* > \varepsilon$ where ε denotes an arbitrarily small positive constant. Based on (4.135), the expression in (4.134) can be rewritten as follows

$$m_i = \underbrace{\frac{z_i^*}{z_i}}_{\alpha_i} \underbrace{\left(R + \frac{x_f}{d^*}n^{*T}\right)}_{H} m_i^* \qquad (4.136)$$

$$m_{di} = \underbrace{\frac{z_i^*}{z_{di}}}_{\alpha_{di}} \underbrace{\left(R_d + \frac{x_{fd}}{d^*}n^{*T}\right)}_{H_d} m_i^*. \qquad (4.137)$$

In (4.136) and (4.137), $\alpha_i(t)$, $\alpha_{di}(t) \in \mathbb{R}$ denote invertible depth ratios, $H(t)$, $H_d(t) \in \mathbb{R}^{3\times3}$ denote Euclidean homographies [23], and $d^* \in \mathbb{R}$ denotes the constant, unknown distance from the origin of $\mathcal{F}^*$ to π. The following projective relationship can also be developed from Fig. 4.14

$$d^* = n^{*T}\bar{m}_i^*. \qquad (4.138)$$

Also from Fig. 4.14, the unknown, time varying distance from the origin of $\mathcal{F}_d$ to π, denoted by $d(t) \in \mathbb{R}$, can be expressed as follows

$$d = n^{*T}R_d^T\bar{m}_{di}. \qquad (4.139)$$

Projective Homography

Each feature point on π has projected pixel coordinates denoted by $u_i(t)$, $v_i(t) \in \mathbb{R}$ in $\mathcal{F}$, $u_{di}(t)$, $v_{di}(t) \in \mathbb{R}$ in $\mathcal{F}_d$, and u_i^*, $v_i^* \in \mathbb{R}$ in $\mathcal{F}^*$, that are defined as follows

$$p_i \triangleq \begin{bmatrix} u_i & v_i & 1 \end{bmatrix}^T \qquad p_{di} \triangleq \begin{bmatrix} u_{di} & v_{di} & 1 \end{bmatrix}^T$$
$$p_i^* \triangleq \begin{bmatrix} u_i^* & v_i^* & 1 \end{bmatrix}^T. \qquad (4.140)$$

In (4.140), $p_i(t)$, $p_{di}(t)$, $p_i^* \in \mathbb{R}^3$ represent the image-space coordinates of the time-varying feature points, the desired time-varying feature point trajectory, and the constant reference feature points, respectively. To calculate the Euclidean homography given in (4.136) and (4.137) from pixel information, the projected pixel coordinates of the target points are related to $m_i(t)$, $m_{di}(t)$, and m_i^* by the following pin-hole lens models [23]

$$p_i = Am_i \qquad p_{di} = Am_{di} \qquad p_i^* = Am_i^* \qquad (4.141)$$

where $A \in \mathbb{R}^{3 \times 3}$ is a known, constant, and invertible intrinsic camera calibration matrix with the following form

$$A = \begin{bmatrix} a_1 & a_2 & a_4 \\ 0 & a_3 & a_5 \\ 0 & 0 & 1 \end{bmatrix} \qquad (4.142)$$

where $a_i \in \mathbb{R}\ \forall i = 1, 2, ..., 5$, denote known, constant calibration parameters (see [23]). After substituting (4.141) into (4.136) and (4.137), the following relationships can be developed

$$p_i = \alpha_i \underbrace{\left(AHA^{-1}\right)}_{G} p_i^* \qquad p_{di} = \alpha_{di} \underbrace{\left(AH_d A^{-1}\right)}_{G_d} p_i^* \qquad (4.143)$$

where $G(t)$, $G_d(t) \in \mathbb{R}^{3 \times 3}$ denote projective homographies. Given the images of the 4 feature points on π expressed in $\mathcal{F}$, $\mathcal{F}_d$, and $\mathcal{F}^*$, a linear system of equations can be developed from (4.143). From the linear system of equations, a decomposition algorithm (e.g., the Faugeras decomposition algorithm in [23]) can be used to compute $\alpha_i(t)$, $\alpha_{di}(t)$, n^*, $R(t)$, and $R_d(t)$ (see [9] for details)[3]. Hence, $\alpha_i(t)$, $\alpha_{di}(t)$, n^*, $R(t)$, and $R_d(t)$ are known signals that can be used in the subsequent development.

Kinematic Model of Vision System

The camera pose, denoted by $\Upsilon(t) \in \mathbb{R}^6$, can be expressed in terms of a hybrid of pixel and reconstructed Euclidean information as follows

$$\Upsilon(t) \triangleq \begin{bmatrix} p_{e1}^T & \Theta^T \end{bmatrix}^T \qquad (4.144)$$

where the extended pixel coordinate $p_{e1}(t) \in \mathbb{R}^3$ is defined as follows

$$p_{e1} = \begin{bmatrix} u_1 & v_1 & -\ln(\alpha_1) \end{bmatrix}^T, \qquad (4.145)$$

and $\Theta(t) \in \mathbb{R}^3$ denotes the following axis-angle representation of $R(t)$ (see [9] for details)

$$\Theta = \mu(t)\theta(t). \qquad (4.146)$$

---

[3] The initial best-guess of n^* can be utilized to resolve the decomposition ambiguity. See [10] for details.

In (4.146), $\mu(t) \in \mathbb{R}^3$ represents the unit axis of rotation, and $\theta(t)$ denotes the rotation angle about that axis. Based on the development in Section B.2.4 of Appendix B, the open-loop dynamics for $\Upsilon(t)$ can be expressed as follows

$$\dot{\Upsilon} = \begin{bmatrix} \dot{p}_{e1} \\ \dot{\Theta} \end{bmatrix} = \begin{bmatrix} -\dfrac{1}{z_1} A_{e1} & A_{e1} [m_1]_\times \\ 0 & -L_\omega \end{bmatrix} \begin{bmatrix} v_c \\ \omega_c \end{bmatrix} \qquad (4.147)$$

where $v_c(t) \in \mathbb{R}^3$ and $\omega_c(t) \in \mathbb{R}^3$ denote the linear and angular velocity of the camera expressed in terms of $\mathcal{F}$, $A_{ei}(u_i, v_i) \in \mathbb{R}^{3 \times 3}$ is a known, invertible matrix defined as follows

$$A_{ei} = A - \begin{bmatrix} 0 & 0 & u_i \\ 0 & 0 & v_i \\ 0 & 0 & 0 \end{bmatrix} \qquad i = 1, 2, 3, 4, \qquad (4.148)$$

and the invertible Jacobian-like matrix $L_\omega(\theta, \mu) \in \mathbb{R}^{3 \times 3}$ is defined as

$$L_\omega = I_3 - \frac{\theta}{2} [\mu]_\times + \left(1 - \frac{sinc(\theta)}{sinc^2 \left(\dfrac{\theta}{2} \right)} \right) [\mu]_\times^2 \qquad (4.149)$$

where

$$sinc(\theta(t)) \triangleq \frac{\sin \theta(t)}{\theta(t)}.$$

Remark 4.8 *As stated in [68], the axis-angle representation of (4.146) is not unique, in the sense that a rotation of $-\theta(t)$ about $-\mu(t)$ is equal to a rotation of $\theta(t)$ about $\mu(t)$. A particular solution[4] for $\theta(t)$ and $\mu(t)$ can be determined as follows [68]*

$$\theta_p = \cos^{-1} \left(\frac{1}{2} (tr(R) - 1) \right) \qquad [\mu_p]_\times = \frac{R - R^T}{2 \sin(\theta_p)} \qquad (4.150)$$

where the notation $tr(\cdot)$ denotes the trace of a matrix, and $[\mu_p]_\times$ denotes the 3×3 skew-symmetric expansion of $\mu_p(t)$. From (4.150), it is clear that

$$0 \leq \theta_p(t) \leq \pi. \qquad (4.151)$$

4.4.2   Image-Based Path Planning

The path planning objective involves regulating the pose of a camera held by the end-effector of a robot manipulator to a desired camera pose along

---

[4]See [8] for further details.

an image-space trajectory while ensuring the target points remain visible. To achieve this objective, a desired camera pose trajectory is constructed in this section so that the desired image feature vector, denoted by $\bar{p}_d(t) \triangleq \begin{bmatrix} u_{d1}(t) & v_{d1}(t) & \ldots & u_{d4}(t) & v_{d4}(t) \end{bmatrix}^T \in \mathbb{R}^8$, remains in a set, denoted by $\mathcal{D} \subset \mathbb{R}^8$, where all four feature points of the target remain visible for a valid camera pose. The constant, goal image feature vector $\bar{p}^* \triangleq \begin{bmatrix} u_1^* & v_1^* & \ldots & u_4^* & v_4^* \end{bmatrix}^T \in \mathbb{R}^8$ is assumed be in the interior of $\mathcal{D}$. To generate the desired camera pose trajectory such that $\bar{p}_d(t) \in \mathcal{D}$, we will use the navigation function [39] as previously defined in Properties 4.2.4 – 4.2.7.

Pose Space to Image Space Relationship

To develop a desired camera pose trajectory that ensures $\bar{p}_d(t) \in \mathcal{D}$, the desired image feature vector is related to the desired camera pose, denoted by $\Upsilon_d(t) \in \mathbb{R}^6$, through the following relationship

$$\bar{p}_d = \Pi(\Upsilon_d) \tag{4.152}$$

where $\Pi(\cdot) : \mathbb{R}^6 \to \mathcal{D}$ denotes a known function that maps the camera pose to the image feature vector[5]. In (4.152), the desired camera pose is defined as follows

$$\Upsilon_d(t) \triangleq \begin{bmatrix} p_{ed1}^T & \Theta_d^T \end{bmatrix}^T \tag{4.153}$$

where $p_{ed1}(t) \in \mathbb{R}^3$ denotes the desired extended pixel coordinates defined as follows

$$p_{ed1} = \begin{bmatrix} u_{d1} & v_{d1} & -\ln(\alpha_{d1}) \end{bmatrix}^T \tag{4.154}$$

where $\alpha_{d1}(t)$ is introduced in (4.137), and $\Theta_d(t) \in \mathbb{R}^3$ denotes the axis-angle representation of $R_d(t)$ as follows

$$\Theta_d = \mu_d(t)\theta_d(t) \tag{4.155}$$

where $\mu_d(t) \in \mathbb{R}^3$ and $\theta_d(t) \in \mathbb{R}$ are defined with respect to $R_d(t)$ in the same manner as $\mu(t)$ and $\theta(t)$ in (4.146) with respect to $R(t)$.

---

[5] The reason we choose four feature points to construct the image feature vector is that the same image of three points can be seen from four different camera poses [33]. A unique camera pose can theoretically be obtained by using at least four points [6]. Therefore, the map $\Pi(\cdot)$ is a unique mapping with the image feature vector corresponding to a valid camera pose.

Desired Image Trajectory Planning

After taking the time derivative of (4.152), the following expression can be obtained

$$\dot{\bar{p}}_d = L_{\Upsilon_d}\dot{\Upsilon}_d \tag{4.156}$$

where $L_{\Upsilon_d}(\bar{p}_d) \triangleq \dfrac{\partial \bar{p}_d}{\partial \Upsilon_d} \in \mathbb{R}^{8\times 6}$ denotes an image Jacobian-like matrix. Based on the development in Section (B.2.5) of Appendix B, a measurable expression for $L_{\Upsilon_d}(t)$ can be developed as follows

$$L_{\Upsilon_d} = \bar{I}T \tag{4.157}$$

where $\bar{I} \in \mathbb{R}^{8\times 12}$ denotes a constant, row-delete matrix defined as follows

$$\bar{I} = \begin{bmatrix} I_2 & 0^2 & 0_2 & 0^2 & 0_2 & 0^2 & 0_2 & 0^2 \\ 0_2 & 0^2 & I_2 & 0^2 & 0_2 & 0^2 & 0_2 & 0^2 \\ 0_2 & 0^2 & 0_2 & 0^2 & I_2 & 0^2 & 0_2 & 0^2 \\ 0_2 & 0^2 & 0_2 & 0^2 & 0_2 & 0^2 & I_2 & 0^2 \end{bmatrix}$$

where $I_n \in \mathbb{R}^{n\times n}$ denotes the $n \times n$ identity matrix, $0_n \in \mathbb{R}^{n\times n}$ denotes an $n \times n$ matrix of zeros, $0^n \in \mathbb{R}^n$ denotes an $n \times 1$ column of zeros, and $T(t) \in \mathbb{R}^{12\times 6}$ is a measurable auxiliary matrix defined as follows

$$T = \begin{bmatrix} I_3 & 0_3 \\ \dfrac{\beta_1}{\beta_2} A_{ed2} A_{ed1}^{-1} & A_{ed2} \left[\dfrac{\beta_1}{\beta_2} m_{d1} - m_{d2} \right]_\times L_{\omega d}^{-1} \\ \dfrac{\beta_1}{\beta_3} A_{ed3} A_{ed1}^{-1} & A_{ed3} \left[\dfrac{\beta_1}{\beta_3} m_{d1} - m_{d3} \right]_\times L_{\omega d}^{-1} \\ \dfrac{\beta_1}{\beta_4} A_{ed4} A_{ed1}^{-1} & A_{ed4} \left[\dfrac{\beta_1}{\beta_4} m_{d1} - m_{d4} \right]_\times L_{\omega d}^{-1} \end{bmatrix}. \tag{4.158}$$

In (4.158), $A_{edi}(u_{di}, v_{di}) \in \mathbb{R}^{3\times 3}$ and the Jacobian-like matrix $L_{\omega d}(\theta_d, \mu_d) \in \mathbb{R}^{3\times 3}$ are defined with respect to $u_{di}(t), v_{di}(t), \mu_d(t)$, and $\theta_d(t)$ in the same manner as $A_{ei}(\cdot)$ and $L_\omega(\cdot)$ in (4.148) and (4.149) with respect to $u_i(t), v_i(t), \mu(t)$, and $\theta(t)$. The auxiliary variable $\beta_i(t) \in \mathbb{R}$ is defined as follows

$$\beta_i \triangleq \dfrac{z_{di}}{d} \qquad i = 1,2,3,4. \tag{4.159}$$

Based on (4.135), (4.139), and (4.141), $\beta_i(t)$ can be rewritten in terms of computed and measurable terms as follows

$$\beta_i = \dfrac{1}{n^{*T} R_d^T A^{-1} p_{di}}. \tag{4.160}$$

Motivated by (4.156) and the definition of the navigation function in Properties 4.2.4 – 4.2.7, the desired camera pose trajectory is designed as follows

$$\dot{\Upsilon}_d = -k_1 L_{\Upsilon_d}^T \bigtriangledown \varphi \qquad (4.161)$$

where $k_1 \in \mathbb{R}$ denotes a positive constant, and $\bigtriangledown\varphi(\bar{p}_d) \triangleq \left(\frac{\partial\varphi(\bar{p}_d)}{\partial\bar{p}_d}\right)^T \in \mathbb{R}^8$ denotes the gradient vector of $\varphi(\bar{p}_d)$. The development of a particular image space NF and its gradient are provided in Section (B.2.6) of Appendix B. After substituting (4.161) into (4.156), the desired image trajectory can be expressed as follows

$$\dot{\bar{p}}_d = -k_1 L_{\Upsilon_d} L_{\Upsilon_d}^T \bigtriangledown \varphi \qquad (4.162)$$

where it is assumed that $\bigtriangledown\varphi(\bar{p}_d)$ is not a member of the null space of $L_{\Upsilon_d}^T (\bar{p}_d)$. Based on (4.156) and (4.161), it is clear that the desired image trajectory generated by (4.162) will satisfy rigid body constraints.

Remark 4.9 *Based on comments in [6] and the current development, it seems that a remaining open problem is to develop a rigorous, theoretical and general approach to ensure that $\bigtriangledown\varphi(\bar{p}_d)$ is not a member of the null space of $L_{\Upsilon_d}^T (\bar{p}_d)$ (i.e., $\bigtriangledown\varphi(\bar{p}_d) \notin NS(L_{\Upsilon_d}^T (\bar{p}_d))$ where $NS(\cdot)$ denotes the null space operator). However, since the approach described here is developed in terms of the desired image-space trajectory (and hence, is an off-line approach), a particular desired image trajectory can be chosen (e.g., by trial and error) a priori to ensure that $\bigtriangledown\varphi(\bar{p}_d) \notin NS(L_{\Upsilon_d}^T (\bar{p}_d))$. Similar comments are provided in [6] and [59] that indicate that this assumption can be readily satisfied in practice for particular cases. Likewise, a particular desired image trajectory is also assumed to be a priori selected to ensure that $\Upsilon_d(t), \dot{\Upsilon}_d(t) \in \mathcal{L}_\infty$ if $\bar{p}_d(t) \in \mathcal{D}$. Based on the structure of (4.153) and (4.154), the assumption that $\Upsilon_d(t), \dot{\Upsilon}_d(t) \in \mathcal{L}_\infty$ if $\bar{p}_d(t) \in \mathcal{D}$ is considered mild in the sense that the only possible alternative case is if the camera could somehow be positioned at an infinite distance from the target while all four feature points remain visible.*

Path Planner Analysis

Lemma 4.8 *Provided the desired feature points can be a priori selected to ensure that $\bar{p}_d(0) \in \mathcal{D}$ and that $\bigtriangledown\varphi(\bar{p}_d) \notin NS(L_{\Upsilon_d}^T (\bar{p}_d))$, then the desired image trajectory generated by (4.162) ensures that $\bar{p}_d(t) \in \mathcal{D}$ and (4.162) has the asymptotically stable equilibrium point $\bar{p}^*$.*

Proof: Let $V_1 (\bar{p}_d) : \mathcal{D} \to \mathbb{R}$ denote a non-negative function defined as follows

$$V_1 (\bar{p}_d) \triangleq \varphi (\bar{p}_d) . \qquad (4.163)$$

After taking the time derivative of (4.163), the following expression can be obtained

$$\dot{V}_1\left(\bar{p}_d\left(t\right)\right) = \left(\nabla\varphi\right)^T \dot{\bar{p}}_d. \tag{4.164}$$

After substituting (4.162) into (4.164) to obtain the following expression

$$\dot{V}_1\left(\bar{p}_d\left(t\right)\right) = -k_1 \left\| L_{\Upsilon_d}^T \nabla\varphi \right\|^2, \tag{4.165}$$

it is clear that $V_1\left(\bar{p}_d\left(t\right)\right)$ is a non-increasing function in the sense that

$$V_1\left(\bar{p}_d\left(t\right)\right) \leq V_1\left(\bar{p}_d(0)\right). \tag{4.166}$$

From (4.163), (4.166), and the development in Section (B.2.6) of Appendix B, it is clear that for any initial condition $\bar{p}_d\left(0\right) \in \mathcal{D}$, that $\bar{p}_d\left(t\right) \in \mathcal{D}$ $\forall t > 0$; therefore, $\mathcal{D}$ is a positively invariant set [35]. Let $E_1 \subset \mathcal{D}$ denote the following set $E_1 \triangleq \{\bar{p}_d\left(t\right)| \ \dot{V}_1\left(\bar{p}_d\right) = 0\}$. Based on (4.165), it is clear that $\left\| L_{\Upsilon_d}^T\left(\bar{p}_d\right) \nabla\varphi\left(\bar{p}_d\right) \right\| = 0$ in E_1; hence, from (4.161) and (4.162) it can be determined that $\left\| \dot{\Upsilon}_d\left(t\right) \right\| = \left\| \dot{\bar{p}}_d\left(t\right) \right\| = 0$ in E_1, and that E_1 is the largest invariant set. By invoking LaSalle's Theorem [35], it can be determined that every solution $\bar{p}_d\left(t\right) \in \mathcal{D}$ approaches E_1 as $t \to \infty$, and hence, $\left\| L_{\Upsilon_d}^T\left(\bar{p}_d\right) \nabla\varphi\left(\bar{p}_d\right) \right\| \to 0$. Since $\bar{p}_d\left(t\right)$ are chosen *a priori* via the off-line path planning routine in (4.162), the four feature points can be *a priori* selected to ensure that $\nabla\varphi(\bar{p}_d) \notin NS(L_{\Upsilon_d}^T\left(\bar{p}_d\right))$. Provided $\nabla\varphi(\bar{p}_d) \notin NS(L_{\Upsilon_d}^T\left(\bar{p}_d\right))$, then $\left\| L_{\Upsilon_d}^T\left(\bar{p}_d\right) \nabla\varphi(\bar{p}_d) \right\| = 0$ implies that $\left\| \nabla\varphi(\bar{p}_d) \right\| = 0$. Based on development given in Section (B.2.6) of Appendix B, since $\nabla\varphi\left(\bar{p}_d\left(t\right)\right) \to 0$ then $\bar{p}_d(t) \to \bar{p}^*$. ∎

4.4.3 *Tracking Control Development*

Based on Lemma 4.8, the desired camera pose trajectory can be generated from (4.161) to ensure that the camera moves along a path generated in the image space such that the desired object features remain visible (i.e., $\bar{p}_d(t) \in \mathcal{D}$). The objective in this section is to develop a controller so that the actual camera pose $\Upsilon\left(t\right)$ tracks the desired camera pose $\Upsilon_d\left(t\right)$ generated by (4.161), while also ensuring that the object features remain visible (i.e., $\bar{p}(t) \triangleq \begin{bmatrix} u_1\left(t\right) & v_1\left(t\right) & ... & u_4\left(t\right) & v_4\left(t\right) \end{bmatrix}^T \in \mathcal{D}$). To quantify this objective, a rotational tracking error, denoted by $e_\omega(t) \in \mathbb{R}^3$, is defined as

$$e_\omega \triangleq \Theta - \Theta_d, \tag{4.167}$$

and a translational tracking error, denoted by $e_v\left(t\right) \in \mathbb{R}^3$, is defined as follows

$$e_v = p_{e1} - p_{ed1}. \tag{4.168}$$

Control Development

After taking the time derivative of (4.167) and (4.168), the open-loop dynamics for $e_\omega(t)$ and $e_v(t)$ can be obtained as follows

$$\dot{e}_\omega = -L_\omega \omega_c - \dot{\Theta}_d \qquad (4.169)$$

$$\dot{e}_v = -\frac{1}{z_1} A_{e1} v_c + A_{e1} [m_1]_\times \omega_c - \dot{p}_{ed1} \qquad (4.170)$$

where (4.147) was utilized. Based on the open-loop error systems in (4.169) and (4.170), $v_c(t)$ and $\omega_c(t)$ are designed as follows

$$\omega_c \triangleq L_\omega^{-1} \left(K_\omega e_\omega - \dot{\Theta}_d \right) \qquad (4.171)$$

$$v_c \triangleq \frac{1}{\alpha_1} A_{e1}^{-1} \left(K_v e_v - \hat{z}_1^* \dot{p}_{ed1} \right) + \frac{1}{\alpha_1} [m_1]_\times \omega_c \hat{z}_1^* \qquad (4.172)$$

where $K_\omega, K_v \in \mathbb{R}^{3\times3}$ denote diagonal matrices of positive constant control gains, and $\hat{z}_1^*(t) \in \mathbb{R}$ denotes a parameter estimate for z_1^* that is designed as follows

$$\dot{\hat{z}}_1^* \triangleq k_2 e_v^T \left(A_{e1} [m_1]_\times \omega_c - \dot{p}_{ed1} \right) \qquad (4.173)$$

where $k_2 \in \mathbb{R}$ denotes a positive constant adaptation gain. After substituting (4.171) and (4.172) into (4.169) and (4.170), the following closed-loop error systems can be developed

$$\dot{e}_\omega = -K_\omega e_\omega \qquad (4.174)$$

$$z_1^* \dot{e}_v = -K_v e_v + \left(A_{e1} [m_1]_\times \omega_c - \dot{p}_{ed1} \right) \tilde{z}_1^* \qquad (4.175)$$

where the parameter estimation error signal $\tilde{z}_1^*(t) \in \mathbb{R}$ is defined as follows

$$\tilde{z}_1^* = z_1^* - \hat{z}_1^*. \qquad (4.176)$$

Controller Analysis

Theorem 4.9 *The control inputs introduced in (4.171) and (4.172), along with the adaptive update law defined in (4.173), ensure that the actual camera pose tracks the desired camera pose trajectory in the sense that*

$$\|e_\omega(t)\| \to 0 \quad \|e_v(t)\| \to 0 \ \text{as} \ t \to \infty. \qquad (4.177)$$

Proof: Let $V_2(t) \in \mathbb{R}$ denote a non-negative function defined as follows

$$V_2 \triangleq \frac{1}{2} e_\omega^T e_\omega + \frac{z_1^*}{2} e_v^T e_v + \frac{1}{2k_2} \tilde{z}_1^{*2}. \qquad (4.178)$$

After taking the time derivative of (4.178) and then substituting for the closed-loop error systems developed in (4.174) and (4.175), the following expression can be obtained

$$
\begin{aligned}
\dot{V}_2 = {}& -e_\omega^T K_\omega e_\omega - e_v^T K_v e_v \\
& + e_v^T \left(A_{e1} \left[m_1 \right]_\times \omega_c - \dot{p}_{ed1} \right) \tilde{z}_1^* - \frac{1}{k_2} \tilde{z}_1^* \dot{\hat{z}}_1^*
\end{aligned}
\tag{4.179}
$$

where the time derivative of (4.176) was utilized. After substituting the adaptive update law designed in (4.173) into (4.179), the following expression can be obtained

$$
\dot{V}_2 = -e_\omega^T K_\omega e_\omega - e_v^T K_v e_v.
\tag{4.180}
$$

Based on (4.176), (4.178), and (4.180), it can be determined that $e_\omega(t)$, $e_v(t)$, $\tilde{z}_1^*(t)$, $\hat{z}_1^*(t) \in \mathcal{L}_\infty$ and that $e_\omega(t)$, $e_v(t) \in \mathcal{L}_2$. Based on the assumption that $\dot{\Theta}_d(t)$ is bounded (see Remark 4.9), the expressions given in (4.167), (4.171), and $L_\omega(t)$ in (4.149) can be used to conclude that $\omega_c(t) \in \mathcal{L}_\infty$. Since $e_v(t) \in \mathcal{L}_\infty$, (4.168), (4.145), (4.141), and $A_{e1}(t)$ in (4.148) can be used to prove that $u_1(t)$, $v_1(t)$, $\alpha_1(t)$, $m_1(t)$, $A_{e1}(t) \in \mathcal{L}_\infty$. Based on the assumption that $\dot{p}_{ed1}(t)$ is bounded (see Remark 4.9), the expressions in (4.172), (4.173), and (4.175) can be used to conclude that $v_c(t)$, $\dot{\hat{z}}_1^*(t)$, $\dot{e}_v(t) \in \mathcal{L}_\infty$. Since $e_\omega(t) \in \mathcal{L}_\infty$, it is clear from (4.174) that $\dot{e}_\omega(t) \in \mathcal{L}_\infty$. Since $e_\omega(t)$, $e_v(t) \in \mathcal{L}_2$ and $e_\omega(t)$, $\dot{e}_\omega(t)$, $e_v(t)$, $\dot{e}_v(t) \in \mathcal{L}_\infty$, Barbalat's Lemma [67] can be used to prove the result given in (4.177). ∎

Remark 4.10 *Based on the result provided in (4.177), it can be proven from the Euclidean reconstruction given in (4.136) and (4.137) that $R(t) \to R_d(t)$, $m_1(t) \to m_{d1}(t)$, and $z_1(t) \to z_{d1}(t)$ (and hence, $x_f(t) \to x_{fd}(t)$). Based on these results, (4.134) can be used to also prove that $\bar{m}_i(t) \to \bar{m}_{di}(t)$. Since $\Pi(\cdot)$ is a unique mapping, we can conclude that the desired camera pose converges to the goal camera pose based on the previous result $\bar{p}_d(t) \to \bar{p}^*$ from Lemma 4.8. Based on the above analysis, $\bar{m}_i(t) \to \bar{m}^*$.*

Remark 4.11 *Based on (4.21) and (4.180), the following inequality can be obtained*

$$
\begin{aligned}
e_\omega^T e_\omega + e_v^T e_v &\leqslant 2 \max \left\{ 1, \frac{1}{z_1^*} \right\} V_2(t) \tag{4.181} \\
&\leqslant 2 \max \left\{ 1, \frac{1}{z_1^*} \right\} V_2(0)
\end{aligned}
$$

where

$$
V_2(0) = \frac{1}{2} e_\omega^T(0) e_\omega(0) + \frac{z_1^*}{2} e_v^T(0) e_v(0) + \frac{1}{2k_2} \tilde{z}_1^{*2}(0).
$$

From (4.144), (4.153), (4.167), (4.168), and the inequality in (4.181), the following inequality can be developed

$$\|\Upsilon - \Upsilon_d\| \leqslant \sqrt{2 \max\left\{1, \frac{1}{z_1^*}\right\} V_2\left(0\right)}. \qquad (4.182)$$

Based on (4.152), the following expression can be developed

$$\bar{p} = \Pi\left(\Upsilon\right) - \Pi\left(\Upsilon_d\right) + \bar{p}_d. \qquad (4.183)$$

After applying the mean-value theorem to (4.183), the following inequality can be obtained

$$\|\bar{p}\| \leqslant \|L_{\Upsilon_d}\| \|\Upsilon - \Upsilon_d\| + \|\bar{p}_d\|. \qquad (4.184)$$

Since all signals are bounded, it can be shown that $L_{\Upsilon_d}^T(\bar{p}_d) \in \mathcal{L}_\infty$; hence, the following inequality can be developed from (4.182) and (4.184)

$$\|\bar{p}\| \leqslant \zeta_b \sqrt{V_2\left(0\right)} + \|\bar{p}_d\| \qquad (4.185)$$

for some positive constant $\zeta_b \in \mathbb{R}$, where $\bar{p}_d\left(t\right) \in \mathcal{D}$ based on Lemma 4.8. To ensure that $\bar{p}\left(t\right) \in \mathcal{D}$, the image space needs to be sized to account for the effects of $\zeta_b\sqrt{V_2\left(0\right)}$. Based on (4.178), $V_2\left(0\right)$ can be made arbitrarily small by increasing k_2 and initializing $\bar{p}_d\left(0\right)$ close or equal to $\bar{p}\left(0\right)$.

4.4.4 Simulation Results

From a practical point of view, we choose a state-related time varying control gain matrix $k_3 \left(L_\Upsilon^T L_\Upsilon\right)^{-1}$ instead of a constant k_1 in (4.161) for the image path planner as follows

$$\dot{\Upsilon}_d = -k_3 \left(L_\Upsilon^T L_\Upsilon\right)^{-1} L_{\Upsilon_d}^T \nabla \varphi \qquad (4.186)$$

where $k_3 \in \mathbb{R}$ is a constant control gain. Through many simulation trials, we conclude that the path planner in (4.186) works better than the path planner in (4.161). Using the path planner in (4.186) instead of the path planner in (4.161) will not affect the proof for Theorem 4.8 as long as $L_\Upsilon^T L_\Upsilon$ is positive definite along the desired image trajectory $\bar{p}_d\left(t\right)$ (It is clear that $L_\Upsilon^T L_\Upsilon$ is positive definite if $L_{\Upsilon_d}\left(\bar{p}_d\right)$ is full rank). Similar to the statement in Remark 4.9, this assumption is readily satisfied for this off-line path planner approach.

To solve the self-occlusion problem (the terminology, self-occlusion, is utilized here to denote the case when the center of the camera is in the plane determined by the feature points) from a practical point of view, we define a distance ratio $\gamma(t) \in \mathbb{R}$ as follows

$$\gamma(t) = \frac{d}{d^*}. \qquad (4.187)$$

From [55], $\gamma(t)$ is measurable. The idea here is to begin by planning a desired image trajectory without self-occlusion. Based on (4.185), we can assume that the actual trajectory is close enough to the desired trajectory such that no self-occlusion occurs for the actual trajectory.

To illustrate the performance of the path planner given in (4.186) and the controller given in (4.171)–(4.173), numerical simulations will performed for four standard visual servo tasks, which are believed to represent the most interesting tasks encountered by a visual servo system [28]:

- Task 1: Optical axis rotation, a pure rotation about the optic axis

- Task 2: Optical axis translation, a pure translation along the optic axis

- Task 3: Camera y-axis rotation, a pure rotation of the camera about the y-axis of the camera coordinate frame.

- Task 4: General camera motion, a transformation that includes a translation and rotation about an arbitrary axis.

For the simulation, the intrinsic camera calibration matrix is given as follows

$$A = \begin{bmatrix} fk_u & -fk_u \cot\phi & u_0 \\ 0 & \dfrac{fk_v}{\sin\phi} & v_0 \\ 0 & 0 & 1 \end{bmatrix} \qquad (4.188)$$

where $u_0 = 257$ [pixels], $v_0 = 253$ [pixels] represent the pixel coordinates of the principal point, $k_u = 101.4$ [pixels·mm^{-1}] and $k_v = 101.4$ [pixels·mm^{-1}] represent camera scaling factors, $\phi = 90$ [degrees] is the angle between the camera axes, and $f = 12.5$ [mm] denotes the camera focal length.

Simulation Results: Optical axis rotation

The initial image-space coordinates and the initial desired image-space coordinates of the 4 target points were selected as follows (in pixels)

$$p_1^T(0) = p_{d1}^T(0) = \begin{bmatrix} 434 & 445 & 1 \end{bmatrix}$$
$$p_2^T(0) = p_{d2}^T(0) = \begin{bmatrix} 56 & 443 & 1 \end{bmatrix}$$
$$p_3^T(0) = p_{d3}^T(0) = \begin{bmatrix} 69 & 49 & 1 \end{bmatrix}$$
$$p_4^T(0) = p_{d4}^T(0) = \begin{bmatrix} 449 & 71 & 1 \end{bmatrix}$$

while the image-space coordinates of the 4 constant reference target points
were selected as follows (in pixels)

$$p_1^* = \begin{bmatrix} 416 & 46 & 1 \end{bmatrix}^T \quad p_2^* = \begin{bmatrix} 479 & 418 & 1 \end{bmatrix}^T$$
$$p_3^* = \begin{bmatrix} 88 & 473 & 1 \end{bmatrix}^T \quad p_4^* = \begin{bmatrix} 45 & 96 & 1 \end{bmatrix}^T.$$

The control parameters were selected as follows

$$K_v = \text{diag}\{1,1,1\} \quad K_\omega = \text{diag}\{0.3, 0.3, 0.3\}$$

$$k_2 = 0.04 \quad k_3 = 400000 \quad \kappa = 8$$

$$K = \text{diag}\{10, 10, 10, 18, 13, 15, 10, 10\}.$$

The desired and actual image trajectories of the feature points are depicted in Figures 4.15 and 4.16, respectively. The translational and rotational tracking errors of the target are depicted in Figures 4.17 and 4.18,
respectively, and the parameter estimate signal is depicted in Figure 4.19.
The control input velocities $\omega_c(t)$ and $v_c(t)$ defined in (4.171) and (4.172)
are depicted in Figures 4.20 and 4.21. From Figures 4.15 and 4.16, it is
clear that the desired feature points and actual feature points remain in
the camera field of view and converge to the goal feature points. Figures
4.17 and 4.18 show that the tracking errors go to zero as $t \to \infty$.

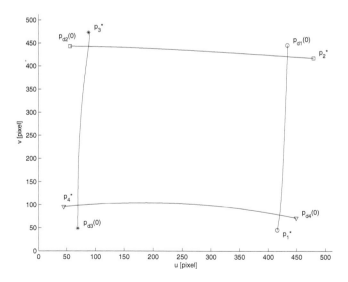

FIGURE 4.15. Task 1: Desired Image Trajectory

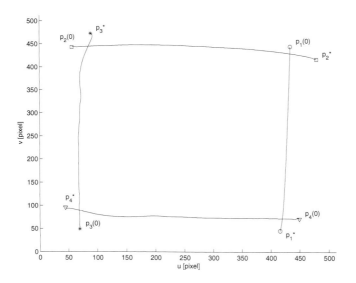

FIGURE 4.16. Task 1: Actual Image Trajectory

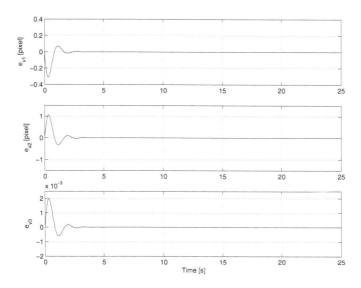

FIGURE 4.17. Task 1: Translational Tracking Error

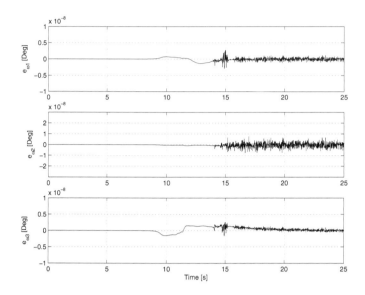

FIGURE 4.18. Task 1: Rotational Tracking Error

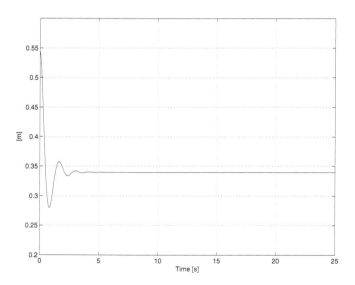

FIGURE 4.19. Task 1: Estimate of z_1^*

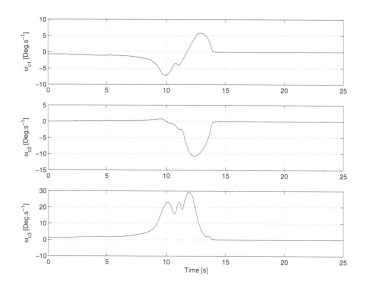

FIGURE 4.20. Task 1: Angular Velocity

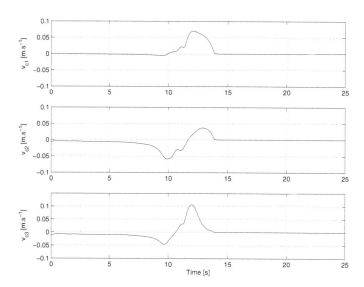

FIGURE 4.21. Task 1: Linear Velocity Input

Simulation Results: Optical axis translation

The initial image-space coordinates and the initial desired image-space coordinates of the 4 target points were selected as follows (in pixels)

$$
\begin{aligned}
p_1(0) &= p_{d1}(0) = \begin{bmatrix} 363 & 115 & 1 \end{bmatrix}^T \\
p_2(0) &= p_{d2}(0) = \begin{bmatrix} 402 & 361 & 1 \end{bmatrix}^T \\
p_3(0) &= p_{d3}(0) = \begin{bmatrix} 147 & 397 & 1 \end{bmatrix}^T \\
p_4(0) &= p_{d4}(0) = \begin{bmatrix} 116 & 148 & 1 \end{bmatrix}^T
\end{aligned}
$$

while the image-space coordinates of the 4 constant reference target points were selected as follows (in pixels)

$$
\begin{aligned}
p_1^* &= \begin{bmatrix} 416 & 46 & 1 \end{bmatrix}^T \quad p_2^* = \begin{bmatrix} 479 & 418 & 1 \end{bmatrix}^T \\
p_3^* &= \begin{bmatrix} 88 & 473 & 1 \end{bmatrix}^T \quad p_4^* = \begin{bmatrix} 45 & 96 & 1 \end{bmatrix}^T.
\end{aligned}
$$

The control parameters were selected as follows

$$
K_v = \operatorname{diag}\{1,1,1\} \quad K_\omega = \operatorname{diag}\{0.3, 0.3, 0.3\}
$$

$$
k_2 = 0.0004 \quad k_3 = 10000 \quad \kappa = 8
$$

$$
K = \operatorname{diag}\{30, 20, 10, 28, 33, 25, 10, 40\}.
$$

The desired and actual image trajectories of the feature points are depicted in Figures 4.22 and 4.23, respectively. The translational and rotational tracking errors of the target are depicted in Figures 4.24 and 4.25, respectively, and the parameter estimate signal is depicted in Figure 4.26. The control input velocities $\omega_c(t)$ and $v_c(t)$ defined in (4.171) and (4.172) are depicted in Figures 4.27 and 4.28. From Figures 4.22 and 4.23, it is clear that the desired feature points and actual feature points remain in the camera field of view and converge to the goal feature points. Figures 4.24 and 4.25 show that the tracking errors go to zero as $t \to \infty$.

Simulation Results: Camera y-axis rotation

The initial image-space coordinates and the initial desired image-space coordinates of the 4 target points were selected as follows (in pixels)

$$
\begin{aligned}
p_1^T(0) &= p_{d1}^T(0) = \begin{bmatrix} 98 & 207 & 1 \end{bmatrix} \\
p_2^T(0) &= p_{d2}^T(0) = \begin{bmatrix} 112 & 288 & 1 \end{bmatrix} \\
p_3^T(0) &= p_{d3}^T(0) = \begin{bmatrix} 29 & 301 & 1 \end{bmatrix} \\
p_4^T(0) &= p_{d4}^T(0) = \begin{bmatrix} 15 & 217 & 1 \end{bmatrix}
\end{aligned}
$$

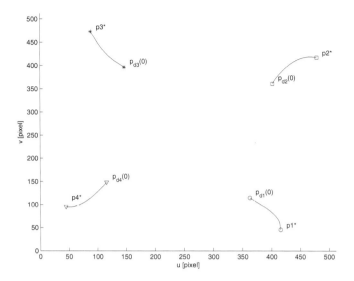

FIGURE 4.22. Task 2: Desired Image Trajectory

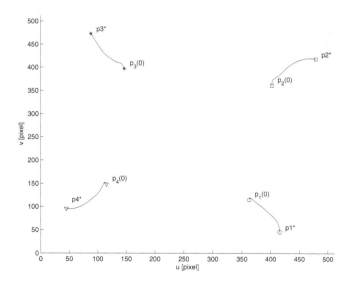

FIGURE 4.23. Task 2: Actual Image Trajectory

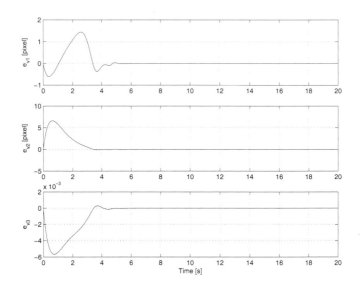

FIGURE 4.24. Task 2: Translational Tracking Error

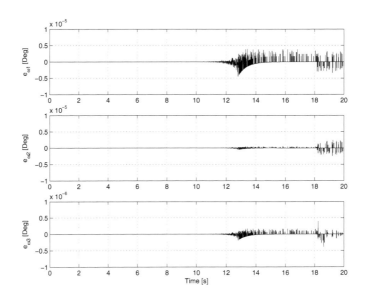

FIGURE 4.25. Task 2: Rotational Tracking Error

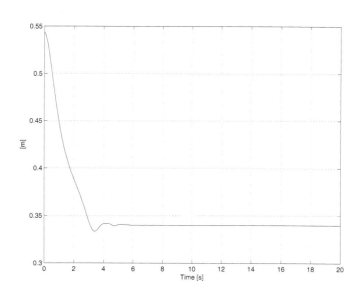

FIGURE 4.26. Task 2: Estimate of z_1^*

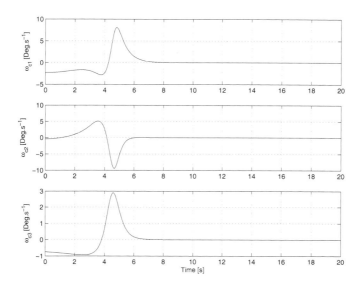

FIGURE 4.27. Task 2: Angular Velocity Input

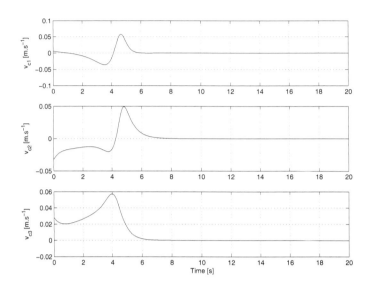

FIGURE 4.28. Task 2: Linear Velocity Input

while the image-space coordinates of the 4 constant reference target points were selected as follows (in pixels)

$$p_1^* = \begin{bmatrix} 478 & 206 & 1 \end{bmatrix}^T \quad p_2^* = \begin{bmatrix} 492 & 289 & 1 \end{bmatrix}^T$$
$$p_3^* = \begin{bmatrix} 408 & 300 & 1 \end{bmatrix}^T \quad p_4^* = \begin{bmatrix} 395 & 218 & 1 \end{bmatrix}^T.$$

The control parameters were selected as follows

$$K_v = \text{diag}\{5, 5, 5\} \quad K_\omega = \text{diag}\{0.3, 0.3, 0.3\}$$

$$k_2 = 0.04 \quad k_3 = 1000000 \quad \kappa = 8$$

$$K = \text{diag}\{30, 20, 10, 28, 33, 25, 10, 40\}.$$

The desired and actual image trajectories of the feature points are depicted in Figures 4.29 and 4.30, respectively. The translational and rotational tracking errors of the target are depicted in Figures 4.31 and 4.32, respectively, and the parameter estimate signal is depicted in Figure 4.33. The control input velocities $\omega_c(t)$ and $v_c(t)$ defined in (4.171) and (4.172) are depicted in Figures 4.34 and 4.35. From Figures 4.29 and 4.30, it is clear that the desired feature points and actual feature points remain in the camera field of view and converge to the goal feature points. Figures 4.31 and 4.32 show that the tracking errors go to zero as $t \to \infty$.

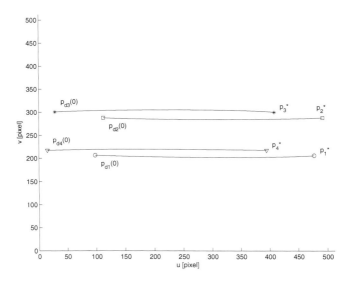

FIGURE 4.29. Task 3: Desired Image Trajectory

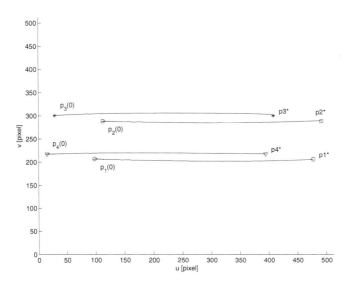

FIGURE 4.30. Task 3: Actual Image Trajectory

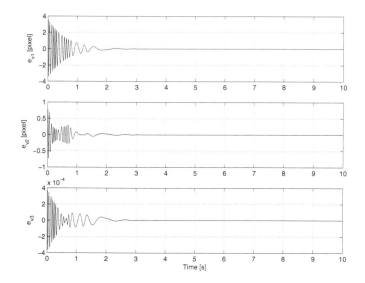

FIGURE 4.31. Task 3: Translational Tracking Error

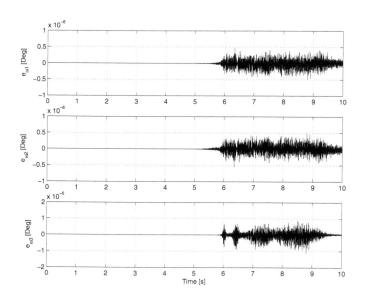

FIGURE 4.32. Task 3: Rotational Tracking Error

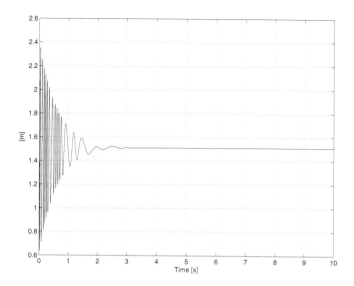

FIGURE 4.33. Task 3: Estimate of z_1^*

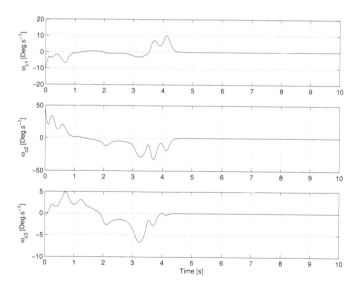

FIGURE 4.34. Task 3: Angular Velocity Input

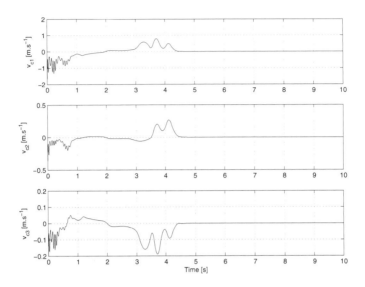

FIGURE 4.35. Task 3: Linear Velocity Input

Simulation Results: General Camera Motion

The initial desired image-space coordinates and the initial desired image-space coordinates of the 4 target points were selected as follows (in pixels)

$$p_1\left(0\right) = p_{d1}\left(0\right) = \left[\begin{array}{ccc} 267 & 428 & 1 \end{array}\right]^T$$
$$p_2\left(0\right) = p_{d2}\left(0\right) = \left[\begin{array}{ccc} 295 & 257 & 1 \end{array}\right]^T$$
$$p_3\left(0\right) = p_{d3}\left(0\right) = \left[\begin{array}{ccc} 446 & 285 & 1 \end{array}\right]^T$$
$$p_4\left(0\right) = p_{d4}\left(0\right) = \left[\begin{array}{ccc} 420 & 449 & 1 \end{array}\right]^T$$

while the image-space coordinates of the 4 constant reference target points were selected as follows (in pixels)

$$p_1^* = \left[\begin{array}{ccc} 416 & 46 & 1 \end{array}\right]^T \quad p_2^* = \left[\begin{array}{ccc} 479 & 418 & 1 \end{array}\right]^T$$

$$p_3^* = \left[\begin{array}{ccc} 88 & 473 & 1 \end{array}\right]^T \quad p_4^* = \left[\begin{array}{ccc} 45 & 96 & 1 \end{array}\right]^T.$$

The control parameters were selected as follows

$$K_v = \mathrm{diag}\left\{1, 1, 1\right\} \quad K_\omega = \mathrm{diag}\left\{0.3, 0.3, 0.3\right\}$$

$$k_2 = 0.004 \quad k_3 = 200000 \quad \kappa = 8$$

$$K = \mathrm{diag}\left\{10, 10, 10, 18, 13, 15, 10, 10\right\}.$$

The desired and actual image trajectories of the feature points are depicted in Figures 4.36 and 4.37, respectively. The translational and rotational errors of the target are depicted in Figures 4.38 and 4.39, respectively, and the parameter estimate signal is depicted in Figure 4.40. The control input velocities $\omega_c(t)$ and $v_c(t)$ defined in (4.171) and (4.172) are depicted in Figures 4.41 and 4.42. From Figures 4.36 and 4.37, it is clear that the desired feature points and actual feature points remain in the camera field of view and converge to the goal feature points. Figures 4.38 and 4.39 show that the tracking errors go to zero as $t \to \infty$.

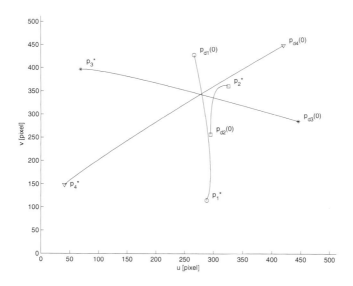

FIGURE 4.36. Task 4: Desired Image Trajectory

4.5   Optimal Navigation and Obstacle Avoidance

As stated in the introduction to this chapter, image-based visual servoing has been a widely used control method owing to the potential for improving robustness of the controller to camera calibration effects. However, some inherent technical problems associated with the use of a non-square image-Jacobian have been the subject of much discussion. In an effort to address these problems, a novel position-based visual servo controller is designed here that works effectively in the presence of uncertain camera calibration

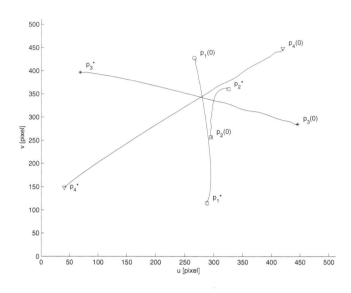

FIGURE 4.37. Task 4: Actual Image Trajectory

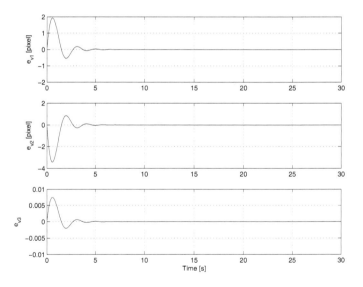

FIGURE 4.38. Task 4: Translational Tracking Error

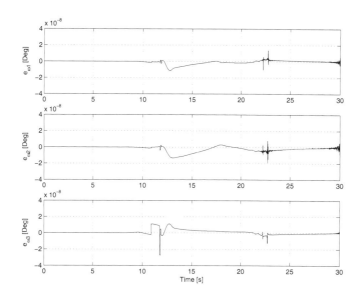

FIGURE 4.39. Task 4: Rotational Tracking Error

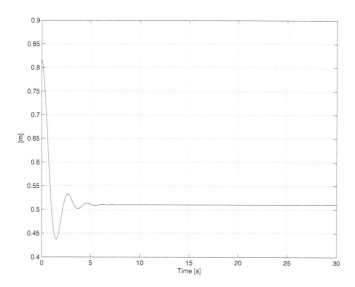

FIGURE 4.40. Task 4: Estimate of z_1^*

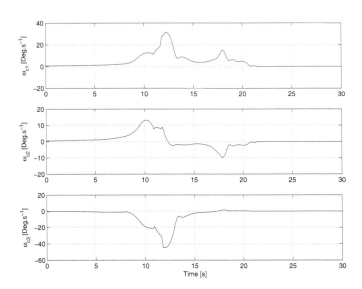

FIGURE 4.41. Task 4: Angular Velocity Input

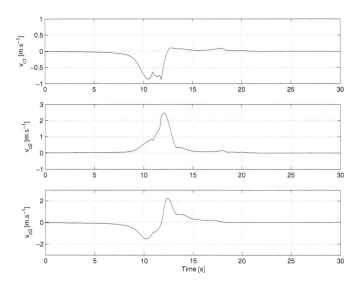

FIGURE 4.42. Task 4: Linear Velocity Input

and nonlinear radial distortion effects. Specifically, an optimization-based, on-line trajectory generator for the image features is fused with a position-based controller to move the kinematic system to a desired setpoint. A planar example based on a fixed camera configuration is used to illustrate the approach. The approach is then extended to the 6-degrees-of-freedom case for the camera-in-hand configuration.

4.5.1 Illustrative Example: Planar PBVS

Manipulator Kinematic Model

The following kinematic equations relate the task-space coordinates, denoted by $x(t) \triangleq \begin{bmatrix} x_1(t) & x_2(t) \end{bmatrix}^T \in \mathbb{R}^2$, of a target point affixed to the tip of a rigid, two-link, revolute, planar robot manipulator to the joint displacements, denoted by $q(t) \triangleq \begin{bmatrix} q_1(t) & q_2(t) \end{bmatrix}^T \in \mathbb{R}^2$

$$x = \Omega(q) \tag{4.189}$$

where $\Omega(q) \in C^2$ is a known function. After taking the time derivative of (4.189), the following expression can be obtained

$$\dot{x} = J(q)\dot{q} \tag{4.190}$$

where the manipulator Jacobian $J(q) \in \mathbb{R}^{2 \times 2}$ is defined as

$$J(q) = \begin{bmatrix} \dfrac{\partial \Omega(q)}{\partial q_1} & \dfrac{\partial \Omega(q)}{\partial q_2} \end{bmatrix} \tag{4.191}$$

and $J(q)$ is assumed to have a bounded first order partial derivative (i.e., $J(q) \in C^1$). The inverse of the manipulator Jacobian, denoted by $J^{-1}(q)$, is assumed to always exist, and all kinematic singularities associated with $J(q)$ are assumed to be always avoided.

Camera Model

As illustrated in Figure 4.43, the visual servo system in this example consists of a planar robot and a single camera mounted in a fixed configuration above the robot workspace with the camera optical axis perpendicular to the robot's plane of motion where the target point on the robot manipulator is assumed to remain in the camera field-of-view throughout the entire range of the manipulator. The task-space position of the target point can be related to the corresponding image-space coordinate, denoted by

$\bar{p}(t) \triangleq \begin{bmatrix} u(t) & v(t) \end{bmatrix}^T \in \mathbb{R}^2$, based on the camera model as follows

$$\bar{p}(t) \triangleq \Pi_1(x) \tag{4.192}$$

$$= \frac{1}{d} f_r(x) A R_r \left(x - \begin{bmatrix} O_{01} \\ O_{02} \end{bmatrix} \right) + \begin{bmatrix} u_0 \\ v_0 \end{bmatrix}$$

where it is assumed that the mapping $\Pi_1(\cdot) : \mathbb{R}^2 \to \mathbb{R}^2$ is a unique mapping (the terminology, unique mapping, is utilized here to denote a one-to-one and invertible mapping). In (4.192), $d \in \mathbb{R}$ denotes a constant unknown distance from the manipulator target point to the camera along the optical axis, $f_r(x) \in \mathbb{R}$ denotes radial distortion effects, $A \in \mathbb{R}^{2 \times 2}$ is a constant and invertible intrinsic camera calibration matrix, $R_r(\theta_r) \in SO(2)$ is a constant rotation offset matrix, $\begin{bmatrix} O_{01} & O_{02} \end{bmatrix}^T \in \mathbb{R}^2$ denotes the projection of the camera's optical center on the task-space plane, and $\begin{bmatrix} u_0 & v_0 \end{bmatrix}^T \in \mathbb{R}^2$ denotes the pixel coordinates of the principal point (i.e., the image center that is defined as the frame buffer coordinates of the intersection of the optical axis with the image plane). In (4.192), the camera calibration matrix has the following form

$$A = \begin{bmatrix} f k_u & -f k_u \cot \phi \\ 0 & \dfrac{f k_v}{\sin \phi} \end{bmatrix} \tag{4.193}$$

where $k_u, k_v \in \mathbb{R}$ represent constant camera scaling factors, $\phi \in \mathbb{R}$ represents the constant angle between the camera axes (i.e., skew angle), and $f \in \mathbb{R}$ denotes the constant camera focal length. The constant offset rotation matrix in (4.192), is defined as follows

$$R_r \triangleq \begin{bmatrix} \cos \theta_r & -\sin \theta_r \\ \sin \theta_r & \cos \theta_r \end{bmatrix}$$

where $\theta_r \in \mathbb{R}$ represents the constant right-handed rotation angle of the image-space coordinate system with respect to the task-space coordinate system that is assumed to be confined to the following regions

$$-\pi < \theta_r < \pi. \tag{4.194}$$

The radial distortion effects in (4.192) are assumed to be modeled by the following nondecreasing polynomial [52], [75]

$$f_r(x) = 1 + c_1 r^2(x) + c_2 r^4(x) \tag{4.195}$$

where c_1 and c_2 are radial distortion coefficients, and the undistorted radial distance $r(x) \in \mathbb{R}$ is defined as follows

$$r = \left\| \frac{1}{d} R_r \left(x - \begin{bmatrix} O_{01} \\ O_{02} \end{bmatrix} \right) \right\|. \tag{4.196}$$

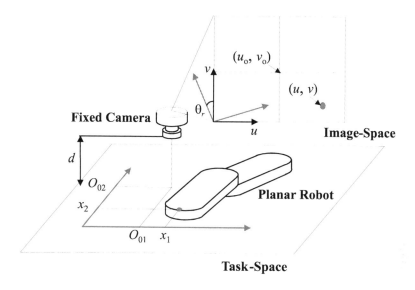

FIGURE 4.43. Planar Visual Servo Control System

Control Objective

The PBVS control objective is to ensure that the task-space position of the manipulator target point is regulated to a desired task-space setpoint, denoted by $x^* \in \mathbb{R}^2$, that corresponds to a given, desired image-space setpoint, denoted by $\bar{p}^* \in \mathbb{R}^2$ (according to (4.192)). The challenging aspect of this problem is that the image-space to task-space relationship is corrupted by uncertainty in the camera calibration, radial distortion effects, and unknown depth information (i.e., x^* is unknown while $\bar{p}^*$ is known). To quantify the control objective, a task-space tracking error signal, denoted by $e(t) \in \mathbb{R}^2$, is defined as follows

$$e(t) \triangleq x(t) - x_d(t) \tag{4.197}$$

where $x_d(t) \in \mathbb{R}^2$ denotes a subsequently designed desired task-space trajectory that is assumed to be designed such that $x_d(t)$, $\dot{x}_d(t) \in \mathcal{L}_\infty$ and that $x_d(t) \to x^*$ as $t \to \infty$.

Closed-Loop Error System Development

By taking the time derivative of (4.197) and utilizing (4.190), the following open-loop error system can be obtained

$$\dot{e}(t) = J(q)\dot{q} - \dot{x}_d(t). \tag{4.198}$$

Based on the previously stated assumption about the existence of the inverse of the manipulator Jacobian, the following kinematic velocity control input is designed

$$\dot{q} = J^{-1}(q)\left(-k_1 e(t) + \dot{x}_d(t)\right) \tag{4.199}$$

where $k_1 \in \mathbb{R}$ denotes an adjustable positive constant. After substituting (4.199) into (4.198), the following closed-loop error system can be obtained

$$\dot{e}(t) = -k_1 e(t). \tag{4.200}$$

Using standard linear analysis techniques, the solution of (4.200) is given by

$$e(t) = e(0)\exp(-k_1 t). \tag{4.201}$$

Image-Based Extremum Seeking Path Planner

As previously described, the simple planar example presents a significant control design challenge because the mapping between the task-space and the image-space is unknown; hence, the unknown setpoint x^* can not be reconstructed from the desired image-space setpoint $\bar{p}^*$. To address this challenge, a numeric optimization routine is utilized to produce a desired trajectory that seeks the unknown setpoint x^* online. Specifically, an objective function, denoted by an $\varphi\left(\bar{p}(t)\right) \in \mathbb{R}$, is defined as follows

$$\varphi\left(\bar{p}\right) \triangleq \frac{1}{2}\left(\bar{p} - \bar{p}^*\right)^T\left(\bar{p} - \bar{p}^*\right). \tag{4.202}$$

Clearly, we can see from (4.202) that $\varphi\left(\bar{p}\right)$ has a unique minimum at $\bar{p}(t) = \bar{p}^*$. By using the mapping defined in (4.192), we can rewrite (4.202) as follows

$$\varphi\left(x\right) = \frac{1}{2}\left(\Pi_1\left(x\right) - \Pi_1\left(x^*\right)\right)^T\left(\Pi_1\left(x\right) - \Pi_1\left(x^*\right)\right). \tag{4.203}$$

From (4.203), it is easy to show that a unique minimum at $\bar{p}(t) = \bar{p}^*$ corresponds to a unique minimum at $x(t) = x^*$ (see Section B.2.7 of Appendix B for the 6 DOF case). If $x(t)$ could be directly manipulated, a standard optimization routine could be utilized to locate the minimum of $\varphi\left(x\right)$ [6]. That is, optimization routines (e.g., Brent's Method [62]) provide a mechanism for numerically searching for the minimum of an objective function whose structure is unknown (i.e., the right-hand side of (4.203) is uncertain because of the presence of $\Pi_1\left(\cdot\right)$) provided that the output of the function

---

[6]There are numerous optimization routines that only require measurement of the objective function; hence, gradient and/or Hessian related information is not required.

can be measured (i.e., $\varphi(x)$ can be measured by using the right-hand side of (4.202)).

While the above optimization procedure is simple to understand conceptually, it can not be used as described above because $x(t)$ can not be directly manipulated; rather, we must indirectly manipulate $x(t)$ through the controller given in (4.199). To illustrate how the optimization routine can be fused with the controller, we utilize (4.197) to rewrite (4.203) as follows

$$\varphi(x_d, e) = \frac{1}{2} (\Pi_1 (x_d + e) - \Pi_1 (x^*))^T \qquad (4.204)$$
$$\cdot (\Pi_1 (x_d + e) - \Pi_1 (x^*)).$$

If $e(t) = 0$ in (4.204), an optimization routine can be used to minimize $\varphi(x_d, e)$ by directly manipulating $x_d(t)$ as explained above. As illustrated by (4.201), $e(t)$ goes to zero very quickly. In fact, $e(t)$ can actually be set to zero for all time by designing $x_d(0) = x(0)$ such that $e(0) = 0$, and hence, $e(t) = 0$ for all time (i.e., at least theoretically). However, it should be noted that we can not utilize an optimization routine for $x_d(t)$ unless we slow down the optimization routine because the kinematic system will not be able to track a desired trajectory that exhibits large desired velocity values (i.e., large values of $\dot{x}_d(t)$). To slow down the desired trajectory generated by the optimization routine, we can utilize a set of low pass filters and some thresholding functions. To illustrate how an optimization routine can be used to generate $x_d(t)$, the following step-by-step procedure is given:

- Step 1. Initialize the optimization routine as follows $\bar{x}_{d(k)}\big|_{k=0} = x(t)\big|_{t=0}$ where $\bar{x}_{d(k)} \in \mathbb{R}^2$ denotes the k-th output of the numeric optimization which is held constant during the k-th iteration. The output of the optimization function, denoted by $\bar{x}_d(t)$, is a discrete signal with the value $\bar{x}_{d(k)}$ at the k-th iteration.

- Step 2. Set the iteration number $k = 1$. Invoke one iteration of an optimization algorithm for the objective function $\varphi(x_d(0), e(0))$.

- Step 3. $\bar{x}_d(t)$ is passed through a set of second order stable and proper low pass filters to generate continuous bounded signals for $x_d(t)$ and $\dot{x}_d(t)$. For example, the following filters could be utilized

$$x_d = \frac{\varsigma_1}{s^2 + \varsigma_2 s + \varsigma_3} \bar{x}_d$$
$$\qquad (4.205)$$
$$\dot{x}_d = \frac{\varsigma_1 s}{s^2 + \varsigma_2 s + \varsigma_3} \bar{x}_d$$

where ς_1, ς_2, and ς_3 denote positive filter constants.

- Step 4. Wait for T seconds (or wait until $\left\|x\left(t\right) - \bar{x}_{d(k)}\right\| \leqslant \varepsilon_1$ where ε_1 is some pre-defined threshold value). Generally, a fixed time delay T can be utilized based on the assumption that $\left\|x\left(t\right) - \bar{x}_{d(k)}\right\| \leqslant \varepsilon_1$ after T seconds.

- Step 5. $k = k + 1$. Invoke one iteration of an optimization algorithm for the objective function $\varphi\left(x_d(t), e(t)\right)$.

- Step 6. If the optimization algorithm has converged (i.e., given a threshold value ε_2, then $\left\|\bar{x}_{d(k)} - \bar{x}_{d(k-1)}\right\| \leqslant \varepsilon_2$), then stop.

- Step 7. Go to Step 3.

Remark 4.12 *The above method does not depend on a specific optimization routine; however, to facilitate real-time implementation, the optimization routine must be capable of running single iterations as indicated in Step 2 and Step 5.*

Remark 4.13 *The unknown mapping from the task-space to the image-space in this example includes unknown camera calibration effects, radial distortion, and unknown constant depth information. The mapping could be further corrupted by additional effects or even the optical axis of the camera need not be perpendicular to the robot motion plan, provided the resulting mapping remains unique. If the mapping is not unique, then it is heuristically evident that it would be nearly impossible to design a visual servo control algorithm to achieve the control objective.*

4.5.2   6D Visual Servoing: Camera-in-Hand

In this section, the basic idea illustrated by the planar example is extended to the full 6-DOF case for the camera-in-hand configuration.

Geometric Model

For the 6-DOF case, four target points, denoted by $O_i \; \forall i = 1, 2, 3, 4$, are assumed to be located on a reference plane π (see Figure 4.44), and are considered to be coplanar and not colinear. Let $\mathcal{I}$ denote a coordinate axis whose $x - y$ axes define the reference plane π. Let $\mathcal{F}$, $\mathcal{F}^*$, and $\mathcal{F}_0$ denote coordinate frames attached to the camera, the goal pose of the camera, and the base frame of the manipulator as depicted in Figure 4.44. To relate the coordinate systems, let $R_t\left(t\right) \in \mathbb{R}^{3\times3}$ and $x_t\left(t\right) \in \mathbb{R}^3$ denote the known, constant rotation and translation from $\mathcal{F}_0$ to $\mathcal{I}$; $R_e\left(t\right) \in \mathbb{R}^{3\times3}$

and $x_e(t) \in \mathbb{R}^3$ denote the measurable rotation and translation from $\mathcal{F}_0$ to $\mathcal{F}$; $R_e^*(t) \in \mathbb{R}^{3 \times 3}$ and $x_e^*(t) \in \mathbb{R}^3$ denote the unknown constant rotation and translation from $\mathcal{F}_0$ to $\mathcal{F}^*$; $R(t) \in \mathbb{R}^{3 \times 3}$ and $x_f(t) \in \mathbb{R}^3$ denote the rotation and translation from $\mathcal{F}$ to $\mathcal{I}$; and $R^*(t) \in \mathbb{R}^{3 \times 3}$ and $x_f^*(t) \in \mathbb{R}^3$ denote the rotation and translation from $\mathcal{F}^*$ to $\mathcal{I}$, respectively. The task-space coordinates of O_i expressed in $\mathcal{F}$ and $\mathcal{F}^*$ are denoted by $\bar{m}_i(t) \triangleq \begin{bmatrix} x_i(t) & y_i(t) & z_i(t) \end{bmatrix}^T$ and $\bar{m}_i^* \triangleq \begin{bmatrix} x_i^* & y_i^* & z_i^* \end{bmatrix}^T$, respectively. Furthermore, the homogeneous coordinates are denoted by $m_i(t)$ and m_i^* and are generated in the manner of (4.135). Here, it is assumed that the distance from the camera to the target along the focal axis remains positive (i.e., $z_i(t)$, $z_i^* > 0$ $\forall i = 1, 2, 3, 4$). From the geometry indicated in Figure 4.44, the following expressions can be obtained

$$R(t) = R_e^T(t) R_t(t) \qquad R^*(t) = R_e^{*T} R_t(t) \tag{4.206}$$

$$\begin{aligned} x_f(t) &= R_e^T(t)\left(x_t(t) - x_e(t)\right) \\ x_f^*(t) &= R_e^{*T}\left(x_t(t) - x_e^*(t)\right). \end{aligned} \tag{4.207}$$

Based on (4.206) and (4.207) and the geometry between the coordinate frames and the feature points located on π, the following relationships can be developed

$$\bar{m}_i(t) = x_f + R s_i \qquad \bar{m}_i^*(t) = x_f^* + R^* s_i \tag{4.208}$$

$\forall i = 1, 2, 3, 4$, where $s_i \in \mathbb{R}$ denotes the constant coordinates of the target points O_i expressed in $\mathcal{I}$. Based on Figure 4.44, it is easy to show that $x_e(t)$ and $R_e(t)$ can be obtained by measuring the manipulator joint angles and utilizing the manipulator Jacobian.

Camera Model

The camera model for the planar manipulator example is based on the assumptions that the camera is fixed and that the depth is constant from the camera to the plane of motion of the target point along the focal axis. Since the camera in this section is mounted in the camera-in-hand configuration and moves with 6-DOF motion, some modifications to the previous model are required. Specifically, the relationship between the image-space coordinates of the ith target point, denoted by $\begin{bmatrix} u_i(t) & v_i(t) \end{bmatrix}^T$, and the corresponding task-space coordinates expressed in $\mathcal{F}$ is given as follows

$$\begin{bmatrix} u_i \\ v_i \end{bmatrix} = f_{ri} A \begin{bmatrix} \dfrac{x_i}{z_i} \\ \dfrac{y_i}{z_i} \\ z_i \end{bmatrix} + \begin{bmatrix} u_0 \\ v_0 \end{bmatrix}. \tag{4.209}$$

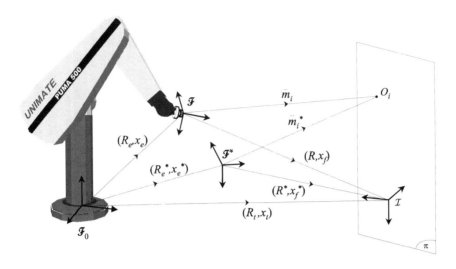

FIGURE 4.44. Coordinate Frame Relationships

In (4.209), u_0, v_0, A, have the same meaning as in Section 4.5.1; $f_{ri}(m_i) \in \mathbb{R}$ denotes the same radial distortion model as given in (4.195), but in this case, the undistorted radial distance for each target point needs to be separately defined as follows

$$r_i(m_i) = \sqrt{\left(\frac{x_i(t)}{z_i(t)}\right)^2 + \left(\frac{y_i(t)}{z_i(t)}\right)^2}.$$

Similar to (4.209), the relationship between the image-space coordinates of the desired position of the ith target point, denoted by $\begin{bmatrix} u_i^* & v_i^* \end{bmatrix}^T$, and the corresponding task-space coordinates expressed in $\mathcal{F}^*$ is given as follows

$$\begin{bmatrix} u_i^* \\ v_i^* \end{bmatrix} = f_{ri}^* A \begin{bmatrix} \frac{x_i^*}{z_i^*} \\ \frac{y_i}{z_i^*} \end{bmatrix} + \begin{bmatrix} u_0 \\ v_0 \end{bmatrix}$$

where f_{ri}^* is defined similar to f_{ri} with respect to m_i^*.

A composite image feature vector comprising four feature point locations, denoted by $\bar{p}(t) \triangleq \begin{bmatrix} u_1(t) & v_1(t) & \dots & u_4(t) & v_4(t) \end{bmatrix}^T \in \mathbb{R}^8$, can be related to the camera pose, denoted by $\Upsilon(t) \in \mathbb{R}^6$, through the following relationship

$$\bar{p} = \Pi(\Upsilon) \tag{4.210}$$

where $\Pi(\cdot) : \mathbb{R}^6 \to \mathcal{D}$ denotes an unknown mapping based on (4.206)–(4.209) and D denotes the space spanned by the image feature vector $\bar{p}(t)$.

In (4.210), the camera pose is defined as

$$\Upsilon(t) \triangleq \left[\begin{array}{cc} x_e^T(t) & \Theta^T(t) \end{array} \right]^T \tag{4.211}$$

where $x_e(t)$ is introduced in Section (4.5.2), and $\Theta(t) = \mu(t)\theta(t) \in \mathbb{R}^3$ denotes the axis-angle representation of $R_e(t)$ in the sense of (4.146).

Remark 4.14 *It is well-known that a unique transformation (R, x_f) can be determined from at least four coplanar but not colinear target points [6], and a unique image of four coplanar but not colinear target points can be obtained by a unique transformation (R, x_f). We assume that the radial distortion will not affect the uniqueness of above mappings. From (4.206), (4.207), and (4.211)–(4.150), it is clear that a unique $\Upsilon(t)$ corresponds to a unique (R, x_f). Therefore a unique $\Upsilon(t)$ corresponds to a unique $\bar{p}(t)$. Similarly, a unique $\bar{p}(t)$ can be shown to correspond to a unique $\Upsilon(t)$. Therefore, it is clear that $\bar{p}(t)$ is constrained to lie on a six-dimensional subspace of the eight-dimensional image feature vector; hence, the rank of $\mathcal{D}$, the span space of $\bar{p}(t)$, is six, and the mapping defined in (4.210) is a unique mapping, where the terminology, unique mapping, is utilized here to denote the one-to-one and invertible mapping between the camera pose and the image feature vector in $\mathcal{D}$.*

Control Objective

In a similar manner as in the planar example, the 6-DOF PBVS control objective is to ensure the task-space pose of the camera is regulated to an unknown desired task-space pose, denoted by $\Upsilon^* \in \mathbb{R}^6$, that corresponds to the known, desired image-space setpoint, denoted by $\bar{p}^* \in \mathbb{R}^8$. To quantify the control objective, a camera pose tracking error signal, denoted by $e(t) \in \mathbb{R}^6$, is defined as follows

$$e(t) \triangleq \Upsilon(t) - \Upsilon_d(t) \tag{4.212}$$

where $\Upsilon_d(t) \in \mathbb{R}^6$ denotes a desired camera pose trajectory that is designed based on an optimization routine such that $\Upsilon_d(t), \dot{\Upsilon}_d(t) \in \mathcal{L}_\infty$ and that $\Upsilon_d(t) \to \Upsilon^*$.

Closed-Loop Error System Development

Similar to (4.190), the 6-DOF robot kinematics can be expressed as follows [46]

$$\dot{\Upsilon}(t) = J(q)\dot{q}(t) \tag{4.213}$$

where $q(t), \dot{q}(t) \in \mathbb{R}^6$ denote the joint displacements and velocities, respectively, and $J(q) \in \mathbb{R}^{6 \times 6}$ denotes the manipulator Jacobian. After taking the time derivative of (4.212) and utilizing (4.213), the following open-loop error system can be obtained

$$\dot{e} = J(q)\dot{q} - \dot{\Upsilon}_d. \tag{4.214}$$

Based on the assumption that the inverse of the manipulator Jacobian, denoted by $J^{-1}(q)$, is assumed to always exist and that all kinematic singularities are always avoided, the kinematic control input is designed as follows

$$\dot{q}(t) = J^{-1}(q)\left(-k_2 e(t) + \dot{\Upsilon}_d(t)\right) \tag{4.215}$$

where $k_2 \in \mathbb{R}$ denotes a positive constant. After substituting (4.215) into (4.214), the following closed-loop error can be obtained

$$\dot{e}(t) = -k_2 e(t). \tag{4.216}$$

Using standard linear analysis techniques, (4.216) can be solved as follows

$$e(t) = e(0)\exp(-k_2 t). \tag{4.217}$$

Image-Based Extremum Seeking Path Planner

To develop the extremum seeking desired trajectory for the 6-DOF case, an objective function, denoted by $\varphi(\bar{p}) \in \mathbb{R}$, is defined as follows

$$\varphi(\bar{p}) \triangleq \frac{1}{2}(\bar{p}(t) - \bar{p}^*)^T(\bar{p}(t) - \bar{p}^*) \tag{4.218}$$

where $\bar{p}^* \in \mathbb{R}^8$ denotes the given image-space coordinates that correspond to the desired camera pose. Based on the previous development given for the planar robot example, $\Upsilon_d(t)$ can be generated with an optimization routine that is modified by using the same steps described for the planar manipulator case. The reader is referred to Section B.2.7 of Appendix B for a discussion related to the global minimum for the 6-DOF case.

4.6 Background and Notes

Researchers have targeted a variety of applications that exploit the ability of a velocity field controller to encode certain contour following tasks. For example, Li and Horowitz [49] utilized a passive VFC approach to control robot manipulators for contour following applications, and more recently, Dee and Li [43] used VFC to achieve passive bilateral teleoperation of robot manipulators. The authors of [47] utilized a passive VFC approach to develop a force controller for robot manipulator contour following applications. Other relevant work utilizing VFC approaches by Li and co-workers can be found in [48] and [50]. Yamakita et al. investigated the application of passive VFC to cooperative mobile robots and cooperative robot manipulators in [73] and [74], respectively. Typically, VFC is based on a nonlinear control approach where exact model knowledge of the system dynamics is required. Motivated by the desire to account for uncertainty in the robot dynamics, Cervantes et al. developed a robust VFC in [4]. Specifically, in [4] a proportional-integral controller was developed that achieved semiglobal practical stabilization of the velocity field tracking errors despite uncertainty in the robot dynamics. From a review of VFC literature, it can be determined that research efforts have focused on ensuring that the robot tracks the velocity field but no development has been provided to ensure that the link position remains bounded. The result in [4] acknowledged the issue of boundedness of the robot position — the issue was addressed by making an assumption that the following norm

$$\left\| q(0) + \int_0^t \vartheta(q(\sigma))d\sigma \right\|$$

yields globally bounded trajectories, where $q(t)$ denotes the position, and $\vartheta(\cdot)$ denotes the velocity field. For a more thorough discussion of the advantages and differences of VFC with respect to traditional trajectory tracking control, the reader is referred to [4], [47], and [49].

Numerous researchers have investigated algorithms to address the path planning and motion control problem when the configuration space of the robot is cluttered with obstacles. A comprehensive summary of techniques that address the classic geometric problem of constructing a collision-free path is provided in Section 9, "Literature Landmarks," of Chapter 1 of [41]. Since the pioneering work by Khatib in [36], it is clear that the construction and use of potential functions has continued to be one of the mainstream approaches to robotic task execution among known obstacles. In short, potential functions produce a repulsive potential field around the boundary of the robot task-space and obstacles and an attractive potential field at the goal configuration. A comprehensive overview of research directed at potential functions is provided in [41]. One criticism of the potential function approach is that local minima can occur that can cause the robot to "get stuck" without reaching the goal position. Several researchers have proposed approaches to address the local minima issue (e.g., see [1], [2], [11], [37], and [71]). Koditschek [38] (see also [39] and [64]) introduced the navigation function (NF) which is a special kind of potential function with a refined mathematical structure which guarantees the existence of a unique minimum.

By leveraging from previous results directed at classic (holonomic) systems, more recent research has focused on the development of potential function-based approaches for nonholonomic systems (e.g., wheeled mobile robots (WMRs)). For example, Laumond et al. [42] used a geometric path planner to generate a collision-free path that ignores the nonholonomic constraints of a WMR, then divided the geometric path into smaller paths that satisfy the nonholonomic constraints, and then applied an optimization routine to reduce the path length. In [30] and [31], Guldner et al. use discontinuous, sliding mode controllers to force the position of a WMR to track the negative gradient of a potential function and to force the orientation to align with the negative gradient. In [3], [40], and [16], continuous potential field-based controllers are developed to also ensure position tracking of the negative gradient of a potential function, and orientation tracking of the negative gradient. More recently, Ge and Cui present a new repulsive potential function approach in [29] to address the case when the goal is nonreachable with obstacles nearby (GNRON). In [69] and [70], Tanner et al. exploit the navigation function research of [64] along with a dipolar potential field concept to develop a navigation function-based controller for a nonholonomic mobile manipulator. Specifically, the results in [69] and [70] use a discontinuous controller to track the negative gradient of the navigation function, where a nonsmooth dipolar potential field causes the WMR to turn in place at the goal position to align with a desired orientation.

Vision based controllers allow the robot to navigate in unstructured environments. IBVS and PBVS controllers have typically been employed by researchers to address the visual servoing problem. For a review of IBVS and PBVS controllers, the user is referred to [32]. To avoid the pitfalls associated with IBVS and PBVS approaches, hybrid approaches such as homography-based visual servoing control techniques (coined 2.5D controllers) have been recently developed in a series of papers by Malis and Chaumette (e.g., see [5], [55], [56]). Motivated by the advantages of the homography-based strategy, several researchers have recently developed various regulation controllers for robot manipulators (see [9], [12], and [15]). A common problem with all the aforementioned approaches is the inability to achieve the control objective while ensuring the visibility of target features. To address this issue, Mezouar and Chaumette developed a path-following IBVS algorithm in [59] where the path to a goal point is generated via a potential function that incorporates motion constraints; however, as stated in [59], local minima associated with traditional potential functions may exist. Using a navigation function, Cowan et al. developed a hybrid position/image-space controller that forces a manipulator to a desired setpoint while ensuring the object remains visible (i.e., the NF ensures no local minima) and by avoiding pitfalls such as self-occlusion [13]. However, as stated in [59], this approach requires a complete knowledge of the space topology and requires an object model. In [27], Gans and Hutchinson developed a strategy that switches between an IBVS and a PBVS controller to ensure asymptotic stability of the position and orientation (i.e., pose) in the Euclidean and image-space. An image-space based follow-the-leader application for mobile robots was developed in [14] that exploits an image-space NF. Specifically, an input/output feedback linearization technique is applied to the mobile robot kinematic model to yield a controller that yields "string stability" [25]. Without a feedforward component, the controller in [14] yields an approximate "input-to-formation" stability (i.e., a local, linear exponential system with a bounded disturbance). A NF based approach to the follow-the-leader problem for a group of fully actuated holonomic mobile robots is considered in [61] where configuration based constraints are developed to ensure the robot edges remain in the sight of an omnidirectional camera. A Lyapunov-based analysis is provided in [61] to ensure that the NF decreases to the goal position, however, the stability of the overall system is not examined.

References

[1] J. Barraquand and J. C. Latombe, "A Monte-Carlo Algorithm for Path Planning with Many Degrees of Freedom," *Proceedings of the IEEE International Conference on Robotics and Automation*, Cincinnati, Ohio, pp. 584–589, 1990.

[2] J. Barraquand, B. Langlois, and J. C. Latombe, "Numerical Potential Fields Techniques for Robot Path Planning," *IEEE Transactions on Systems, Man, and Cybernetics*, Vol. 22, pp. 224–241, 1992.

[3] A. Bemporad, A. De Luca, and G. Oriolo, "Local Incremental Planning for a Car-Like Robot Navigating Among Obstacles," *Proceedings of the IEEE International Conference on Robotics and Automation*, Minneapolis, Minnesota, pp. 1205–1211, April 1996.

[4] I. Cervantes, R. Kelly, J. Alvarez-Ramirez, and J. Moreno, "A Robust Velocity Field Control," *IEEE Transactions on Control Systems Technology*, Vol. 10, No. 6, pp. 888–894, 2002.

[5] F. Chaumette, E. Malis, and S. Boudet, "2D 1/2 Visual Servoing with Respect to a Planar Object," *Proceedings of the Workshop on New Trends in Image-Based Robot Servoing*, pp. 45–52, 1997.

[6] F. Chaumette, "Potential Problems of Stability and Convergence in Image-Based and Position-Based Visual Servoing," *The Confluence of Vision and Control*, ser. LNCIS, D. Kriegman, G. Hager, and A. Morse, Eds. New York: Springer Verlag, Vol. 237, pp. 66–78, 1998.

[7] F. Chaumette and E. Malis, "2 1/2 D Visual Servoing: A Possible Solution to Improve Image-based and Position-based Visual Servoing," *Proceedings of the IEEE International Conference on Robotics and Automation*, San Francisco, California, pp. 630–635, April 2000.

[8] J. Chen, A. Behal, D. Dawson, and Y. Fang, "2.5D Visual Servoing with a Fixed Camera," *Proceedings of the American Control Conference*, Denver, Colorado, pp. 3442–3447, June 2003.

[9] J. Chen, D. M. Dawson, W. E. Dixon, and A. Behal, "Adaptive Homography-based Visual Servo Tracking," *Proceedings of the IEEE/RSJ International Conference on Intelligent Robots and Systems*, Las Vegas, Nevada, pp. 230–235, October 2003.

[10] J. Chen, W. E. Dixon, D. M. Dawson, and M. McIntire, "Homography-based Visual Servo Tracking Control of a Wheeled Mobile Robot," *IEEE Transactions on Robotics*, Vol. 22, No. 2, pp. 407–416, 2006.

[11] C. I. Connolly, J. B. Burns, and R. Weiss, "Path Planning Using Laplace's Equation," *Proceedings of the IEEE International Conference on Robotics and Automation*, Cincinnati, Ohio, pp. 2102–2106, 1990.

[12] P. I. Corke and S. A. Hutchinson, "A New Hybrid Image-Based Visual Servo Control Scheme," *Proceedings of the IEEE Conference on Decision and Control*, Las Vegas, Nevada, pp. 2521–2527, Dec. 2000.

[13] N. J. Cowan, J. D. Weingarten, and D. E. Koditschek, "Visual Servoing via Navigation Function," *IEEE Transactions on Robotics and Automation*, Vol. 18, No. 4, pp. 521–533, 2002.

[14] N. J. Cowan, O. Shakernia, R. Vidal, and S. Sastry, "Vision-Based Follow-the-Leader," *Proceedings of the International Conference on Intelligent Robots and Systems*, Las Vegas, Nevada, pp. 1796–1801, October 2003.

[15] K. Deguchi, "Optimal Motion Control for Image-Based Visual Servoing by Decoupling Translation and Rotation," *Proceedings of the International Conference on Intelligent Robots and Systems*, Victoria, B.C., Canada, pp. 705–711, Oct. 1998.

[16] A. De Luca and G. Oriolo, "Local Incremental Planning for Nonholonomic Mobile Robots," *Proceedings of the IEEE International Conference on Robotics and Automation*, San Diego, California, pp. 104–110, 1994.

[17] M. S. de Queiroz, D. M. Dawson, S. P. Nagarkatti, and F. Zhang, *Lyapunov-Based Control of Mechanical Systems*, Birkhäuser, 1999.

[18] W. E. Dixon, D. M. Dawson, E. Zergeroglu, and F. Zhang, "Robust Tracking and Regulation Control for Mobile Robots," *International Journal of Robust and Nonlinear Control*, Vol. 10, pp. 199–216, 2000.

[19] W. E. Dixon, D. M. Dawson, E. Zergeroglu, and A. Behal, *Nonlinear Control of Wheeled Mobile Robots*, Springer-Verlag London Limited, 2001.

[20] W. E. Dixon, A. Behal, D. M. Dawson, and S. Nagarkatti, *Nonlinear Control of Engineering Systems: A Lyapunov-Based Approach*, Birkhäuser Boston, 2003.

[21] Y. Fang, *Lyapunov-based Control for Mechanical and Vision-based System*, Ph.D. dissertation, Department of Electrical and Computer Engineering, Clemson University, Clemson, SC, 2002.

[22] Y. Fang, W. E. Dixon, D. M. Dawson and J. Chen, "Robust 2.5D Visual Servoing for Robot Manipulators," *Proceedings of the American Control Conference*, Denver, Colorado, pp. 3311–3316, June 2003.

[23] O. Faugeras, *Three-Dimensional Computer Vision*, The MIT Press, Cambridge Massachusetts, 2001.

[24] C. A. Felippa, *A Systematic Approach to the Element-Independent Corotational Dynamics of Finite Elements*, Center for Aerospace Structures Document Number CU-CAS-00-03, College of Engineering, University of Colorado, January 2000.

[25] R. Fierro, P. Song, A. Das, and V. Kumar, "Cooperative Control of Robot Formations," in *Cooperative Control and Optimization*, Vol. 66, Chapter 5, pp. 73–93, Kluwer Academic Press, 2002.

[26] M. A. Fischler and R. C. Bolles, "Random Sample Consensus: A Paradigm for Model Fitting with Applications to Image Analysis and Automated Cartography," *Communications ACM*, Vol. 44, pp. 381–395, 1981.

[27] N. R. Gans and S. A. Hutchinson, "An Asymptotically Stable Switched System Visual Controller for Eye in Hand Robots," *Proceedings of the IEEE/RSJ International Conference on Intelligent Robots and Systems*, Las Vegas, Nevada, pp. 735–742, Oct. 2003.

[28] N. R. Gans, S. A. Hutchinson, and P. I. Corke, "Performance Tests for Visual Servo Control Systems, with Application to Partitioned Approaches to Visual Servo Control," *International Journal of Robotics Research*, Vol. 22, No. 10-11, pp. 955–981, 2003.

[29] S. S. Ge and Y. J. Cui, "New Potential Functions for Mobile Robot Path Planning," *IEEE Transactions on Robotics and Automation*, Vol. 16, No. 5, pp. 615–620, 2000.

[30] J. Guldner and V. I. Utkin, "Sliding Mode Control for Gradient Tracking and Robot Navigation Using Artificial Potential Fields," *IEEE Transactions on Robotics and Automation*, Vol. 11, No. 2, pp. 247–254, 1995.

[31] J. Guldner, V. I. Utkin, H. Hashimoto, and F. Harashima, "Tracking Gradients of Artificial Potential Field with Non-Holonomic Mobile Robots," *Proceedings of the American Control Conference*, Seattle, Washington, pp. 2803–2804, 1995.

[32] G. D. Hager and S. Hutchinson (guest editors), Special Section on Vision-Based Control of Robot Manipulators, *IEEE Transactions on Robotics and Automation*, Vol. 12, No. 5, 1996.

[33] R. Horaud, "New Methods for Matching 3-D Objects with Single Perspective View," *IEEE Transactions. on Pattern Analysis and Machine Intelligence*, Vol. PAMI-9, No. 3, pp. 401–412, 1987.

[34] S. Hutchinson, G. D. Hager, and P. I. Corke, "A tutorial on Visual Servo Control," *IEEE Transactions on Robotics and Automation*, Vol. 12, No. 5, pp. 651–670, 1996.

[35] H. K. Khalil, *Nonlinear Systems*, Third edition, Prentice Hall, 2002.

[36] O. Khatib, *Commande dynamique dans l'espace opérational des robots manipulateurs en présence d'obstacles*, Ph.D. Dissertation, École Nationale Supéieure de l'Aeéronatique et de l'Espace (ENSAE), France, 1980.

[37] O. Khatib, "Real-Time Obstacle Avoidance for Manipulators and Mobile Robots," *International Journal of Robotics Research*, Vol. 5, No. 1, pp. 90–99, 1986.

[38] D. E. Koditschek, "Exact Robot Navigation by Means of Potential Functions: Some Topological Considerations," *Proceedings of the IEEE International Conference on Robotics and Automation*, Raleigh, North Carolina, pp. 1–6, 1987.

[39] D. E. Koditschek and E. Rimon, "Robot Navigation Functions on Manifolds with Boundary," *Advances in Applied Math.*, Vol. 11, pp. 412–442, 1990.

[40] K. J. Kyriakopoulos, H. G. Tanner, and N. J. Krikelis, "Navigation of Nonholonomic Vehicles in Complex Environments with Potential Fields and Tracking," *International Journal of Intelligent Control Systems*, Vol. 1, No. 4, pp. 487–495, 1996.

[41] J. C. Latombe, *Robot Motion Planning*, Kluwer Academic Publishers: Boston, Massachusetts, 1991.

[42] J. P. Laumond, P. E. Jacobs, M. Taix, and R. M. Murray, "A Motion Planner for Nonholonomic Mobile Robots," *IEEE Transactions on Robotics and Automation*, Vol. 10, No. 5, pp. 577–593, 1994.

[43] D. Lee and P. Li, "Passive Bilateral Feedforward Control of Linear Dynamically Similar Teleoperated Manipulators," *IEEE Transactions on Robotics and Automation*, Vol. 19, No. 3, pp. 443–456, 2003.

[44] F. L. Lewis, *Optimal Control*, John Wiley & Sons, Inc: New York, NY, 1986.

[45] F. Lewis, C. Abdallah, and D. Dawson, *Control of Robot Manipulators*, New York: MacMillan Publishing Co., 1993.

[46] F. L. Lewis, D.M. Dawson, and C.T. Abdallah, *Robot Manipulator Control: Theory and Practice*, 2nd edition, revised and expanded, Marcel Dekker, Inc: New York, NY, 2004.

[47] J. Li and P. Li, "Passive Velocity Field Control (PVFC) Approach to Robot Force Control and Contour Following," *Proceedings of the Japan/USA Symposium on Flexible Automation*, Ann Arbor, Michigan, 2000.

[48] P. Li, "Adaptive Passive Velocity Field Control," *Proceedings of the American Controls Conference*, San Diego, California, 1999, pp. 774–779.

[49] P. Li and R. Horowitz, "Passive Velocity Field Control of Mechanical Manipulators," *IEEE Transactions on Robotics and Automation*, Vol. 15, No. 4, pp. 751–763, 1999.

[50] P. Li and R. Horowitz, "Passive Velocity Field Control (PVFC): Part II — Application to Contour Following," *IEEE Transactions on Automatic Control*, Vol. 46, No. 9, pp. 1360–1371, 2001.

[51] M. Loffler, N. Costescu, and D. Dawson, "QMotor 3.0 and the QMotor Robotic Toolkit — An Advanced PC-Based Real-Time Control Platform," *IEEE Control Systems Magazine*, Vol. 22, No. 3, pp. 12–26, June 2002.

[52] L. L. Ma, Y. Q. Chen, and K. L. Moore, "Flexible Camera Calibration Using a New Analytical Radial Undistortion Formula with Application to Mobile Robot Localization," *Proceedings of the IEEE International Symposium on Intelligent Control*, Houston, Texas, October 2003, pp. 799–804.

[53] B. Nelson and N. Papanikolopoulos (guest editors), Special Issue of Visual Servoing, *IEEE Robotics and Automation Mag.*, Vol. 5, No. 4, Dec. 1998.

[54] E. Malis, *Contributions à la Modélisation et à la Commande en Asservissement Visuel*, Ph.D. Dissertation, University of Rennes I, IRISA, France, Nov. 1998.

[55] E. Malis, F. Chaumette, and S. Bodet, "2 1/2 D Visual Servoing," *IEEE Transactions on Robotics and Automation*, Vol. 15, No. 2, pp. 238–250, 1999.

[56] E. Malis and F. Chaumette, "2 1/2 D Visual Servoing with Respect to Unknown Objects Through a New Estimation Scheme of Camera Displacement," *International Journal of Computer Vision*, Vol. 37, No. 1, pp. 79–97, June 2000.

[57] E. Malis and F. Chaumette, "Theoretical Improvements in the Stability Analysis of a New Class of Model-Free Visual Servoing Methods," *IEEE Transactions on Robotics and Automation*, Vol. 18, No. 2, pp. 176–186, 2002.

[58] *Optimization Toolbox Documentation*, http://www.mathworks.com.

[59] Y. Mezouar and F. Chaumette, "Path Planning for Robust Image-Based Control," *IEEE Transactions on Robotics and Automation*, Vol. 18, No. 4, pp. 534–549, 2002.

[60] Y. Mezouar and F. Chaumette, "Optimal Camera Trajectory with Image-Based Control," *International Journal of Robotics Research*, Vol. 22, No. 10-11, pp. 781–803, 2003.

[61] G. A. Pereira, A. K. Das, V. Kumar, and M. F. Campos, "Formation Control with configuration Space Constraints," *Proceedings of the International Conference on Intelligent Robots and Systems*, Las Vegas, Nevada, pp. 2755–2760, October 2003.

[62] W. H. Press, S. A. Teukosky, W. T. Vetterly and B. P. Flamneny, *Numerical Recipes in Fortran, the Art of Scientific Computing*, 2nd edition, Cambridge University Press, 1992.

[63] Z. Qu, *Robust Control of Nonlinear Uncertain Systems*, New York: John Wiley & Sons, 1998.

[64] E. Rimon and D. E. Koditschek, "Exact Robot Navigation Using Artificial Potential Function," *IEEE Transactions on Robotics and Automation*, Vol. 8, No. 5, pp. 501–518, 1992.

[65] A. A. Rizzi and D. E. Koditschek, "An Active Visual Estimator for Dexterous Manipulation," *IEEE Transactions on Robotics and Automation*, Vol. 12, No. 5, pp. 697–713, 1996.

[66] S. Sastry and M. Bodson, *Adaptive Control: Stability, Convergence, and Robustness*, Englewood Cliffs, NJ, Prentice Hall Co. 1989.

[67] J. J. E. Slotine and W. Li, *Applied Nonlinear Control*, Englewood Cliffs, NJ: Prentice Hall, Inc., 1991.

[68] M. W. Spong and M. Vidyasagar, *Robot Dynamics and Control*, New York: John Wiley and Sons, Inc., 1989.

[69] H. G. Tanner and K. J. Kyriakopoulos, "Nonholonomic Motion Planning for Mobile Manipulators," *Proceedings of the IEEE International Conference on Robotics and Automation*, San Francisco, California, 2000, pp. 1233–1238.

[70] H. G. Tanner, S. G. Loizou, and K. J. Kyriakopoulos, "Nonholonomic Navigation and Control of Cooperating Mobile Manipulators," *IEEE Transactions on Robotics and Automation*, Vol. 19, No. 1, pp. 53–64, 2003.

[71] R. Volpe and P. Khosla, "Artificial Potential with Elliptical Isopotential Contours for Obstacle Avoidance," *Proceedings of the IEEE Conference on Decision and Control*, Los Angeles, California, pp. 180–185, 1987.

[72] E. W. Weisstein, *CRC Concise Encyclopedia of Mathematics*, Second Edition, CRC Press, 2002.

[73] M. Yamakita, T. Yazawa, X.-Z. Zheng, and K. Ito, "An Application of Passive Velocity Field Control to Cooperative Multiple 3-Wheeled Mobile Robots," *Proceedings of the IEEE/RJS International Conference on Intelligent Robots and Systems*, Victoria, B. C., Canada, pp. 368–373, 1998.

[74] M. Yamakita, K. Suzuki, X.-Z. Zheng, M. Katayama, and K. Ito, "An Extension of Passive Velocity Field Control to Cooperative Multiple Manipulator Systems," *Proceedings of the IEEE/RJS International Conference on Intelligent Robots and Systems*, Grenoble, France, pp. 11–16, 1997.

[75] Z. Zhang, "Flexible Camera Calibration by Viewing a Plane from Unknown Orientation," *Proceedings of the IEEE International Conference on Computer Vision*, pp. 666–673, September 1999.

5

Human Machine Interaction

5.1 Introduction

Typically, machines are used for simple, repetitive tasks in structured environments isolated from humans. However, the last decade has seen a surge in active research in the area of human machine interaction. Smart exercise machines [37, 38], steer-by-wire applications [58, 59], bilateral teleoperated robots [15, 30, 34, 65], rehabilitation robots [11, 28, 40, 41], and human assist gantry cranes [63] are among the multitude of application areas that drive this research. As a specific example, the teleoperation or remote control of robotic manipulators is of considerable interest as it permits the introduction of human intelligence and decision making capabilities into a possibly hostile remote environment. Even though the primary control objective varies from one application to the next, a common thread that runs through all application areas is the need to rigorously ensure user safety. Approaches based on passivity ensure that the net flow of energy during the human robot interaction is from the user to the machine [2, 37].

The first application area that will be addressed in this chapter concerns smart exercise machines. Generally, exercise machines are classified according to characteristics such as the source of exercise resistance, exercise motions, and exercise objectives [3], [37]. Traditional exercise machines (e.g., [24]) do not incorporate user specific information in the machine functionality. Typically, traditional exercise machines either rely on manual adjustment of the machine parameters (e.g., altering resistance levels) or automatic adjustment based on an open-loop approach. Exercise based on manual adjustments by the user are affected by the psychological state of the user, resulting in suboptimal performance (e.g., quantified by the power output by the user). Motivated by the desire to maximize the user's power output, recent research has focused on closed-loop, actuated exercise equipment that incorporates feedback from the user. That is, next generation exercise machines will incorporate user performance information to actively change the resistance. In addition to maximizing the user's power output, an additional challenge for actuated exercise machines is to maintain passivity with respect to the user.

The second application area that will be addressed in this chapter is the steer-by-wire control of vehicles. In recent years, engineers and scientists from specialized fields such as information technology, advanced materials, defense systems, and aerospace have collaborated with the automotive industry to introduce advanced technologies for large vehicle production volumes. Examples such as hybrid electric vehicles (HEV) featuring hydrogen, fuel cells, electric motors, solar cells, and/or internal combustion engines are common. Although the concepts of electric and specifically steer-by-wire steering systems have been explored in vehicular research, attention must be focused on the haptic interface. The concept of force feedback follows directly and its advantage in drive-by-wire vehicles is very evident. An operator functioning within a remote driving environment primarily depends on visual feedback to make meaningful maneuvers. The "feel" of the road, due to both the vehicle acceleration forces (i.e., G forces) and the tire/road forces, plays a very prominent role in recreating the driving experience [39]. The physiological and psychological effect of these forces has been documented [13]. An appropriate magnitude is important for force feedback to be valuable to the driver. For instance, excessive feedback results in the need for large driver forces to steer the system, which defeats the purpose of easing the driving experience. Hence, it is essential for the control strategy to ensure that the road "feel" provided by the force feedback can be adjusted. The control design pursued here rigorously ensures global asymptotic regulation of the "locked tracking error" and the "driver experience tracking error."

Another fascinating area which involves human-machine interaction is teleoperator systems. A teleoperator system consists of a user interacting with some type of input device (i.e., a master manipulator) with the intention of imparting a predictable response by an output system (i.e., a slave manipulator). Practical applications of teleoperation are motivated by the

need for task execution in hazardous environments (e.g., contaminated fa-
cilities, space, underwater), the need for remote manipulation due to the
characteristics of the object (e.g., size and mass of an object, hazardous nature of the object), or the need for precision beyond human capacity (e.g.,
robotic assisted medical procedures). In the past few years, significant research has been aimed at the development and control of teleoperator systems due to both the practical importance and the challenging theoretical
nature of the human-robot interaction problem. The teleoperator problem
is theoretically challenging due to issues that impact the user's ability to
impart a desired motion and a desired force on the remote environment
through the coupled master-slave system. Some difficult issues include the
presence of uncertainty in the master and slave dynamics, the ability to
accurately model or measure environmental and user inputs to the system,
the ability to safely reflect desired forces back to the user while mitigating
other forces, and the stability of the overall system (e.g., as stated in [34],
a stable teleoperator system may be destabilized when interacting with a
stable environment due to coupling between the systems).

The final topic that would be addressed in this chapter deals with a
rehabilitation robot. The framework created here is inspired by the desire
to provide passive resistance therapy to patients affected by dystrophies
in the muscles of the upper extremities — these patients need to target
specific groups of muscles in order to regain muscle tone [4]. As stated in [4],
moderate (submaximal) resistance weight lifting, among other treatments,
may improve strength in slowly progressive NMDs such as Lou Gehrig's
Disease (ALS), Spinal Muscular Atrophy, etc. The idea being pursued here
attempts to cast the robot as a reconfigurable passive exercise machine —
along any desired curve of motion in 3D space that satisfies a criterion
of merit, motion is permitted against a programmable apparent inertia
[29] when the user "pushes" at the end-effector; force applied in all other
directions is penalized. As with any other application of human-machine
interaction, safety of the user is a prime consideration and is rigorously
ensured by maintaining the net flow of energy during the interaction from
the user toward the manipulator.

5.2   Exercise Machine

While a variety of machine configurations are available to facilitate different exercises, many configurations can be reduced to a user torque input
to an actuated motor. With that in mind, the specific problem that will be
addressed here is the design of a next generation exercise machine controller

for a single degree of freedom system. As previously stated, one goal of the exercise machine controller is to maximize the user's power expenditure. To attain this goal, a desired trajectory signal is designed at first to seek the optimal velocity setpoint that will maximize the user's power output. A controller is then designed to ensure that the exercise machine tracks the resulting desired trajectory. To generate the desired trajectory, two different algorithms are presented (e.g., [27], [50]) to seek the optimal velocity while ensuring that the trajectory remains sufficiently differentiable. In contrast to the linear approximation of the user force input required in previous research (e.g., [37] and [60]), the development being presented here is based on a general form of the user torque input. As previously stated, another goal of the controller is to ensure that the exercise machine remains passive with respect to the user's power input. To ensure this while also achieving trajectory tracking, two different controllers are developed. The first controller is developed based on the assumption that the user's torque input can be measured. Based on the desire to eliminate the need for force/torque sensors, a second controller is designed that estimates the user's torque input. Both controllers are proven to remain passive with respect to the user's power output and yield semi-global tracking through Lyapunov-based analyses provided that mild assumptions are satisfied for the machine dynamics and the user input. Proof-of-concept experimental results are provided that illustrate the performance of the torque estimation controller.

5.2.1   Exercise Machine Dynamics

The model for a one-degree-of-freedom (DOF) exercise machine is assumed to be as follows[1]

$$J\ddot{q}(t) = \tau(\dot{q}) + u(t) \tag{5.1}$$

where $J \in \mathbb{R}$ denotes the constant inertia of the machine, $q(t)$, $\dot{q}(t)$, $\ddot{q}(t) \in \mathbb{R}$ denote the angular position, velocity, and acceleration of the machine, respectively, $\tau(\dot{q}) \in \mathbb{R}$ denotes a velocity dependent user torque input, and $u(t) \in \mathbb{R}$ denotes the motor control input. The user input is assumed to exhibit the following characteristics that are exploited in the subsequent development.

---

[1] Additional dynamic effects (e.g., friction) can be incorporated in the exercise machine model and subsequent control design. These terms have been neglected in the control development for simplicity.

- Assumption 5.1.1: The user input is a function of the machine velocity (i.e., $\tau(\dot{q})$)

- Assumption 5.1.2: The user input is a second order differentiable function (i.e., $\tau(\dot{q}) \in C^2$).

- Assumption 5.1.3: The user input is unidirectional (i.e., assumed to be positive w.l.o.g.) and satisfies the following inequalities

$$0 \leq \tau(\dot{q}) \leq \tau_{\max} \tag{5.2}$$

where $\tau_{\max} \in \mathbb{R}$ is a positive constant that denotes the maximum possible torque applied by the user.

- Assumption 5.1.4: The desired trajectory is assumed to be designed such that $\dot{q}_d(t)$, $\ddot{q}_d(t)$, $\dddot{q}_d(t) \in \mathcal{L}_\infty$, where the desired velocity, denoted by $\dot{q}_d(t) \in \mathbb{R}$, is assumed to be in the same direction as the user input (i.e., assumed positive w.l.o.g.).[2]

Remark 5.1 *In biomechanics literature, a user's joint torque is typically expressed as a function of position, velocity, and time (i.e., $\tau(q, \dot{q}, t)$). The position dependence is related to the configuration of the limbs attached to the joint. As in [37], the user is assumed to be able to exert the same amount of torque throughout the required range-of-motion for the exercise, and hence, the position dependence can be neglected. The time dependence of the user's joint torque is due to the effects of fatigue (i.e., the amount of maximum torque diminishes as the user fatigues). As also described in [37], the user is assumed to maintain a constant level of fatigue during the exercise session, and hence, the time dependence can be neglected.*

5.2.2 Control Design with Measurable User Input

Control Objectives

One objective of the exercise machine controller is to ensure that the exercise machine tracks a desired velocity. To quantify this objective, a velocity tracking error, denoted by $e(t) \in \mathbb{R}$, is defined as follows

$$e(t) \triangleq \dot{q}(t) - \dot{q}_d(t) \tag{5.3}$$

---

[2]The assumption that $\dot{q}_d(t)$ is assumed to be positive is a similar assumption that is exploited in [37] and [38]. The assumption is considered to be mild since the trajectory generation algorithm can easily be restricted (e.g., a projection algorithm) to produce a positive value.

where $\dot{q}_d(t) \in \mathbb{R}$ denotes a desired velocity that is assumed to be designed such that $\dot{q}_d(t)$, $\ddot{q}_d(t)$, $\dddot{q}_d(t) \in \mathcal{L}_\infty$. Another objective is to maximize the modified user power output, denoted by $p(\dot{q})$, that is defined as follows [37]

$$p(\dot{q}) = \tau(\dot{q})\dot{q}^\rho(t) \tag{5.4}$$

where $\rho \in \mathbb{R}$ is a positive constant[3]. To achieve this objective, the desired trajectory must also be designed to ensure that $\dot{q}_d(t) \to \dot{q}_d^*$ as $t \to \infty$ where $\dot{q}_d^* \in \mathbb{R}$ is a positive constant that denotes an unknown, user-dependent optimal velocity setpoint. A final objective for the exercise machine controller is to ensure the safety of the user by guaranteeing that the machine remains passive with respect to the user's power input. The exercise machine is passive with respect to the user's power input provided the following integral inequality is satisfied [37]

$$\int_{t_0}^t \tau(\sigma)\dot{q}(\sigma)d\sigma \geq -c^2 \tag{5.5}$$

where c is a bounded positive constant.

Remark 5.2 *In contrast to the linear approximation of the user force input required in [60] and [37], the subsequent development is based on a general form of the user torque input. Specifically, Assumptions 5.1.1–5.1.3 should be satisfied and $p(\dot{q})$ of (5.4) should have a global maximum for some value of $\dot{q}(t)$ (i.e., $\dot{q}_d^*$).*

Control Development and Analysis

The open-loop error system is determined by taking the time derivative of (5.3) and multiplying the result by J as follows

$$J\dot{e}(t) = \tau(\dot{q}) + u(t) - J\ddot{q}_d(t) \tag{5.6}$$

where (5.1) has been utilized. In this section, the user torque input is assumed to be measurable. Based on this assumption, the structure of (5.6), and the subsequent stability analysis, the following controller is developed

$$u(t) = -ke(t) + J\ddot{q}_d(t) - \tau(\dot{q}) \tag{5.7}$$

where $k \in \mathbb{R}$ is a positive constant control gain. After substituting (5.7) into (5.6), the following closed-loop error system can be determined

$$J\dot{e}(t) = -ke(t). \tag{5.8}$$

---

[3] A discussion of the physical interpretation of ρ is provided in [37].

The linear differential equation in (5.8) can be directly solved to obtain the following solution

$$e(t) = e(0) \exp(-\frac{k}{J}t). \tag{5.9}$$

Based on (5.9), it is clear that $e(t) \in \mathcal{L}_\infty \cap \mathcal{L}_1$. The expression in (5.3) and the assumption that $\dot{q}_d(t) \in \mathcal{L}_\infty$ can be used to conclude that $\dot{q}(t) \in \mathcal{L}_\infty$; hence, (5.2), (5.7), and the assumption that $\ddot{q}_d(t) \in \mathcal{L}_\infty$ can be used to determine that $u(t) \in \mathcal{L}_\infty$. Thus, it can be stated that the exercise machine controller in (5.7) ensures that all system signals are bounded under closed-loop operation, and the velocity tracking error is exponentially stable in the sense of (5.9). Furthermore, one can substitute (5.3) into (5.5) to obtain the following expression

$$\int_{t_0}^t \tau(\sigma)\dot{q}(\sigma)d\sigma = \int_{t_0}^t \tau(\sigma)e(\sigma)d\sigma + \int_{t_0}^t \tau(\sigma)\dot{q}_d(\sigma)d\sigma. \tag{5.10}$$

Based on Assumptions 5.1.3 and 5.1.4, it is clear that the right-most term in (5.10) is always positive; hence, since $e(t) \in \mathcal{L}_1$, (5.10) can be lower bounded as follows

$$\int_{t_0}^t \tau(\sigma)e(\sigma)d\sigma \geq -\tau_{\max}\int_{t_0}^t |e(\sigma)|\,d\sigma = -c^2. \tag{5.11}$$

Based on (5.11), it is clear that the passivity condition given in (5.5) is satisfied. Thus, one can conclude that the controller in (5.7) ensures that the exercise machine is passive with respect to the user's power input.

5.2.3   Desired Trajectory Generator

In the previous development, it is assumed that a desired trajectory can be generated such that $\dot{q}_d(t)$, $\ddot{q}_d(t)$, $\dddot{q}_d(t) \in \mathcal{L}_\infty$ and that $\dot{q}_d(t) \to \dot{q}_d^*$ where $\dot{q}_d^*$ is an unknown constant that maximizes the user power output. From (5.3) and (5.4), the user power output can be expressed as follows (where $\rho = 1$ w.l.o.g.)

$$p(e,t) = \tau(\dot{q}_d(t) + e(t))(\dot{q}_d(t) + e(t)). \tag{5.12}$$

Since $e(t) \to 0$ exponentially fast, (5.12) can be approximated as follows

$$p(t) \cong \tau(\dot{q}_d)\dot{q}_d(t). \tag{5.13}$$

From (5.13), it is clear that if $\dot{q}_d(t) \to \dot{q}_d^*$ then $p(t) \to \tau(\dot{q}_d^*)\dot{q}_d^*$, and hence, the user power output will be maximized. To generate a desired trajectory that ensures $\dot{q}_d(t)$, $\ddot{q}_d(t)$, $\dddot{q}_d(t) \in \mathcal{L}_\infty$ and that $\dot{q}_d(t) \to \dot{q}_d^*$, several extremum seeking algorithms can be utilized. Two algorithms that can be used to generate the trajectory are described in the following sections.

Perturbation-Based Extremum Generation

For brevity, the extremum seeking algorithm is simply presented along with a heuristic commentary on the internal workings of the algorithm as opposed to extensive mathematical efforts to prove convergence of the scheme. Specifically, following the work presented in [27], a saturated extremum algorithm for generating $\dot{q}_d(t)$ can be designed as follows

$$
\begin{aligned}
\dot{q}_d(t) &= a_e \sin(\omega t) + \hat{\theta}(t) \\
\dot{\hat{\theta}}(t) &= -\alpha_f \hat{\theta}(t) + \kappa(t) \\
\dot{\kappa}(t) &= -\alpha_0 \kappa(t) + \alpha_0 (k_{f1}(sat(p) - \eta(t)) a_e \sin(\omega t)) \\
\dot{\eta}(t) &= -k_{f2} \eta(t) + k_{f2} sat(p)
\end{aligned}
\tag{5.14}
$$

where a_e, ω, α_0, α_f, k_{f1} and $k_{f2} \in \mathbb{R}$ are constant design parameters, $\hat{\theta}(t)$, $\kappa(t)$ and $\eta(t)$ are filtered signals, and $sat(\cdot)$ denotes a continuous saturation function. The algorithm given in (5.14) reduces to the algorithm presented in [27] when the saturation functions are removed and $\alpha_f = 0$. These modifications to the algorithm are incorporated to ensure that $\dot{q}_d(t)$, $\ddot{q}_d(t)$, $\dddot{q}_d(t) \in \mathcal{L}_\infty$. The design parameters a_e, ω, α_0, α_f, k_{f1}, and k_{f2} must be selected sufficiently small because the convergence analysis associated with (5.14) utilizes averaging techniques. Specifically, the convergence analysis requires that the cut-off frequency of the $\eta(t)$ filter used in (5.14) be lower than the frequency of the perturbation signal (i.e., ω). In fact, the convergence analysis requires that the closed-loop system exhibit three distinct time scales: i) high speed — the convergence of $e(t)$, ii) medium speed — the periodic perturbation parameter ω, and iii) slow speed — the filter parameter k_{f2} in the $\eta(t)$ dynamics. As presented in [27], the convergence analysis illustrates that an extremum algorithm similar to (5.14) finds a near-optimum solution (i.e., $\dot{q}_d(t)$ goes to some value very close to $\dot{q}_d^*$). With regard to the periodic terms in (5.14) (i.e., $\sin(\omega t)$ and $\cos(\omega t)$), an extremum-seeking scheme must "investigate" the neighborhood on both sides of the maximum. This "investigation" motivates the use of slow periodic terms in the algorithm.

Numerically-Based Extremum Generation

As previously described, (5.13) can be used to show that if $\dot{q}_d(t) \to \dot{q}_d^*$ then the user power output will be maximized. An extremum algorithm for generating $\dot{q}_d(t)$ was presented in (5.14); however, this algorithm can be slow to find $\dot{q}_d^*$. As an alternative to the approach given by (5.14), several numerically-based extremum search algorithms (e.g., Brent's Method [50], Simplex Method [50], etc.) can be utilized for the online computation of $\dot{q}_d(t)$. For example, Brent's Method only requires measurement of the output function (i.e., $p(t)$ in (5.4)) and two initial guesses that enclose the unknown value for $\dot{q}_d^*$ (the two initial guesses are not required to be close to the value of $\dot{q}_d^*$). Brent's Method then uses an inverse parabolic interpolation algorithm and measurements of $p(t)$ to generate estimates for $\dot{q}_d^*$ until the estimates converge. Specifically, the filter-based algorithm for computing $\dot{q}_d(t)$ is described in Section B.3.1 of Appendix B.

5.2.4   Control Design without Measurable User Input

The control development in the previous section requires that the user torque input be measurable. To measure the user input, an additional sensor (i.e., a force/torque sensor) has to be included in the exercise machine design. Inclusion of the additional sensor results in additional cost and complexity of the system. Motivated by the desire to eliminate the additional sensor, the controller in this section is crafted by developing a nonlinear integral feedback term that produces a user torque input estimate.

Open-Loop Error System

To facilitate the subsequent development, a filtered tracking error, denoted by $r(t) \in \mathbb{R}$, is defined as follows

$$r(t) \triangleq \dot{e}(t) + \alpha_r e(t) \tag{5.15}$$

where $\alpha_r \in \mathbb{R}$ denotes a positive constant parameter. After differentiating (5.15) and multiplying both sides of the resulting equation by J, the following expression can be obtained

$$J\dot{r}(t) = -e(t) + N(\dot{q}, \ddot{q}) + \dot{u}(t) - J\dddot{q}_d(t) \tag{5.16}$$

where the time derivative of (5.1) and (5.3) have been utilized, and the auxiliary function $N(\dot{q}, \ddot{q}) \in \mathbb{R}$ is defined as follows

$$N(\dot{q}, \ddot{q}) \triangleq \frac{d}{dt}[\tau(\dot{q})] + e(t) + J\alpha_r \dot{e}(t). \tag{5.17}$$

To further facilitate the subsequent analysis, an auxiliary signal $N_d(t) \in \mathbb{R}$ is defined as follows

$$N_d(t) \triangleq N(\dot{q}, \ddot{q})|_{\dot{q}(t)=\dot{q}_d(t),\ \ddot{q}(t)=\ddot{q}_d(t)} \tag{5.18}$$

where (5.17) can be utilized to prove that $N_d(t)$, $\dot{N}_d(t) \in \mathcal{L}_\infty$ based on the assumptions that $\dot{q}_d(t)$, $\ddot{q}_d(t)$, $\dddot{q}_d(t) \in \mathcal{L}_\infty$ and $\tau(\dot{q}) \in \mathcal{C}^2$. After adding and subtracting $N_d(t)$ to the right side of (5.16) the following expression can be obtained

$$J\dot{r}(t) = -e(t) + \dot{u}(t) - J\ddot{q}_d(t) + \tilde{N}(\dot{q}, \ddot{q}) + N_d(t) \tag{5.19}$$

where $\tilde{N}(\dot{q}, \ddot{q}) \in \mathbb{R}$ is defined as follows

$$\tilde{N}(\dot{q}, \ddot{q}) \triangleq N(\dot{q}, \ddot{q}) - N_d(t). \tag{5.20}$$

Remark 5.3 *Since $N(\dot{q}, \ddot{q})$ defined in (5.17) is continuously differentiable, $\tilde{N}(\dot{q}, \ddot{q})$ introduced in (5.20) can be upper bounded as indicated by the following inequality [64]*

$$\left| \tilde{N}(\dot{q}, \ddot{q}, t) \right| \leq \rho(\|z(t)\|) \, \|z(t)\| \tag{5.21}$$

where $z(t) \in \mathbb{R}^2$ is defined as

$$z(t) \triangleq [e(t) \quad r(t)]^T \tag{5.22}$$

and $\rho(\|z(t)\|) \in \mathbb{R}$ is a positive bounding function that is non-decreasing in $\|z(t)\|$.

Closed-Loop Error System

Based on the structure of (5.19) and the subsequent stability analysis, the following controller is developed

$$\begin{aligned} u(t) &= J\ddot{q}_d(t) - [J\ddot{q}_d(t_0) - (k_s + 1)e(t_0)] \tag{5.23} \\ &\quad -(k_s + 1)e(t) - \int_{t_0}^{t} (k_s + 1)\alpha_r e(\sigma) d\sigma \\ &\quad - \int_{t_0}^{t} (\beta_1 + \beta_2) sgn(e(\sigma)) d\sigma \end{aligned}$$

where $sgn(\cdot)$ represents the standard signum function, k_s, β_1, $\beta_2 \in \mathbb{R}$ are positive control gains, and the bracketed terms in (5.23) ensure that $u(0) = 0$. The time derivative of (5.23) is given by the following expression

$$\dot{u}(t) = J\ddot{q}_d(t) - (k_s + 1)r(t) - (\beta_1 + \beta_2) sgn(e). \tag{5.24}$$

After substituting (5.24) into (5.19), the closed-loop dynamics for $r(t)$ can be determined as follows

$$\begin{aligned} J\dot{r}(t) &= -e(t) - (k_s + 1)r(t) \tag{5.25} \\ &\quad -(\beta_1 + \beta_2) sgn(e) + \tilde{N}(\dot{q}, \ddot{q}) + N_d(t). \end{aligned}$$

Stability Analysis

Before we proceed with the statement and proof of the main result, two preliminary Lemmas are presented.

Lemma 5.1 *Let $L_1(t), L_2(t) \in \mathbb{R}$ be defined as follows*

$$
\begin{aligned}
L_1(t) &\triangleq r(t)\left(N_d(t) - \beta_1 sgn(e)\right) \qquad &(5.26)\\
L_2(t) &\triangleq -\beta_2 \dot{e}(t) sgn(e).
\end{aligned}
$$

If β_1 and β_2 introduced in (5.23) are selected to satisfy the following sufficient conditions

$$
\beta_1 > |N_d(t)| + \frac{1}{\alpha_r}\left|\dot{N}_d(t)\right| \qquad \beta_2 > 0 \qquad (5.27)
$$

then

$$
\int_{t_0}^{t} L_1(\sigma)d\sigma \leq \zeta_{b1} \qquad \int_{t_0}^{t} L_2(\sigma)d\sigma \leq \zeta_{b2} \qquad (5.28)
$$

where the positive constants $\zeta_{b1}, \zeta_{b2} \in \mathbb{R}$ are defined as

$$
\zeta_{b1} \triangleq \beta_1 |e(t_0)| - e(t_0)N_d(t_0) \qquad \zeta_{b2} \triangleq \beta_2 |e(t_0)|. \qquad (5.29)
$$

Proof. See Section B.3.2 of Appendix B. ■

Lemma 5.2 *Consider the system $\dot{\xi} = f(\xi, t)$ where $f : \mathbb{R}^m \times \mathbb{R} \to \mathbb{R}^m$ for which a solution exists. Let $\mathcal{D} := \{\xi \in \mathbb{R}^m \mid \|\xi\| < \varepsilon\}$ where ε is some positive constant, and let $V : \mathcal{D} \times \mathbb{R} \to \mathbb{R}$ be a continuously differentiable function such that*

$$
W_1(\xi) \leq V(\xi, t) \leq W_2(\xi) \quad and \quad \dot{V}(\xi, t) \leq -W(\xi) \qquad (5.30)
$$

$\forall t \geq 0$ and $\forall \xi \in \mathcal{D}$ where $W_1(\xi), W_2(\xi)$ are continuous positive definite functions, and $W(\xi)$ is a uniformly continuous positive semi-definite function. Provided (5.30) is satisfied and $\xi(0) \in \mathcal{S}$, we have

$$
W(\xi(t)) \to 0 \ as \ t \to \infty \qquad (5.31)
$$

where the region denoted by $\mathcal{S}$ is defined as follows

$$
\mathcal{S} := \{\xi \in \mathcal{D} \mid W_2(\xi) \leq \delta\} \quad where \quad \delta < \min_{\|\xi\|=\varepsilon} W_1(\xi) \qquad (5.32)
$$

where δ denotes some positive constant.

Proof. Direct application of Theorem 8.4 in [22]. ■

The stability of the exercise machine controller can now be proven by the following Theorem.

Theorem 5.3 *The exercise machine controller introduced in (5.23) ensures all signals are bounded under closed-loop operation and that*

$$e(t), \dot{e}(t) \to 0 \ as \ t \to \infty \tag{5.33}$$

provided the control gains β_1 and β_2 are selected according to the sufficient conditions given in (5.27), and the control gain k_s is selected sufficiently large with respect to the initial conditions of the system.

Proof. Let $P_1(t)$, $P_2(t) \in \mathbb{R}$ denote the following auxiliary functions

$$\begin{aligned}
P_1(t) &= \zeta_{b1} - \int_{t_0}^t L_1(\sigma)d\sigma \\
P_2(t) &= \zeta_{b2} - \int_{t_0}^t L_2(\sigma)d\sigma
\end{aligned} \tag{5.34}$$

where ζ_{b1}, ζ_{b2}, $L_1(t)$, and $L_2(t)$ are defined in Lemma 5.1. The results from Lemma 5.1 can be used to show that $P_1(t)$ and $P_2(t)$ are non-negative. Let $V(y,t) \colon \mathbb{R}^2 \times \mathbb{R} \times \mathbb{R} \times \mathbb{R}$ denote the following non-negative function

$$V(y,t) \triangleq \frac{1}{2}e^2(t) + \frac{1}{2}Jr^2(t) + P_1(t) + P_2(t) \tag{5.35}$$

where $y(t) \in \mathbb{R}^2 \times \mathbb{R} \times \mathbb{R}$ is defined as

$$y(t) \triangleq \left[\ z^T(t) \quad \sqrt{P_1(t)} \quad \sqrt{P_2(t)} \ \right]^T \tag{5.36}$$

and $z(t)$ was defined in (5.21). Since J is a positive constant, (5.35) can be lower and upper bounded by the following inequalities

$$W_1(y) \leq V(y,t) \leq W_2(y) \tag{5.37}$$

where

$$W_1(y) = \lambda_1 \|y(t)\|^2 \qquad W_2(y) = \lambda_2 \|y(t)\|^2 \tag{5.38}$$

where $\lambda_1 \triangleq \frac{1}{2}\min\{1, J\}$ and $\lambda_2 \triangleq \max\{1, \frac{1}{2}J\}$.

After differentiating (5.35) and utilizing (5.15), (5.25), (5.26), and the time derivative of (5.34), the following expression can be obtained

$$\begin{aligned}
\dot{V}(y,t) \ &= \ -\alpha_r e^2(t) - r^2(t) - k_s r^2(t) + r(t)\tilde{N}(\cdot) \\
&\quad -\beta_2(\dot{e}(t) + \alpha_r e(t))sgn(e) + \beta_2\dot{e}(t)sgn(e) \\
&\leq \ -\lambda_3 \|z(t)\|^2 - k_s r^2(t) + r(t)\tilde{N}(\cdot) - \alpha_r\beta_2 |e(t)|
\end{aligned} \tag{5.39}$$

where $\lambda_3 \triangleq \min\{1, \alpha_r\}$. By utilizing (5.21), the following inequality can be developed

$$\begin{aligned}
\dot{V}(y,t) \ \leq \ &-\lambda_3 \|z(t)\|^2 - \alpha_r\beta_2 |e(t)| \\
&+ \left[|r(t)| \rho(\|z(t)\|) \|z(t)\| - k_s r^2(t)\right].
\end{aligned} \tag{5.40}$$

Completing the squares on the bracketed term in (5.40) yields the following inequality

$$\dot{V}(y,t) \le -(\lambda_3 - \frac{\rho^2(\|z(t)\|)}{4k_s}) \|z(t)\|^2 - \alpha_r \beta_2 |e(t)|. \qquad (5.41)$$

Based on (5.41), the following inequality can be developed

$$\dot{V}(y,t) \le W(y) - \alpha_r \beta_2 |e(t)| \quad \text{for } k_s > \frac{\rho^2(\|z(t)\|)}{4\lambda_3} \qquad (5.42)$$
$$\text{or } \|z(t)\| < \rho^{-1}(2\sqrt{\lambda_3 k_s})$$

where

$$W(y) = -\gamma \|z\|^2 \qquad (5.43)$$

and $\gamma \in \mathbb{R}$ is some positive constant. From (5.42) and (5.43), the regions $\mathcal{D}$ and $\mathcal{S}$ can be defined as follows

$$\mathcal{D} \triangleq \{ y \in \mathbb{R}^2 \times \mathbb{R} \times \mathbb{R} \,\big|\, \|y\| \le \rho^{-1}(2\sqrt{\lambda_3 k_s}) \} \qquad (5.44)$$

$$\mathcal{S} \triangleq \{ y \in \mathcal{D} \,\big|\, W_2(y) < \lambda_1(\rho^{-1}(2\sqrt{\lambda_3 k_s}))^2 \}. \qquad (5.45)$$

The region of attraction in (5.45) can be made arbitrarily large to include any initial conditions by increasing the control gain k_s (i.e., a semi-global stability result). Specifically, (5.38) and the region defined in (5.45) can be used to calculate the region of attraction as follows

$$W_2(y(t_0)) < \lambda_1(\rho^{-1}(2\sqrt{\lambda_3 k_s}))^2$$
$$\implies \|y(t_0)\| < \sqrt{\frac{\lambda_1}{\lambda_2}} \, \rho^{-1}(2\sqrt{\lambda_3 k_s}) \qquad (5.46)$$

which can be rearranged as

$$k_s > \frac{1}{4\lambda_3} \rho^2(\sqrt{\frac{\lambda_2}{\lambda_1}} \|y(t_0)\|). \qquad (5.47)$$

By using (5.15) and (5.36), the following explicit expression for $\|y(t_0)\|$ can be obtained

$$\|y(t_0)\| = \sqrt{e^2(t_0) + (\dot{e}(t_0) + \alpha_r e(t_0))^2 + P_1(t_0) + P_2(t_0)} \qquad (5.48)$$

where (5.1), (5.3), and the fact that $u(t_0) = 0$ can be used to determine that

$$\dot{e}(0) = J^{-1}\tau(t_0) - \ddot{q}_d(t_0).$$

Hereafter, we restrict the analysis to be valid for all initial conditions $y(t_0) \in \mathcal{S}$. From (5.35), (5.41), (5.42), and (5.45), it is clear that $V(y,t) \in$

$\mathcal{L}_\infty$; hence, $e(t)$, $r(t)$, $z(t)$, $y(t) \in \mathcal{L}_\infty$. From (5.42), it is also clear that $e(t) \in \mathcal{L}_1$. From (5.15), it can be shown that $\dot{e}(t) \in \mathcal{L}_\infty$. Since $\ddot{q}_d(t)$ is assumed to be bounded, (5.23) can be used to prove that $u(t) \in \mathcal{L}_\infty$. The previous boundedness statements can also be used along with (5.21), (5.25) and (5.43) to prove that $\dot{W}(y(t)) \in \mathcal{L}_\infty$; hence, $W(y(t))$ is uniformly continuous. Lemma 5.2 can now be invoked to prove that $\|z(t)\| \to 0$ as $t \to \infty$ $\forall y(t_0) \in \mathcal{S}$. Hence, (5.15) can be used to show that $e(t)$, $\dot{e}(t)$, $r(t) \to 0$ as $t \to \infty$. ∎

Remark 5.4 *Since $e(t) \in \mathcal{L}_1$, similar arguments as provided in Equations 5.10 and 5.11 can be utilized to conclude that the exercise machine controller in (5.23) is passive with respect the user power input.*

5.2.5 Desired Trajectory Generator

The perturbation and numeric trajectory generators described previously could be used to generate a reference trajectory that ensures that $\dot{q}_d(t)$, $\ddot{q}_d(t)$, $\dddot{q}_d(t) \in \mathcal{L}_\infty$ with the exception that both methods depend on measurement of the user's power input $p(t)$. As indicated by (5.4), $p(t)$ is computed based on the assumption that the user torque input is measurable. Since the development in this section is based on the assumption that the user torque input is not measurable, a torque estimator, denoted by $\hat{\tau}(t) \in \mathbb{R}$, is constructed as follows

$$\hat{\tau}(t) = -u(t) + J\ddot{q}_d(t) \tag{5.49}$$

where $u(t)$ is introduced in (5.23). Based on (5.49) the following Lemma can be stated.

Lemma 5.4 *The torque observer in (5.49) ensures that $\hat{\tau}(t) \in \mathcal{L}_\infty$ and $\tau(t) - \hat{\tau}(t) \to 0$ as $t \to \infty$ provided the control gains k_s, β_1 and β_2 are selected according to Theorem 5.3.*

Proof. Theorem 5.3 indicates that $u(t)$, $e(t)$, $\dot{e}(t) \in \mathcal{L}_\infty$ and $\dot{e}(t) \to 0$ as $t \to \infty$. The assumption that $\ddot{q}_d(t) \in \mathcal{L}_\infty$ and the facts that $u(t)$, $\dot{e}(t) \in \mathcal{L}_\infty$ can be used along with (5.49) to show that $\hat{\tau}(t) \in \mathcal{L}_\infty$. After taking the time derivative of (5.3) and multiplying the result by J, the following expression is obtained

$$\begin{aligned} J\dot{e}(t) &= J\ddot{q}(t) - J\ddot{q}_d(t) \\ &= \tau(t) - \hat{\tau}(t) \end{aligned} \tag{5.50}$$

where (5.1) and (5.49) have been utilized. By integrating both sides of (5.50) as follows

$$\int_0^t (\tau(\sigma) - \hat{\tau}(\sigma))d\sigma = J(e(t) - e(0)), \qquad (5.51)$$

the facts that $e(t_0)$, $e(t) \in \mathcal{L}_\infty$ can be used to show that $\tau(t) - \hat{\tau}(t) \in \mathcal{L}_1$. Based on the fact that $\dot{e}(t) \to 0$ as $t \to \infty$, (5.50) can also be used to conclude that $\tau(t) - \hat{\tau}(t) \to 0$ as $t \to \infty$. ∎

Based on Lemma 5.4, the perturbation and numerically-based extremum seeking algorithms can be rewritten where $p(t)$ is replaced by $\hat{\tau}(t)\dot{q}_d(t)$.

5.2.6   Experimental Results and Discussion

The exercise machine testbed illustrated in Figure 5.1 was constructed and used to complete experiments that illustrate the feasibility of using the proposed control strategy for maximizing the power expenditure of the user. As illustrated in Figure 5.1, the exercise machine consisted of a handle that a user grasps that is connected to a rotating assembly that is mounted on the rotor of a switched reluctance motor. The exercise machine testbed can be modeled by the single-input single-output nonlinear system introduced in (5.1). The inertia of the motor assembly was experimentally determined to be $J = 0.1 \ kg \cdot m^2$. A resolver mounted on the motor is used to measure the rotor position while rotor velocity was calculated using a standard backward difference algorithm. The motor was interfaced with a Pentium IV personal computer (PC) operating under Microsoft Windows 2000. The control algorithm given in (5.23) was implemented in SIMULINK and converted to an executable file via the Real-Time Workshop and the dSPACE Target. The executable file was loaded in the dSPACE ControlDesk user interface for control parameter tuning and data logging and plotting.

To demonstrate the performance of the control algorithm given in (5.23), two experiments were conducted. For each experiment, a user held the handle of the exercise machine shown in Figure 5.1 and rotated the motor shaft. Based on the desired angular velocity generated by the numerical-based extremum generation (Brent's Method) algorithm, the controller given in (5.23) modifies the resistive torque output of the motor to maximize the user's power expenditure. Quantifying the ability of the exercise machine to find the maximum power expenditure of a user requires that the maximum be known. Since the maximum power output for some user is unknown, the first experiment exploits an artificial power function with a known maximum. Specifically, the following surrogate parabolic power function was utilized in the first experiment to generate $\dot{q}_d(t)$

$$p(t) = 3 - \frac{1}{3}(\dot{q}_d(t) - 3)^2 \tag{5.52}$$

where it is clear that (5.52) is maximized at $\dot{q}_d^* = 3\ rad/\sec$. That is, by using the surrogate input described in (5.52) to generate $\dot{q}_d(t)$, the ability of the extremum seeking trajectory generator to accurately determine the maximum can be quantitatively tested.

To generate $\dot{q}_d(t)$ via Brent's Method (see Section B.3.1 of Appendix B), an initial estimate of the maximum $\dot{q}_d(t)$ is required (i.e., γ_2) along with lower and upper bounds (i.e., γ_1 and γ_3, respectively). For the first experiment, γ_1, γ_2, and γ_3 were selected as follows

$$\gamma_1 = 1 \quad \gamma_2 = 2.5 \quad \gamma_3 = 4.$$

To generate continuous bounded signals for $q_d^{(i)}(t)\ \forall i = 1, 2, 3$, the following stable and proper fourth order filters were utilized

$$q_d^{(i)} = \frac{81 s^{i-1}}{s^4 + 12 s^3 + 54 s^2 + 108 s + 81}. \tag{5.53}$$

A 1.5 second time delay was utilized to allow for the torque estimate $\hat{\tau}(t)$ to converge to $\tau(t)$ before Brent's Method is invoked. Figure 5.2 illustrates that the desired exercise machine velocity converges to the optimal velocity setpoint (i.e., $\dot{q}_d^* = 3\ rad/\sec$).

The control gains in (5.23) were adjusted to the following values

$$k_s = 1 \quad \beta_1 + \beta_2 = 0.05 \quad \alpha_r = 0.05.$$

Figure 5.3 depicts the tracking error $e(t)$. Figure 5.4 depicts the control torque input $u(t)$.

In the first experiment, the desired exercise machine trajectory was generated via Brent's Method where $\tau(\dot{q}_d)$ was provided by a surrogate signal with a known maximum as a means to illustrate the ability of the extremum seeking trajectory generator to converge to the desired maximum. In the second experiment, the surrogate signal was eliminated from the trajectory generator, allowing the desired trajectory to seek the maximum power expenditure of the user. For the second experiment, γ_1, γ_2, and γ_3 were selected as follows

$$\gamma_1 = 1 \quad \gamma_2 = 3.5 \quad \gamma_3 = 6.$$

The desired trajectory was constructed using the same filters given in (5.53), and a 1.5 second time delay was utilized to allow for the torque estimate $\hat{\tau}(t)$ to converge to $\tau(t)$. Figure 5.5 depicts $\dot{q}_d(t)$.

The control gains in (5.23) were adjusted to the following values

$$k_s = 2 \quad \beta_1 + \beta_2 = 0.1 \quad \alpha_r = 0.1 \ .$$

Figure 5.6 depicts the tracking error $e(t)$. Figure 5.7 depicts the control torque input $u(t)$.

In [67], the control algorithm given in (5.23) has been simulated for cases where the desired trajectory was generated by the perturbation-based extremum seeking algorithm and Brent's Method. The results shown in that work indicate the controller's performance in the ideal case. However, as can be seen in Figures 5.3 and 5.6, the tracking error signals contain high frequency components in practice and exhibit steady state tracking errors of ±0.5 [rad/sec] and 1.0 to −0.5 [rad/sec], respectively. The magnitude of the tracking errors may not be acceptable for typical tracking applications; however, this application is atypical since a human is directly interacting with the system in real-time. That is, the human input, denoted by $\tau(\dot{q})$ in (5.1), can be viewed as an additive bounded disturbance that corrupts the tracking performance. While the high-frequency estimator given in (5.49) should theoretically compensate for this additive bounded disturbance, measurement noise as well as the limited bandwidth of the actuator used in the experimental hardware resulted in some degradation in tracking performance.

5.3   Steer-by-Wire

The general concept of the proposed steer-by-wire haptic-interface control architecture is presented in Figure 5.8. Signal flow for a typical hybrid vehicle is shown in this figure. It can be seen that the flow of information in a steering system is bidirectional. The input to the Primary System from the operator/driver has to be translated to the Secondary System. At the same time, reaction forces at the Secondary System have to be fed back to the Primary System even though no mechanical linkage exists between these two systems. Hence, providing force feedback only handles one of the two issues that arise out of decoupling the driver interface and the directional control assembly. The other equally important piece of this system involves the actuation of the directional control assembly to translate the driver's intentions into actions. In this section, a full state feedback controller is designed to provide the desired force feedback on the steering wheel to reflect the tire/road interface forces and simultaneously synchronize the motion of the directional control assembly with the motion of the steering wheel. For

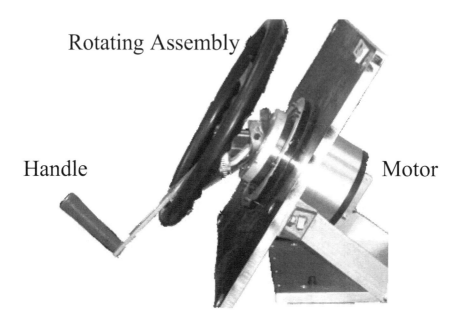

FIGURE 5.1. Exercise Machine Testbed (Side View)

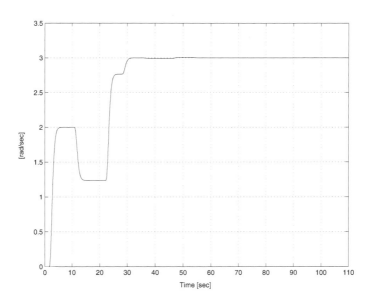

FIGURE 5.2. Desired Velocity

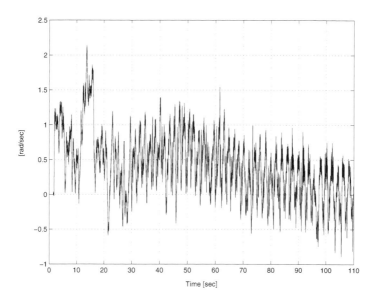

FIGURE 5.3. Tracking Error

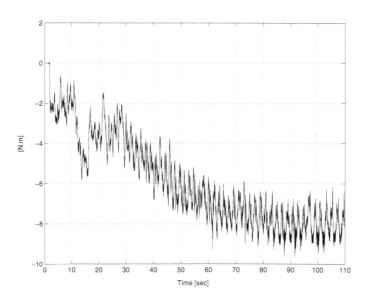

FIGURE 5.4. Computed Motor Torque

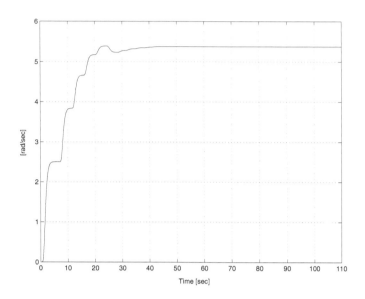

FIGURE 5.5. Desired Velocity

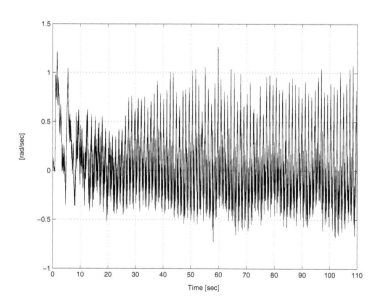

FIGURE 5.6. Tracking Error

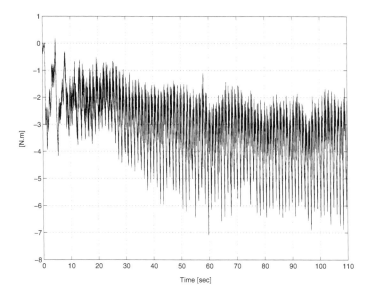

FIGURE 5.7. Computed Motor Torque

the force feedback control design, a target system is used to generate the reference signal for the displacement of the primary system. This type of approach is motivated by the impedance control concept detailed in [35]. The controller adapts for parametric uncertainties in the system while ensuring global asymptotic tracking for the "driver experience error" and the "locked error"; however, torque measurements are required.

To make this concept economically feasible, it is essential to avoid the use of sensors and actuators that are either expensive or require frequent maintenance. Traditionally force/torque sensors have been used to measure the forces that need to be fed back to the driver. However, sensors that can provide the quality of signals necessary for satisfactory performance are both expensive and unreliable. To eliminate torque measurements, a recent idea found in [51] has been modified to develop torque observers for the design of an exact model knowledge tracking controller. This controller ensures that the torque observation error converges asymptotically to zero while also ensuring global asymptotic tracking for the driver experience error and the locked tracking error. Roughly speaking, the torque observer design borrows concepts from robust control techniques that only impose boundedness and smoothness restrictions on the unmeasurable torque signals.

The design of the adaptive, nonlinear tracking controller being presented here ensures that: i) the directional control assembly follows the driver commanded input, and ii) the dynamics of the driver input device follows that of a target system. As previously stated in the introduction, a complete stability analysis, using Lyapunov-based techniques, will be used to demonstrate that the proposed control law guarantees global asymptotic regulation of the locked tracking and the driver experience tracking errors. Furthermore, an extension is presented to study the controller design under elimination of torque measurements. Representative numerical simulation results are presented to validate the performance of the proposed control law. Finally, a detailed description of an experimental test setup is provided along with experimental verification of the control algorithm.

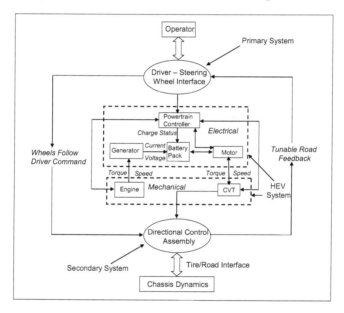

FIGURE 5.8. Steer-by-Wire Interface System Architecture

5.3.1 Control Problem Statement

The steer-by-wire haptic interface control objective is twofold. First, the driver's steering angle commands must be accurately followed; this requires the torque control input provided by the drive motor to be designed so that the angular position of the directional control assembly accurately tracks the input. Second, the driver must be given a realistic "virtual driving experience." To this end, a reference model, or target dynamics for the driver input device, should be designed to generate the desired angular position of the driver input device. The control torque provided by the feedback motor must then be designed to ensure that the response of the driver input device follows that of the reference system. The reader is referred to Figure 5.9 for definition of the driver interface and the directional control assembly. The figure depicts the essential components present in a steer-by-wire system. The reader is referred to the nomenclature included in the figure as an introduction to the various signals considered in the ensuing development.

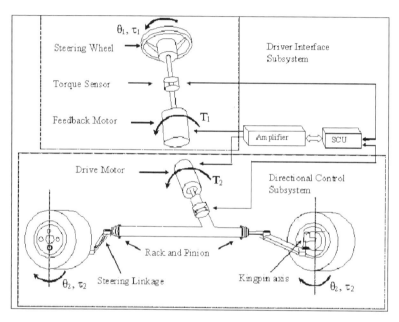

FIGURE 5.9. Driver Interface and Directional Control Subsystems in a Steer-by-Wire Assembly

5.3.2   Dynamic Model Development

Detailed models for conventional and power assisted steering systems can be found in many works in this research area (e.g., [44], [45]). The steer-by-wire system involves the removal of the steering column present in a conventional steering system and the introduction of two servo motors. The steering system is separated into two subsystems: the primary and the secondary subsystems. The primary system consists of the driver input device (e.g., steering wheel or joystick) and a servo motor to provide the driver with force feedback. The secondary subsystem is composed of the directional control assembly (e.g., rack and pinion system) and a servo motor that provides the necessary torque input to drive the assembly and steer the vehicle.

Steering System Model Formulation

In general, the steering system dynamics may be expressed in a simplified form as

$$I_1 \ddot{\theta}_1 + N_1 \left(\theta_1, \dot{\theta}_1 \right) = \alpha_1 \tau_1 + T_1 \tag{5.54}$$

$$I_2 \ddot{\theta}_2 + N_2 \left(\theta_2, \dot{\theta}_2 \right) = \alpha_2 \tau_2 + T_2 \tag{5.55}$$

where $\theta_1(t)$, $\dot{\theta}_1(t)$, $\ddot{\theta}_1(t) \in \mathbb{R}$ denote the angular position, velocity, and acceleration, respectively, of the driver input device, I_1, $I_2 \in \mathbb{R}$ represent the inertias of the driver input device and the vehicle directional control assembly, respectively. $N_1 \left(\theta_1, \dot{\theta}_1 \right) \in \mathbb{R}$ is an auxiliary nonlinear function that describes the dynamics on the driver side, $\tau_1(t) \in \mathbb{R}$ denotes the driver input torque, $T_1(t) \in \mathbb{R}$ represents a control input torque applied to the driver input device, $\theta_2(t)$, $\dot{\theta}_2(t)$, $\ddot{\theta}_2(t) \in \mathbb{R}$ denote the angular position, velocity, and acceleration, respectively, of the vehicle directional control assembly, $N_2 \left(\theta_2, \dot{\theta}_2 \right) \in \mathbb{R}$ is an auxiliary nonlinear function that is used to describe the dynamics of the vehicle directional control assembly, $\tau_2(t) \in \mathbb{R}$ represents the reaction torque between the actuator on the directional control assembly and mechanical subsystem actuated by the directional control assembly, and $T_2(t) \in \mathbb{R}$ denotes a control input torque applied to the directional control assembly. The constants $\alpha_1, \alpha_2 \in \mathbb{R}$ are scaling factors that could arise due to gearing in the system.

Remark 5.5 *The damping and friction effects modeled by $N_1 \left(\cdot \right)$ and $N_2 \left(\cdot \right)$ are assumed to be linearly parameterizable as*

$$N_1 \left(\theta_1, \dot{\theta}_1 \right) = Y_{N1} \left(\theta_1, \dot{\theta}_1 \right) \phi_{N1} \tag{5.56}$$

$$N_2 \left(\theta_2, \dot{\theta}_2 \right) = Y_{N2} \left(\theta_2, \dot{\theta}_2 \right) \phi_{N2} \tag{5.57}$$

where $Y_{N1} \left(\cdot \right) \in \mathbb{R}^{1 \times p}$, $Y_{N2} \left(\cdot \right) \in \mathbb{R}^{1 \times q}$ are regression matrices containing the measurable signals, and $\phi_{N1} \in \mathbb{R}^{p \times 1}$, $\phi_{N2} \in \mathbb{R}^{q \times 1}$ are constant matrices containing the unknown parameters in the model $N_1 \left(\cdot \right)$ and $N_2 \left(\cdot \right)$. Further, it is also assumed that if $\theta_1 \left(t \right), \dot{\theta}_1 \left(t \right) \in \mathcal{L}_\infty$ then $N_1 \left(\theta_1, \dot{\theta}_1 \right) \in \mathcal{L}_\infty$ and if $\theta_2 \left(t \right), \dot{\theta}_2 \left(t \right) \in \mathcal{L}_\infty$ then $N_2 \left(\theta_2, \dot{\theta}_2 \right) \in \mathcal{L}_\infty$.

Reference Model Development

The second control objective is the provision of road "feel" to the driver. To satisfy this goal, impedance control concepts [35] used for robot manipulator position/force control problems have been applied to the problem. Specifically, a reference model is designed as

$$I_T \ddot{\theta}_{d1} + N_T \left(\theta_{d1}, \dot{\theta}_{d1} \right) = \alpha_{T1} \tau_1 + \alpha_{T2} \tau_2 \tag{5.58}$$

where $\theta_{d1}(t)$, $\dot{\theta}_{d1}(t)$, $\ddot{\theta}_{d1}(t) \in \mathbb{R}$ denote the desired angular position, velocity, and acceleration, respectively, of the driver input device, $N_T \left(\theta_{d1}, \dot{\theta}_{d1} \right) \in \mathbb{R}$ represents an auxiliary target dynamic function for the driver input device, and α_{T1}, $\alpha_{T2} \in \mathbb{R}$ are scaling constants. The structure for (5.58) is motivated by the following philosophy: If $T_1(t)$ was exactly equal to $\alpha_2 \tau_2(t)$ in (5.54), then the dynamics given by (5.54) would give the driver a realistic experience provided the auxiliary term $N_1 (\cdot)$ in (5.54) could be designed or constructed according to some desirable mechanical response. Thus, the $N_T (\cdot)$ term in (5.58) is designed to simulate the desired driving experience, and hence, the dynamics given by (5.58) serve as a desired trajectory generator for control design purposes for (5.54).

Remark 5.6 *The target dynamic function $N_T (\cdot)$ must be selected to ensure that the desired driver input device trajectory and its first two derivatives remain bounded at all times (i.e., $\theta_{d1}(t)$, $\dot{\theta}_{d1}(t)$, $\ddot{\theta}_{d1}(t) \in \mathcal{L}_\infty$). Suppose $N_T (\cdot)$ is selected as*

$$N_T = B_T \dot{\theta}_{d1} + K_T \theta_{d1} \tag{5.59}$$

where B_T, $K_T \in \mathbb{R}$ are some positive design constants. If $N_T (\cdot)$ is selected according to (5.59), then standard linear arguments show that $\theta_{d1}(t)$, $\dot{\theta}_{d1}(t)$, $\ddot{\theta}_{d1}(t) \in \mathcal{L}_\infty$. Furthermore, $N_T (\cdot)$ can be constructed as a nonlinear damping function by utilizing Lyapunov-type arguments.

Open-Loop Error System Development

To quantify the mismatches between the target system and the primary system or driver experience tracking error, as well as the primary and the secondary system or locked tracking error, filtered error signals, $r_1(t)$, $r_2(t) \in \mathbb{R}$ are defined as

$$r_1 = \dot{e}_1 + \mu_1 e_1 \tag{5.60}$$

$$r_2 = \dot{e}_2 + \mu_2 e_2 \tag{5.61}$$

where $\mu_1, \mu_2 \in \mathbb{R}$ represent positive control gains, and the target following error signal $e_1(t)$ and the locked tracking error signal $e_2(t) \in \mathbb{R}$ are defined as

$$e_1 = \theta_{d1} - \theta_1 \tag{5.62}$$

$$e_2 = \theta_1 - \theta_2. \tag{5.63}$$

After taking the first time derivative of (5.60) and (5.61), and substituting the dynamics in (5.54), (5.55) and (5.58), the open loop error systems are

$$I_1 \dot{r}_1 = Y_1 \phi_1 - T_1 \tag{5.64}$$

$$I_2 \dot{r}_2 = Y_2 \phi_2 - T_2 \tag{5.65}$$

where $Y_1(\cdot) \in \mathbb{R}^{1 \times r}$, $Y_2(\cdot) \in \mathbb{R}^{1 \times s}$ are regression matrices consisting of measurable quantities, and $\phi_1 \in \mathbb{R}^{r \times 1}$, $\phi_2 \in \mathbb{R}^{s \times 1}$ are constant unknown vectors. The reader is referred to Section B.3.3 of Appendix B for explicit definitions of $Y_1(\cdot), Y_2(\cdot), \phi_1$ and ϕ_2.

Remark 5.7 *Based on the definition of $r_1(t)$ and $r_2(t)$ given in (5.60) and (5.61), standard arguments [9] can be used to prove that: (i) if $r_1(t)$, $r_2(t) \in \mathcal{L}_\infty$, then $e_1(t)$, $e_2(t)$, $\dot{e}_1(t)$, $\dot{e}_2(t) \in \mathcal{L}_\infty$, and (ii) if $r_1(t)$ and $r_2(t)$ are asymptotically regulated, then $e_1(t)$ and $e_2(t)$ are asymptotically regulated.*

5.3.3   Control Development

The first control objective requires the target following and the locked tracking error signals to approach zero asymptotically, while adapting for the system parameters that are assumed to be unknown. Further, the signals $\theta_1(t)$, $\theta_2(t)$, $\dot{\theta}_1(t)$, $\dot{\theta}_2(t)$, $\tau_1(t)$, and $\tau_2(t) \in \mathcal{L}_\infty$ must be available for measurement.

Control Formulation

Based on the subsequent stability analysis in Section 5.3.4 and the structure of the open-loop error system given in (5.64) and (5.65), the control inputs $T_1(t)$ and $T_2(t)$ are designed as

$$T_1 = k_1 r_1 + Y_1 \hat{\phi}_1 \tag{5.66}$$

$$T_2 = k_2 r_2 + Y_2 \hat{\phi}_2 \tag{5.67}$$

where $k_1, k_2 \in \mathbb{R}$ are constant positive control gains, and $\hat{\phi}_1(t) \in \mathbb{R}^{r \times 1}$, $\hat{\phi}_2(t) \in \mathbb{R}^{s \times 1}$ are adaptive estimates for the unknown parameter matrices.

The adaptive update laws are designed based on the subsequent stability analysis as

$$\dot{\hat{\phi}}_1 = \Gamma_1 Y_1^T r_1 \tag{5.68}$$

$$\dot{\hat{\phi}}_2 = \Gamma_2 Y_2^T r_2 \tag{5.69}$$

where $\Gamma_1 \in \mathbb{R}^{r \times r}, \Gamma_2 \in \mathbb{R}^{s \times s}$ are positive constant diagonal gain matrices.

Closed-Loop Error System Development

After substituting the control torques in (5.66) and (5.67) into the open-loop dynamics in (5.64) and (5.65), the closed-loop error system becomes

$$I_1 \dot{r}_1 = -k_1 r_1 + Y_1 \tilde{\phi}_1 \tag{5.70}$$

$$I_2 \dot{r}_2 = -k_2 r_2 + Y_2 \tilde{\phi}_2 \tag{5.71}$$

where the parameter estimation error signals, $\tilde{\phi}_1(t) \in \mathbb{R}^{r \times 1}, \tilde{\phi}_2(t) \in \mathbb{R}^{s \times 1}$ are defined as

$$\tilde{\phi}_1 = \phi_1 - \hat{\phi}_1 \tag{5.72}$$

$$\tilde{\phi}_2 = \phi_2 - \hat{\phi}_2. \tag{5.73}$$

5.3.4 Stability Analysis

Theorem 5.5 *Given the closed-loop system of (5.70) and (5.71), the tracking error signals defined in (5.62) and (5.63) are globally asymptotically regulated in the sense that*

$$\lim_{t \to \infty} e_1(t), e_2(t) = 0. \tag{5.74}$$

Proof: A non-negative, scalar function, denoted by $V(t) \in \mathbb{R}$, is defined as

$$V = \frac{1}{2} I_1 r_1^2 + \frac{1}{2} I_2 r_2^2 + \frac{1}{2} \tilde{\phi}_1^T \Gamma_1^{-1} \tilde{\phi}_1 + \frac{1}{2} \tilde{\phi}_2^T \Gamma_2^{-1} \tilde{\phi}_2. \tag{5.75}$$

After taking the time derivative of (5.75) and making the appropriate substitutions from (5.70), (5.71), (5.72), and (5.73), the following expression is obtained

$$\begin{aligned} \dot{V} &= r_1 \left[-k_1 r_1 + Y_1 \tilde{\phi}_1 \right] + r_2 \left[-k_2 r_2 + Y_2 \tilde{\phi}_2 \right] \\ &\quad - \tilde{\phi}_1^T \left[Y_1^T r_1 \right] - \tilde{\phi}_2^T \left[Y_2^T r_2 \right] \end{aligned} \tag{5.76}$$

where the fact that Γ_1, Γ_2 are constant diagonal gain matrices has been utilized along with the following equalities: $\dot{\tilde{\phi}}_1 = -\dot{\hat{\phi}}_1$ and $\dot{\tilde{\phi}}_2 = -\dot{\hat{\phi}}_2$.

After cancelling common terms, it is easy to see that we can upper bound $\dot{V}(t)$ as follows

$$\dot{V} \leq -k_1 r_1^2 - k_2 r_2^2. \tag{5.77}$$

From (5.77) and (5.75), it is straightforward to see that $r_1(t)$, $r_2(t)$, $\tilde{\phi}_1(t)$, $\tilde{\phi}_2(t) \in \mathcal{L}_\infty$. After utilizing (5.72), (5.73), and Remark 5.7, we can conclude that $e_1(t)$, $\dot{e}_1(t)$, $e_2(t)$, $\dot{e}_2(t)$, $\hat{\phi}_1(t)$, $\hat{\phi}_2(t) \in \mathcal{L}_\infty$. Using Remark 5.6, (5.62), (5.63) and their first derivatives, it is clear that $\theta_1(t)$, $\theta_2(t)$, $\dot{\theta}_1(t)$ $\dot{\theta}_2(t) \in \mathcal{L}_\infty$. This implies (using 5.56 and 5.57) that $N_1(\cdot)$ and $N_2(\cdot)$ are bounded. From the explicit definition for $Y_1(\cdot)$ given in Section B.3.3 of Appendix B and using the fact that $\tau_1(t)$, $\tau_2(t) \in \mathcal{L}_\infty$, it is easy to see that $Y_1(\cdot) \in \mathcal{L}_\infty$. From (5.66), it is clear that the control torque $T_1(t) \in \mathcal{L}_\infty$. Again, from the definition of $Y_2(\cdot)$ in Section B.3.3 of Appendix B and from the above facts, $Y_2(\cdot) \in \mathcal{L}_\infty$. From (5.67), it is clear that $T_2(t) \in \mathcal{L}_\infty$. Using standard signal chasing arguments, it can be shown that all the signals in the closed-loop system remain bounded. In particular, from (5.70) and (5.71), $\dot{r}_1(t)$, $\dot{r}_2(t) \in \mathcal{L}_\infty$. After employing a corollary to Barbalat's Lemma [61], it is easy to show that

$$\lim_{t \to \infty} r_1(t), r_2(t) = 0.$$

Finally, Remark 5.7 can be used to prove the result stated in (5.74).

5.3.5   Elimination of Torque Measurements: Extension

For the steering system in (5.54) and (5.55), a controller can be designed to eliminate the need for torque sensors. Here the assumption is that the driver input torque, denoted by $\tau_1(t)$, and the reaction torque between the actuator on the directional control assembly and mechanical subsystem actuated by the directional control assembly, denoted by $\tau_2(t)$, are not available for measurement. It is further assumed that the driver input torque $\tau_1(t)$ and the reaction torque $\tau_2(t)$ and their first two derivatives remain bounded at all times. Another assumption is that the dynamics are exactly known with the exception of the torque signals $\tau_1(t)$ and $\tau_2(t)$, and that the signals $\theta_1(t)$, $\theta_2(t)$, $\dot{\theta}_1(t)$, and $\dot{\theta}_2(t)$ are available for measurement. To account for the lack of torque measurements, the torque observation errors, denoted by $\tilde{\tau}_1(t)$, $\tilde{\tau}_2(t) \in \mathbb{R}$, are defined as

$$\tilde{\tau}_1 = \tau_1 - \hat{\tau}_1 \tag{5.78}$$

$$\tilde{\tau}_2 = \tau_2 - \hat{\tau}_2 \tag{5.79}$$

where $\hat{\tau}_1(t)$, $\hat{\tau}_2(t) \in \mathbb{R}$ denote the driver input and the reaction observer torques, respectively. The observer torques are subsequently designed in

Section 5.3.5. The target dynamics are now generated as

$$I_T \ddot{\theta}_{d1} + N_T \left(\theta_{d1}, \dot{\theta}_{d1} \right) = \alpha_{T1} \hat{\tau}_1 + \alpha_{T2} \hat{\tau}_2. \tag{5.80}$$

Open-Loop Error System Development

After taking two time derivatives of (5.62) and (5.63), and substituting the dynamics given in (5.54), (5.55), and the target dynamics given in (5.80), the open loop error systems for the two systems can be written as

$$\ddot{e}_1 = \left(\frac{1}{I_T} \right) \left(-N_T \left(\cdot \right) + \alpha_{T1} \hat{\tau}_1 + \alpha_{T2} \hat{\tau}_2 \right) - \left(\frac{1}{I_1} \right) \left(-N_1 \left(\cdot \right) + \alpha_1 \tau_1 + T_1 \right)$$
$$\tag{5.81}$$

$$\ddot{e}_2 = \left(\frac{1}{I_1} \right) \left(-N_1 \left(\cdot \right) + \alpha_1 \tau_1 + T_1 \right) - \left(\frac{1}{I_2} \right) \left(-N_2 \left(\cdot \right) + \alpha_2 \tau_2 + T_2 \right). \tag{5.82}$$

To simplify further analysis, two auxiliary signals $p_1 \left(t \right)$, $p_2 \left(t \right) \in \mathbb{R}$ are defined as

$$p_1 = \dot{e}_1 + \beta_1 e_1 \tag{5.83}$$

$$p_2 = \dot{e}_2 + \beta_1 e_2 \tag{5.84}$$

where $\beta_1 \in \mathbb{R}$ is a constant positive control gain. Furthermore, two filtered tracking error signals, denoted by $s_1 \left(t \right)$, $s_2 \left(t \right) \in \mathbb{R}$, are defined as

$$s_1 = \dot{p}_1 + p_1 \tag{5.85}$$

$$s_2 = \dot{p}_2 + p_2. \tag{5.86}$$

Control Development

Based on the subsequent control design/stability analysis and the structure of the open-loop error system given by (5.81) and (5.82), the driver input torque observer and the reaction torque observer are

$$\dot{\hat{\tau}}_1 = - \left(\beta_1 + K_s + 1 \right) \hat{\tau}_1 - \frac{I_1}{\alpha_1} \left[\left(\beta_1 + K_s \left(\beta_1 + 1 \right) \right) \dot{e}_1 + K_s \beta_1 e_1 + \rho_1 sgn \left(p_1 \right) \right]$$
$$\tag{5.87}$$

$$\dot{\hat{\tau}}_2 = \ -\left(\beta_1 + K_s + 1 \right) \hat{\tau}_2 - \frac{I_2}{\alpha_2} \left[\left(\beta_1 + K_s \left(\beta_1 + 1 \right) \right) \dot{e}_2 + K_s \beta_1 e_2 \right]$$
$$- \frac{I_2}{\alpha_2} \left[\frac{\alpha_1}{I_1} \left(\dot{\hat{\tau}}_1 + \left(\beta_1 + K_s + 1 \right) \hat{\tau}_1 \right) + \rho_2 sgn \left(p_2 \right) \right]$$
$$\tag{5.88}$$

where $sgn \left(\cdot \right)$ denotes the standard signum function, K_s, ρ_1, $\rho_2 \in \mathbb{R}$ are positive control gains while p_1, p_2, and β were introduced in (5.83) and

(5.84). Based on the subsequent stability analysis, the control inputs $T_1(t)$ and $T_2(t)$ are designed as

$$T_1 = N_1(\cdot) + \left(\frac{I_1}{I_T}\right)(-N_T(\cdot) + \alpha_{T1}\hat{\tau}_1 + \alpha_{T2}\hat{\tau}_2) - \alpha_1\hat{\tau}_1 \tag{5.89}$$

$$T_2 = N_2(\cdot) + \left(\frac{I_2}{I_1}\right)(-N_1(\cdot) + T_1 + \alpha_1\hat{\tau}_1) - \alpha_2\hat{\tau}_2. \tag{5.90}$$

To facilitate the subsequent stability analysis, two auxiliary "disturbance signals," denoted by $\eta_1(t)$, $\eta_2(t) \in \mathbb{R}$, are defined as

$$\eta_1 = \left(\frac{\alpha_1}{I_1}\right)(\dot{\tau}_1 + (\beta_1 + K_s + 1)\tau_1) \tag{5.91}$$

$$\eta_2 = \left(\frac{\alpha_2}{I_2}\right)(\dot{\tau}_2 + (\beta_2 + K_s + 1)\tau_2). \tag{5.92}$$

Based on the fact that $\tau_1(t), \tau_2(t)$ and its derivatives are bounded, $\eta_1(t)$, $\eta_2(t)$, $\dot{\eta}_1(t)$, $\dot{\eta}_2(t) \in \mathcal{L}_\infty$. To facilitate the subsequent closed-loop error system development and the stability analysis, one is required to select the control gains ρ_1 and ρ_2, introduced in (5.87) and (5.88), as follows

$$\rho_1 \geq |\eta_1| + \left|\frac{d\eta_1}{dt}\right| \tag{5.93}$$

$$\rho_2 \geq |\eta_2| + \left|\frac{d\eta_2}{dt}\right| + \rho_1. \tag{5.94}$$

Closed-Loop Error System Development

The control torques in (5.89) and (5.90) can be substituted into the open-loop dynamics in (5.81) and (5.82), so that the closed-loop error system becomes

$$\ddot{e}_1 = -\left(\frac{\alpha_1}{I_1}\right)(\tau_1 - \hat{\tau}_1) \tag{5.95}$$

$$\ddot{e}_2 = \left(\frac{\alpha_1}{I_1}\right)(\tau_1 - \hat{\tau}_1) - \left(\frac{\alpha_2}{I_2}\right)(\tau_2 - \hat{\tau}_2). \tag{5.96}$$

After taking the first time derivative of (5.85) and (5.86), and using (5.95), (5.96), and their first derivatives, the closed-loop error system can be written as

$$\dot{s}_1 = -K_s s_1 - \eta_1 - \rho_1 sgn(p_1) \tag{5.97}$$

$$\dot{s}_2 = -K_s s_2 + \eta_1 - \eta_2 - \rho_2 sgn(p_2) \tag{5.98}$$

where (5.91), (5.92), (5.87), and (5.88) have been utilized.

Remark 5.8 *Based on the definition of $s_1(t)$ and $s_2(t)$ given in (5.85) and (5.86), extensions of the arguments made in [9] can be used to prove that: (i) if $s_1(t)$, $s_2(t) \in \mathcal{L}_\infty$, then $e_1(t)$, $e_2(t)$, $\dot{e}_1(t)$, $\dot{e}_2(t)$, $\ddot{e}_1(t)$, $\ddot{e}_2(t) \in \mathcal{L}_\infty$, and (ii) if $s_1(t)$ and $s_2(t)$ are asymptotically regulated, then $e_1(t)$, $e_2(t)$, $\dot{e}_1(t)$, $\dot{e}_2(t)$, $\ddot{e}_1(t)$, $\ddot{e}_2(t)$ are also asymptotically regulated [10].*

Stability Proof

Theorem 5.6 *Given the closed-loop system of (5.97) and (5.98), the tracking error signals defined in (5.62) and (5.63) along with the torque observation errors defined in (5.78) and (5.79) are globally asymptotically regulated in the sense that*

$$\lim_{t \to \infty} e_1(t), \ e_2(t), \ \tilde{\tau}_1(t), \ \tilde{\tau}_2(t) = 0. \tag{5.99}$$

Proof: A non-negative, scalar function, denoted by $V_{a1}(t) \in \mathbb{R}$, is defined as

$$V_{a1} = \frac{1}{2}s_1^2. \tag{5.100}$$

After taking the time derivative of (5.100) and making appropriate substitutions from (5.97), then $\dot{V}_{a1}(t)$ maybe expressed as

$$\dot{V}_{a1} = -K_s s_1^2 + (\dot{p}_1 + p_1)(-\eta_1 - \rho_1 sgn(p_1)) \tag{5.101}$$

where the definitions of $s_1(t)$ and $p_1(t)$ given in (5.85) and (5.83) have been utilized. After integrating both sides of (5.101), and performing some mathematical operations (refer to Section B.3.4 of Appendix B), the following inequality can be obtained

$$
\begin{aligned}
V_{a1}(t) - V_{a1}(t_0) \leq \ & -K_s \int_{t_0}^{t} s_1^2(\sigma)\, d\sigma \\
& + \left[\int_{t_0}^{t} |p_1(\sigma)| \left(|\eta_1| + \left| \frac{d\eta_1(\sigma)}{d\sigma} \right| - \rho_1 \right) d\sigma \right] \\
& + [|p_1|(|\eta_1| - \rho_1)] + \eta_1(t_0) p_1(t_0) + \rho_1 |p_1(t_0)|
\end{aligned} \tag{5.102}
$$

After applying (5.93) to the bracketed terms in (5.102), $V_{a1}(t)$ can be upper bounded as

$$V_{a1}(t) \leq V_{a1}(t_0) - K_s \int_{t_0}^{t} s_1^2(\sigma)\, d\sigma + \zeta_{01} \tag{5.103}$$

where $\zeta_{01} \in \mathbb{R}$ is a positive constant defined as

$$\zeta_{01} = \eta_1(t_0) p_1(t_0) + \rho_1 |p_1(t_0)|. \tag{5.104}$$

For the error system given in (5.98), a second non-negative, scalar function, denoted by $V_{a2}(t) \in \mathbb{R}$, is

$$V_{a2} = \frac{1}{2} s_2^2. \tag{5.105}$$

Following a similar analysis to that presented above, the upper bound for $V_{a2}(t)$ becomes

$$V_{a2}(t) \leq V_{a2}(t_0) - K_s \int_{t_0}^{t} s_2^2(\sigma) \, d\sigma + \zeta_{02} \tag{5.106}$$

where $\zeta_{02} \in \mathbb{R}$ is a positive constant defined as

$$\zeta_{02} = \eta_2(t_0) p_2(t_0) - \eta_1(t_0) p_2(t_0) + \rho_2 |p_2(t_0)| . \tag{5.107}$$

To complete the analysis, the following composite non-negative, scalar function, $V_a(t) \in \mathbb{R}$, is selected

$$V_a = V_{a1} + V_{a2} = \frac{1}{2} s_1^2 + \frac{1}{2} s_2^2. \tag{5.108}$$

After using the bounds for $V_{a1}(t)$ and $V_{a2}(t)$ obtained in (5.103) and (5.106), the upper bound for $V_a(t)$ becomes

$$V_a(t) \leq V_a(t_0) - K_s \int_{t_0}^{t} s_1^2(\sigma) \, d\sigma - K_s \int_{t_0}^{t} s_2^2(\sigma) \, d\sigma + \zeta_0 \tag{5.109}$$

where $\zeta_0 = \zeta_{01} + \zeta_{02}$. Clearly, from the inequality in (5.109), $V_a(t) \in \mathcal{L}_\infty$. From the definition of $V_a(t)$ in (5.108), $s_1(t)$, $s_2(t) \in \mathcal{L}_\infty$. From Remark 5.8, $e_1(t)$, $e_2(t)$, $\dot{e}_1(t)$, $\dot{e}_2(t)$, $\ddot{e}_1(t)$, $\ddot{e}_2(t) \in \mathcal{L}_\infty$. From (5.97), (5.98), (5.83) and (5.84) and the boundedness of $\eta_1(t)$ and $\eta_2(t)$, it is easy to see that $\dot{s}_1(t)$, $\dot{s}_2(t)$, $p_1(t)$, $p_2(t) \in \mathcal{L}_\infty$. From (5.95) and (5.96), it is now clear that $\hat{\tau}_1(t)$, $\hat{\tau}_2(t) \in \mathcal{L}_\infty$. From equation (5.89), it can be shown that $T_1(t) \in \mathcal{L}_\infty$. Using this fact, it is easy to see from (5.90) that $T_2(t) \in \mathcal{L}_\infty$. Standard signal chasing arguments can now be employed to show that all signals in the system remain bounded. If (5.109) is rewritten as

$$K_s \int_{t_0}^{t} s_1^2(\sigma) \, d\sigma + K_s \int_{t_0}^{t} s_2^2(\sigma) \, d\sigma \leq V_a(t) - V_a(t_0) + \zeta_0, \tag{5.110}$$

then it is clear that $s_1(t)$, $s_2(t) \in \mathcal{L}_2$. Further, Barbalat's Lemma [61] can be applied to show that

$$\lim_{t \to \infty} s_1(t), \, s_2(t) = 0, \tag{5.111}$$

and hence, from Remark 5.8,

$$\lim_{t \to \infty} e_1(t), e_2(t), \dot{e}_1(t), \dot{e}_2(t), \ddot{e}_1(t), \ddot{e}_2(t) = 0. \tag{5.112}$$

Since $\ddot{e}_1(t)$ and $\ddot{e}_2(t)$ are regulated to zero as indicated by (5.112), then (5.95) and (5.96) and the definitions provided in (5.78) and (5.79) show that

$$\lim_{t \to \infty} \tilde{\tau}_1(t), \tilde{\tau}_2(t) = 0. \quad \blacksquare \tag{5.113}$$

Remark 5.9 *The structure of the torque observers given by (5.87) and (5.88) contain discontinuous terms; however, the control inputs that are applied to the plant are not discontinuous. That is, after a close examination of (5.87) and (5.88), it clear that signals $\hat{\tau}_1(t)$ and $\hat{\tau}_2(t)$ are low pass filtered outputs of a discontinuous control signal. Therefore, the control strategy only utilizes the signals $\hat{\tau}_1(t)$ and $\hat{\tau}_2(t)$, so that the control signals applied to the plant will not be discontinuous.*

5.3.6   Numerical Simulation Results

Two sets of simulations were performed to study the performance of the control algorithms: (i) System with an adaptive controller as developed in Section 5.3.3, and (ii) System with a exact model knowledge controller (EMK controller) presented in Section 5.3.5. The simulated vehicle steering system was assumed to have the dynamics described by (5.54) and (5.55). The nonlinear damping and friction functions were chosen as [44]

$$N_i\left(\theta_i, \dot{\theta}_i\right) = B_i \dot{\theta}_i + K_i \theta_i \qquad i = 1, 2 \tag{5.114}$$

The system parameters were chosen to be the same as that of the actual experimental setup described in Section 5.3.7. These values are listed in Table 5.1.

The reference trajectory was generated according to (5.58). The user defined function $N_T\left(\theta_{d1}, \dot{\theta}_{d1}\right)$ was chosen to have the same form as in (5.114). The reaction torque applied on the directional control assembly (due to the tire-road interface forces), was assumed to be related to the angular deflection of the directional control assembly in the following manner [58]

$$\tau_2 = -C_d \tanh(\gamma \theta_2) \tag{5.115}$$

where $C_d, \gamma \in \mathbb{R}$ are constant tire-dependent parameters listed in Table 5.1. Each set of simulations were performed for two driver input torque profiles: *Case 1*: $\tau_1(t) = 0.8 \sin(5t)(1 - \exp(-3t))$ which represents the input to perform a standard slalom maneuver; and *Case 2*: $\tau_1(t) = 0.9(1 - \exp(-3t))$

Symbol	Value	Units	Symbol	Value	Units
I_1	$1.16e-2$	$kg.m^2$	I_T	$1.5e-2$	$kg.m^2$
B_1	$1.9e-2$	$kg.m^2/s$	B_T	$2e-2$	$kg.m^2/s$
K_1	0	$N.m$	K_T	0	$N.m$
α_1	1	$-$	α_{T1}	1	$-$
I_2	$2.35e-2$	$kg.m^2$	α_{T2}	0.15	$-$
B_2	0.6	$kg.m^2/s$	C_d	150	$N.m$
K_2	0	$N.m$	γ	0.02	$-$
α_2	1	$-$			

TABLE 5.1. List of Simulation Parameters and Corresponding Values

which is the input that the driver would apply to follow a circular trajectory (refer to Figure 5.10).

For the first set of simulations, the torque measurements are directly available for use in the control algorithm as opposed to torque observer estimation. All adaptive estimates were initialized to zero in this simulation. The driver experience and locked tracking errors, $e_1(t)$ and $e_2(t)$, are presented in Figure 5.11. The driver experience tracking error corresponds to the differences between the target and the primary subsystem of the haptic interface steer-by-wire system. As shown, the error $e_1(t)$ approaches zero after $t = 5$s for both inputs, which implies that the driver experiences the desired "feel" as specified by the target parameters (which corresponds to a conventional steering system with the target parameters). The locked tracking error, $e_2(t)$, also approaches zero for both inputs. This demonstrates that the driver's steering commands are followed by the directional control assembly. These two facts prove that the control algorithm achieves the two goals outlined in the control objective (refer to Section 5.3.1 for details). Selected gains can be increased for faster convergence rates at the cost of larger control effort. All the parameter estimates were observed to settle down at constant values. The plots for adaptive estimates have been left out for brevity. The corresponding motor control torques are displayed in Figure 5.12. Due to the significantly higher dynamic friction parameter, B_2, of the secondary subsystem (refer to Table 5.1), the directional control assembly requires a larger magnitude of control effort as compared to the primary subsystem for both steering profiles. The performance of the control algorithm is further evaluated in terms of three performance measures: (i) Peak error (in %), (ii) Steady-state error (in %), and (iii) Settling

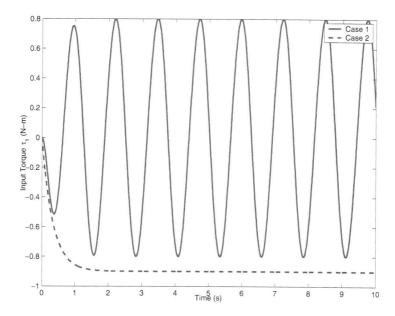

FIGURE 5.10. Driver Input Torque Profiles for Steering Input for a Slalom Maneuver and a Circular Trajectory

time (in seconds). These are tabulated as shown in the Table 5.2. Here, 5% settling time is considered for Case 2 only.

For the second set of simulations, the control algorithm proposed in Section 5.3.5 was simulated on the same system as described in (5.54) and (5.55) having the same parameters as given in Table 5.1. However, the estimated torques are used to generate the target dynamics per (5.80) as opposed to "measured" torques that were available in the previous set of simulations. The driver experience and locked tracking errors, $e_1(t)$ and $e_2(t)$, are shown in Figure 5.13. The errors $e_1(t)$ and $e_2(t)$ are within ± 0.01 rad for both cases. The corresponding control torques, $T_1(t)$ and $T_2(t)$, are

Performance	Case 1		Case 2	
Measure	e_1	e_2	e_1	e_2
Peak Error (%)	0.0577	0.9755	0.9351	2.4913
Steady-State Error (%)	0	0.001	0.0002	0.0025
Settling time (s)	–	–	2.297	2.285

TABLE 5.2. Summary of Results for Adaptive Controller Simulation

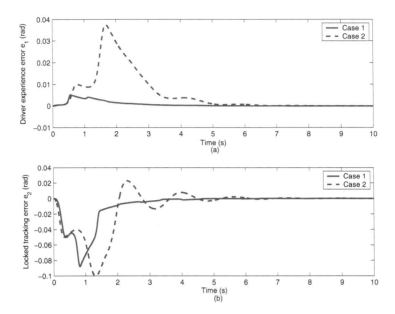

FIGURE 5.11. Adaptive Control Simulation Results: Tracking Errors (a) $e_1(t)$, and (b) $e_2(t)$

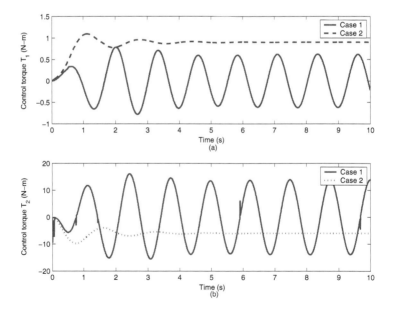

FIGURE 5.12. Adaptive Control Simulation Results: Control Torques, (a) $T_1(t)$, and (b) $T_2(t)$

| Performance | Case 1 | | Case 2 | |
Measure	e_1	e_2	e_1	e_2
Peak Error (%)	0.0222	0.0644	0.1521	0.0698
Steady-state Error (%)	0.0222	0.0444	0.1521	0.0698

TABLE 5.3. Summary of Results for EMK Controller Simulation

shown in Figure 5.14. The torque estimate values $\hat{\tau}_1$ and $\hat{\tau}_2$ are shown in Figure 5.15. Here it is observed that the driver input torque, $\tau_1(t)$ shown in Figure 5.10 is accurately observed by the signal $\hat{\tau}_1(t)$ shown in Figure 5.15. These numerical results indicate the feasibility of using torque observation values in the control algorithm thus eliminating the need for the introduction of torque transducers in the steering system hardware. As with the previous set of simulations, the performance is evaluated using peak error and steady-state error performance measures, and the results are tabulated in Table 5.3

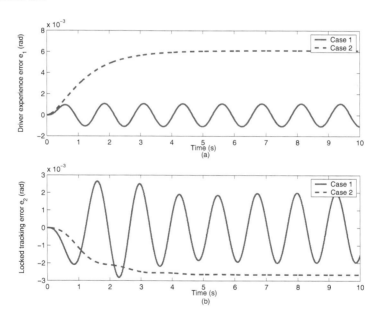

FIGURE 5.13. EMK Controller Simulation Results: Tracking Errors (a) $e_1(t)$, and (b) $e_2(t)$.

To evaluate the steering system behavior, a standard steering mechanism has been simulated whose dynamics have the form as (5.58). Furthermore,

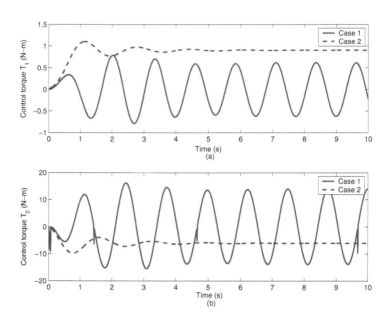

FIGURE 5.14. EMK Controller Simulation Results: Control Torques (a) $T_1(t)$, and (b) $T_2(t)$

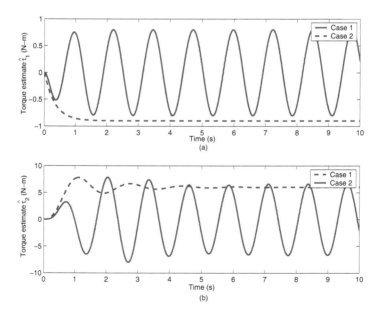

FIGURE 5.15. EMK Controller Results: Estimated torques, (a) $\hat{\tau}_1(t)$, and (b) $\hat{\tau}_2(t)$

the lumped model parameters are assumed to be the sum of the primary and secondary system parameters given in Table 5.1 so that

$$
\begin{aligned}
(I_1 + I_2)\ddot{\theta}_a + N_a\left(\theta_a, \dot{\theta}_a\right) &= \alpha_1 \tau_1 + \alpha_2 \tau_2 \\
N_a\left(\theta_a, \dot{\theta}_a\right) &= (B_1 + B_2)\dot{\theta}_a + (K_1 + K_2)\theta_a.
\end{aligned}
\tag{5.116}
$$

To demonstrate that the driver can be provided with a tunable force feedback, the simulation was executed with a different set of target parameters. In this case, the parameter B_T was increased an order of magnitude to 0.2 $kg - m^2/s$ while all other parameters and inputs were unchanged for Case 1. The results are shown in Figure (5.16) which shows the target system displacement, $\theta_d(t)$ for the original system and the system with increased damping along with the response of a standard steering system, $\theta_a(t)$. As can be observed, the magnitude of $\theta_d(t)$ for the system with the original parameters is greater than $\theta_a(t)$ showing that the values α_{T1} and α_{T2} can be used to provide variable power assist. Further, the magnitude of $\theta_d(t)$ for the system with the original parameters is greater than that with increased damping, showing that the nature of the system response can also be varied by adjusting the target system parameters I_T, B_T, and K_T. Thus, the driver can be provided with a customized "feel" at the steering wheel (i.e., the steering wheel can be made lighter or heavier depending on the operator's need and comfort) by appropriately changing the target parameters for a given steer-by-wire system.

5.3.7 Experimental Results

To address some of the practical issues involved in the implementation of the control algorithm on a prototype steer-by-wire system, the proposed control laws were tested on an experimental testbench.

Experimental Setup

The experimental configuration is shown in Figure 5.17. It consisted of two switched-reluctance motors (SRMs) controlled using NSK drives, a steering wheel, a rack and pinion system, a hydraulic damper, and a spring. One of the SRMs provided the road feedback to the operator by means of the steering wheel, while the second motor actuated the directional control assembly. The SRMs had inbuilt resolvers that provided high resolution angular displacement measurements. Additionally, a Linear Variable Differential Transformer (LVDT) was also used to measure the rack displacement. Optionally, the hydraulic damper and/or the spring could be connected to

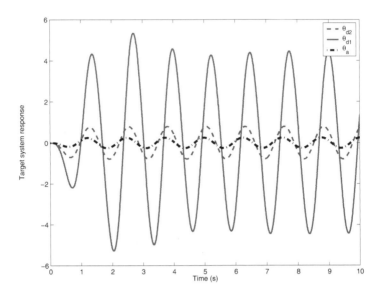

FIGURE 5.16. EMK Controller Simulation Results: Response of System with (a) $B_T = 0.2 \ kg - m^2/s$, (b) $B_T = 0.02 \ kg - m^2/s$, and (c) Standard Steering System Response

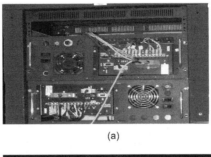

(a)

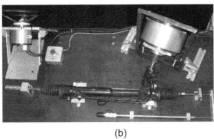

(b)

FIGURE 5.17. Laboratory Experimental Setup: (a) NSK Motor Drives, (b) Steering Wheel, Rack and Pinion System

the rack to simulate road reaction forces. This setup created a steer-by-wire steering system similar to one that can be used in a ground vehicle to experimentally validate the proposed control algorithm. Notice that the essential components are quite the same as shown in Figure 5.9. Precision torque sensors that are required for the accurate implementation of the adaptive control algorithm were prohibitively expensive to incorporate in this study, while attempts with low quality sensors were unsuccessful due to the inherent drift and noise in the sensor signals. Hence, in this experimental section, the algorithms presented in Section 5.3.5 (which do not require the measurement of torque) are implemented.

An AMD Athlon 1.2 GHz PC running QNX 6.2 RTP (Real Time Platform), a real-time micro-kernel based operating system, provided the computational power to implement the control algorithms. An in-house graphical user interface program, Qmotor 3.0, ensured real-time execution of the control algorithms written in C++. This also facilitated real time graphing, data logging and on-line gain tuning. Data acquisition and control implementation were performed at a frequency of 1.0 kHz using the ServoToGo I/O board.

To study and evaluate the effect of this haptic interface feedback system, the system was also equipped with a virtual reality environment. This consisted of a 60″ × 80″ screen along with a high capacity projector which provided the visual feedback for driver-in-loop experiments. A MATLAB-based system was built, both to simulate the vehicle chassis dynamics as well as to render a VRML scene in real-time. Alternate input devices such as a joystick incorporating force feedback were also considered to provide assistance to handicapped drivers.

Tests and Results

Preliminary tests were performed to determine the system parameters listed in Table 5.1. The target dynamics were generated as shown in (5.80). Again, experiments were performed for the two cases as specified in the previous section. For each case, the gains were tuned until the best system performance was obtained. The values of the target system parameters, I_T, B_T, and K_T were chosen in the previous section and listed in Table 5.1. The experimental results for driver experience error $e_1(t)$, and locked tracking error $e_2(t)$ are shown in Figure 5.18. As with the simulations, the performance of the control algorithm in an experimental test is evaluated using the same performance indices as the EMK controller simulation. The results are shown in Table 5.4.

Performance	Case 1		Case 2	
Measure	e_1	e_2	e_1	e_2
Peak Error (%)	1.5	3.2	5.2	2.5
Steady-state Error (%)	1.5	2.0	5.2	2.39

TABLE 5.4. Summary of Results for Experimental Evaluation

As can be viewed from Figure 5.18, the errors $e_1(t)$ and $e_2(t)$, have small magnitudes for both the maximum and steady-state values in each case. The steady-state errors for the case of the step input (case 2) are mainly due to the large static friction present in the physical system, which was very hard to compensate for. As in the simulations, these error values show that $\theta_1(t)$ tracks the reference trajectory $\theta_d(t)$ and that $\theta_2(t)$ follows $\theta_1(t)$ (and hence achieves the two control objectives outlined). The values for the control torques $T_1(t)$ and $T_2(t)$ are shown in Figure 5.19. It can be seen that the control input torques follow a similar pattern as in the simulations. One of the differences, though is, the inherent noise in the experimental system. The actual driver input torques, $\tau_1(t)$, for both cases, and their estimates, $\hat{\tau}_1(t)$, are shown in Figure 5.20. The estimated values shown here and small magnitude of error values in Figure 5.18 prove the efficacy of the controller and the ability of torque observers to accurately estimate input torque values.

The target system response for all the simulation and experimental tests has been plotted in Figure 5.21. During the experiment, the operator could feel the steering wheel become lighter or heavier and the system response differ for different sets of target parameters. Since $e_1(t)$ and $e_2(t)$ are small in magnitude in comparison to the target trajectories, in effect, the target system response is indeed the actual response of the system.

5.4 Robot Teleoperation

As stated in the introduction to the chapter, the practical importance as well as the challenging theoretical nature of the teleoperator problem has spurred a significant amount of research activity in the past few years. The teleoperation problem that is being solved here comprises of two parts: (a) ensuring the coordination of a master and a slave manipulator, and (b) ensuring passivity of the overall system. Through the use of transformations, dynamic trajectory generations, and continuous nonlinear integral feedback

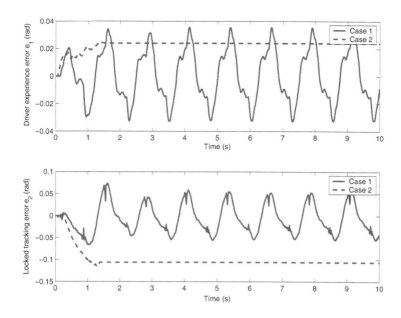

FIGURE 5.18. Experimental Results: Tracking errors $e_1(t)$, and $e_2(t)$

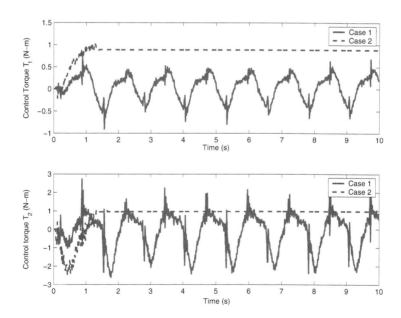

FIGURE 5.19. Experimental Results: Control Torques $T_1(t)$, and $T_2(t)$

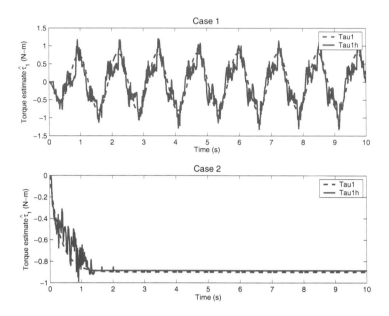

FIGURE 5.20. Experimental Results: Actual and estimated driver input torques, $\tau_1(t)$ and $\hat{\tau}_1(t)$ for (a) Case 1, and (b) Case 2

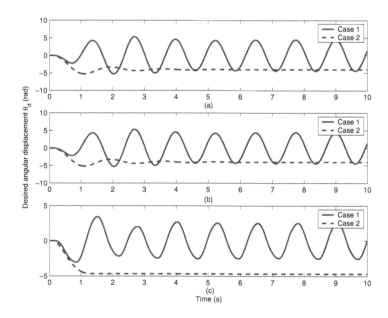

FIGURE 5.21. Target System Responses: (a) Adaptive Controller, (b) EMK Controller, and (c) Experimental Results

terms, two controllers are proven through Lyapunov-based techniques to passively coordinate the master and slave manipulators with respect to the scaled user and environmental power despite incomplete system knowledge. Implementing either controller would provide the user of the closed loop teleoperator system with a power scalable, coordinated master-slave tool that provides safe and stable user feedback. The first controller is proven to yield a semi-global asymptotic result in the presence of parametric uncertainty in the master and slave manipulator dynamic models provided the user and environmental input forces are measurable; henceforth, referred to as the MIF (Measurable Input Force) controller. The second controller yields a global asymptotic result despite unmeasurable user and environmental input forces (UMIF) provided the dynamic models of the respective manipulators are known. The development of each controller consists of the following three steps. In the first step, a transformation is utilized which encodes both the coordination and passivity objectives within the closed loop system. Next, a dynamic trajectory generating system is designed which assists in achieving overall system passivity as well as keeping all signals bounded in the closed loop system. Finally, a continuous nonlinear integral feedback observer (see [51] and [64]) is exploited to compensate for the lack of system dynamics information or user and environmental force measurements. Simulation results provided at the end of the section demonstrate for both controllers that the coordination and tracking control objectives are met.

5.4.1 System Model

The dynamic model for a $2n$-DOF nonlinear teleoperator consisting of a revolute n-DOF master and a revolute n-DOF slave revolute robot are described by the following expressions [32]

$$\gamma \left\{ M_1(q_1(t))\ddot{q}_1(t) + C_1(q_1(t), \dot{q}_1(t))\dot{q}_1(t) + B_1\dot{q}_1(t) = T_1(t) + F_1(t) \right\} \tag{5.117}$$

$$M_2(q_2(t))\ddot{q}_2(t) + C_2(q_2(t), \dot{q}_2(t))\dot{q}_2(t) + B_2\dot{q}_2(t) = T_2(t) + F_2(t). \tag{5.118}$$

In (5.117) and (5.118), $\gamma \in \mathbb{R}$ denotes a positive adjustable power scaling term, $q_i(t), \dot{q}_i(t), \ddot{q}_i(t) \in \mathbb{R}^n$ denote the link position, velocity, and acceleration, respectively, $\forall i = 1, 2$ where $i = 1$ denotes the master manipulator and $i = 2$ denotes the slave manipulator, $M_i(q_i) \in \mathbb{R}^{n \times n}$ represents the inertia effects, $C_i(q_i, \dot{q}_i) \in \mathbb{R}^{n \times n}$ represents centripetal-Coriolis effects, $B_i \in \mathbb{R}^{n \times n}$ represents the constant positive definite, diagonal dynamic frictional effects, $T_i(t) \in \mathbb{R}^n$ represents the torque input control vector, $F_1(t) \in \mathbb{R}^n$

represents the user input force, and $F_2(t) \in \mathbb{R}^n$ represents the input force from the environment. The subsequent development is based on the property that the master and slave inertia matrices are positive definite and symmetric in the sense that [36]

$$m_{1i} \left\| \xi \right\|^2 \le \xi^T M_i(q_i) \xi \le m_{2i} \left\| \xi \right\|^2 \qquad (5.119)$$

$\forall \xi \in \mathbb{R}^n$ and $i = 1, 2$ where m_{1i}, $m_{2i} \in \mathbb{R}$ are positive constants, and $\left\| \cdot \right\|$ denotes the Euclidean norm. The subsequent development is also based on the assumption that $q_i(t)$, $\dot{q}_i(t)$ are measurable, and that the inertia and centripetal-Coriolis matrices are second order differentiable.

5.4.2   MIF Control Development

For the MIF controller development, the subsequent analysis will prove a semi-global asymptotic result in the presence of parametric uncertainty in the master and slave manipulator dynamic models provided the user and environmental input forces are measurable. This development requires the assumption that $F_i(t)$, $\dot{F}_i(t)$, $\ddot{F}_i(t) \in L_\infty$ $\forall i = 1, 2$ (precedence for this type of assumption is provided in [32] and [34]).

Objective and Model Transformation

One of the two primary objectives for the bilateral teleoperator system is to ensure coordination between the master and the slave manipulators in the following sense

$$q_2(t) \to q_1(t) \text{ as } t \to \infty. \qquad (5.120)$$

The other primary objective is to ensure that the system remains passive with respect to the scaled user and environmental power in the following sense [32]

$$\int_{t_0}^{t} (\gamma \dot{q}_1^T(\tau) F_1(\tau) + \dot{q}_2^T(\tau) F_2(\tau)) d\tau \ge -c_1^2 \qquad (5.121)$$

where c_1 is a bounded positive constant, and γ was introduced in (5.117). The passivity objective is included in this section to ensure that the human can interact with the robotic system in a stable and safe manner, and that the robot can also interact with the environment in a stable and safe manner. To facilitate the passivity objective in (5.121), an auxiliary control objective is utilized. Specifically, the coordinated master and slave manipulators are forced to track a desired bounded trajectory, denoted by $q_d(t) \in \mathbb{R}^n$, in the sense that [33]

$$q_1(t) + q_2(t) \to q_d(t). \qquad (5.122)$$

An additional objective is that all signals are required to remain bounded within the closed loop system.

To facilitate the subsequent development, a globally invertible transformation is defined that encodes both the coordination and passivity objectives as follows

$$x \triangleq Sq \tag{5.123}$$

where $x(t) \triangleq [x_1^T(t)\ x_2^T(t)]^T \in \mathbb{R}^{2n}$, $q(t) \triangleq [q_1^T(t)\ q_2^T(t)]^T \in \mathbb{R}^{2n}$, and $S \in \mathbb{R}^{2n \times 2n}$ is defined as follows

$$S \triangleq \begin{bmatrix} I & -I \\ I & I \end{bmatrix} \qquad S^{-1} = \frac{1}{2} \begin{bmatrix} I & I \\ -I & I \end{bmatrix} \tag{5.124}$$

where $I \in \mathbb{R}^{n \times n}$ denotes the identity matrix. Based on (5.123), the dynamic models given in (5.117) and (5.118) can be expressed as follows

$$\bar{M}(x)\ddot{x} + \bar{C}(x,\dot{x})\dot{x} + \bar{B}\dot{x} = \bar{T}(t) + \bar{F}(t) \tag{5.125}$$

where

$$\bar{M}(x) = S^{-T} \begin{bmatrix} \gamma M_1 & 0_{2n} \\ 0_{2n} & M_2 \end{bmatrix} S^{-1} \in \mathbb{R}^{2n \times 2n} \tag{5.126}$$

$$\bar{C}(x,\dot{x}) = S^{-T} \begin{bmatrix} \gamma C_1 & 0_{2n} \\ 0_{2n} & C_2 \end{bmatrix} S^{-1} \in \mathbb{R}^{2n \times 2n} \tag{5.127}$$

$$\bar{B} = S^{-T} \begin{bmatrix} \gamma B_1 & 0_{2n} \\ 0_{2n} & B_2 \end{bmatrix} S^{-1} \in \mathbb{R}^{2n \times 2n} \tag{5.128}$$

$$\bar{T}(t) = S^{-T} \begin{bmatrix} \gamma T_1^T & T_2^T \end{bmatrix}^T \in \mathbb{R}^{2n} \tag{5.129}$$

$$\bar{F}(t) \triangleq \begin{bmatrix} \bar{F}_1(t) \\ \bar{F}_2(t) \end{bmatrix} = S^{-T} \begin{bmatrix} \gamma F_1 \\ F_2 \end{bmatrix} \in \mathbb{R}^{2n} \tag{5.130}$$

and $0_{2n} \in \mathbb{R}^{n \times n}$ denotes an $n \times n$ matrix of zeros. The subsequent development is based on the property that $\bar{M}(x)$, as defined in (5.126), is a positive definite and symmetric matrix in the sense that [36]

$$\bar{m}_1 \|\xi\|^2 \le \xi^T \bar{M}(x)\xi \le \bar{m}_2 \|\xi\|^2 \tag{5.131}$$

$\forall \xi \in \mathbb{R}^{2n}$ where $\bar{m}_1, \bar{m}_2 \in \mathbb{R}$ are positive constants. It is also noted that $\bar{M}(x)$ is second order differentiable by assumption.

To facilitate the subsequent development and analysis, the control objectives can be combined through a filtered tracking error signal, denoted by $r(t) \in \mathbb{R}^{2n}$, that is defined as follows

$$r \triangleq \dot{e}_2 + \alpha_1 e_2 \tag{5.132}$$

where $e_2(t) \in \mathbb{R}^{2n}$ is defined as follows

$$e_2 \triangleq \dot{e}_1 + \alpha_2 e_1 \qquad (5.133)$$

where α_1, $\alpha_2 \in \mathbb{R}$ are positive control gains, and $e_1(t) \in \mathbb{R}^{2n}$ is defined as follows

$$e_1 \triangleq x_d - x \qquad (5.134)$$

where $x_d(t) \in \mathbb{R}^{2n}$ is defined as follows

$$x_d \triangleq \begin{bmatrix} 0_n^T & q_d^T(t) \end{bmatrix}^T \qquad (5.135)$$

where $0_n \in \mathbb{R}^n$ denotes an $n \times 1$ vector of zeros. Based on the definition of $x(t)$ in (5.123) and $e_1(t)$ in (5.134), it is clear that if $\|e_1(t)\| \to 0$ as $t \to \infty$ then $q_2(t) \to q_1(t)$ and that $q_1(t) + q_2(t) \to q_d(t)$ as $t \to \infty$.

The desired trajectory $q_d(t)$ introduced in (5.122) and (5.135) is generated by the following expression

$$M_T \ddot{q}_d + B_T \dot{q}_d + K_T q_d = \bar{F}_2. \qquad (5.136)$$

In (5.136), M_T, B_T, $K_T \in \mathbb{R}^{n \times n}$ represent constant positive definite, diagonal matrices, and $\bar{F}_2(t)$ was introduced in (5.130). Based on the assumption that $\bar{F}_2(t) \in \mathcal{L}_\infty$, standard linear analysis techniques can be used to prove that $q_d(t)$, $\dot{q}_d(t)$, $\ddot{q}_d(t) \in \mathcal{L}_\infty$. The time derivative of (5.136) is given by the following expression

$$M_T \dddot{q}_d + B_T \ddot{q}_d + K_T \dot{q}_d = \dot{\bar{F}}_2. \qquad (5.137)$$

From (5.137), the fact that $\dot{q}_d(t)$, $\ddot{q}_d(t) \in \mathcal{L}_\infty$, and the assumption that $\dot{\bar{F}}_2(t) \in \mathcal{L}_\infty$, it is clear that $\dddot{q}_d(t) \in \mathcal{L}_\infty$. By taking the time derivative of (5.137), and utilizing the assumption that $\ddot{\bar{F}}_2(t) \in \mathcal{L}_\infty$, we can also show that $\ddddot{q}_d(t) \in \mathcal{L}_\infty$.

Closed-Loop Error System

Based on the assumption that the user and environmental forces are measurable, the control input $\bar{T}(t)$ of (5.129) is designed as follows

$$\bar{T} \triangleq \bar{u} - \bar{F} \qquad (5.138)$$

where $\bar{u}(t) \in \mathbb{R}^{2n}$ is an auxiliary control input. Substituting (5.138) into (5.125) yields the following simplified system

$$\bar{M}\ddot{x} + \bar{C}\dot{x} + \bar{B}\dot{x} = \bar{u}. \qquad (5.139)$$

After taking the time derivative of (5.132) and premultiplying by $\bar{M}(x)$, the following expression can be obtained

$$\bar{M}\dot{r} = \bar{M}\dddot{x}_d + \dot{\bar{M}}\ddot{x} + \frac{d}{dt}\left[\bar{C}\dot{x} + \bar{B}\dot{x}\right] - \dot{\bar{u}} + \alpha_2\bar{M}\dot{e}_1 + \alpha_1\bar{M}\dot{e}_2 \qquad (5.140)$$

where (5.132)–(5.134), and the time derivative of (5.139) were utilized. To facilitate the subsequent analysis, the expression in (5.140) is rewritten as follows

$$\bar{M}\dot{r} = \tilde{N} + N_d - e_2 - \dot{\bar{u}} - \frac{1}{2}\dot{\bar{M}}r \qquad (5.141)$$

where the auxiliary signal $\tilde{N}(x, \dot{x}, \ddot{x}, t) \in \mathbb{R}^{2n}$ is defined as

$$\tilde{N} \triangleq N - N_d \qquad (5.142)$$

where $N(x, \dot{x}, \ddot{x}, t) \in \mathbb{R}^{2n}$ is defined as

$$N \triangleq \bar{M}\dddot{x}_d + \dot{\bar{M}}\ddot{x} + \alpha_2\bar{M}\dot{e}_1 + \alpha_1\bar{M}\dot{e}_2 + e_2 + \frac{d}{dt}\left[\bar{C}\dot{x} + \bar{B}\dot{x}\right] + \frac{1}{2}\dot{\bar{M}}r \quad (5.143)$$

and $N_d(t) \in \mathbb{R}^{2n}$ is defined as

$$\begin{aligned} N_d &\triangleq N|_{x=x_d, \ \dot{x}=\dot{x}_d, \ \ddot{x}=\ddot{x}_d} \qquad (5.144) \\ &= \bar{M}(x_d)\dddot{x}_d + \dot{\bar{M}}(x_d, \dot{x}_d)\ddot{x}_d + \frac{d}{dt}\left[\bar{C}(x_d, \dot{x}_d)\dot{x}_d + \bar{B}\dot{x}_d\right]. \end{aligned}$$

After defining an augmented error vector $z(t)$ as follows

$$z \triangleq \left[\ e_1^T \quad e_2^T \quad r^T\ \right]^T, \qquad (5.145)$$

the subsequent analysis can be facilitated by developing the following upper bound

$$\left\|\tilde{N}\right\| \leq \rho(\|z\|)\|z\| \qquad (5.146)$$

where the positive function $\rho(\|z\|)$ is non-decreasing in $\|z\|$ (see Section B.3.5 of Appendix B for further details).

Based on (5.141), the auxiliary control input $\bar{u}(t)$ introduced in (5.138) is designed as follows

$$\bar{u} \triangleq (k_s + 1)\left[e_2(t) - e_2(t_0) + \alpha_1\int_{t_0}^t e_2(\tau)d\tau\right] + (\beta_1 + \beta_2)\int_{t_0}^t sgn(e_2(\tau))d\tau \qquad (5.147)$$

where $k_s, \beta_1, \beta_2 \in \mathbb{R}$ are positive control gains, and $sgn(\cdot)$ denotes the vector signum function. The term $e_2(t_0)$ in (5.147) is included so that $\bar{u}(t_0) = 0$. The time derivative of (5.147) is given by the following expression

$$\dot{\bar{u}} = (k_s + 1)r + (\beta_1 + \beta_2) sgn(e_2). \qquad (5.148)$$

Substituting (5.148) into (5.141) yields the following closed-loop error system

$$\bar{M}\dot{r} = -(k_s + 1)r - (\beta_1 + \beta_2)\,sgn(e_2) + \tilde{N} + N_d - e_2 - \frac{1}{2}\dot{\bar{M}}r. \quad (5.149)$$

Remark 5.10 *Based on the expressions in (5.135), (5.144) and the fact that $q_d(t)$, $\dot{q}_d(t)$, $\ddot{q}_d(t)$, $\dddot{q}_d(t)$, and $\ddddot{q}_d(t) \in \mathcal{L}_\infty$, then $\|N_d(t)\|$ and $\left\|\dot{N}_d(t)\right\|$ can be upper bounded by known positive constants ς_1, $\varsigma_2 \in \mathbb{R}$ as follows*

$$\|N_d(t)\| \le \varsigma_1 \qquad \left\|\dot{N}_d(t)\right\| \le \varsigma_2. \quad (5.150)$$

Stability Analysis

Theorem 5.7 *The controller given in (5.138) and (5.147), ensures that all closed-loop signals are bounded and that coordination between the master and slave manipulators is achieved in the sense that*

$$q_2(t) \to q_1(t) \quad as \quad t \to \infty \quad (5.151)$$

provided the control gain β_1 introduced in (5.147) is selected to satisfy the following sufficient condition

$$\beta_1 > \varsigma_1 + \frac{1}{\alpha_1}\varsigma_2 \quad (5.152)$$

where ς_1 and ς_2 are given in (5.150), the control gains α_1 and α_2 are selected greater than 2, and k_s is selected sufficiently large with respect to the initial conditions of the system.

Proof. See Section B.3.6 of Appendix B. ∎

Theorem 5.8 *The controller given in (5.138) and (5.147) ensures that the teleoperator system is passive with respect to the scaled user and environmental power.*

Proof. See Section B.3.7 of Appendix B. ∎

Simulation Results

A numerical simulation was performed to demonstrate the performance of the controller given in (5.138) and (5.147). The following 2-link, revolute robot dynamic model was utilized for both the master and slave manipulators [61]

$$\begin{bmatrix} \tau_{i_1} \\ \tau_{i_2} \end{bmatrix} + \begin{bmatrix} F_{i_1} \\ F_{i_2} \end{bmatrix} = A(q_{i_1}, q_{i_2})\begin{bmatrix} \ddot{q}_{i_1} \\ \ddot{q}_{i_2} \end{bmatrix} + + \begin{bmatrix} f_{d1_i} & 0 \\ 0 & f_{d2_i} \end{bmatrix}\begin{bmatrix} \dot{q}_{i_1} \\ \dot{q}_{i_2} \end{bmatrix}$$
$$+ B(q_{i_1}, q_{i_2}, \dot{q}_{i_1}, \dot{q}_{i_2})\begin{bmatrix} \dot{q}_{i_1} \\ \dot{q}_{i_2} \end{bmatrix}$$

$$(5.153)$$

where

$$A\left(\cdot\right) \triangleq \begin{bmatrix} p_{1_i} + 2p_{3_i}c(q_{i_2}) + 2p_{4_i}s(q_{i_2}) & p_{2_i} + p_{3_i}c(q_{i_2}) + p_{4_i}s(q_{i_2}) \\ p_{2_i} + p_{3_i}c(q_{i_2}) + p_{4_i}s(q_{i_2}) & p_{2_i} \end{bmatrix}$$

$$B\left(\cdot\right) \triangleq \begin{bmatrix} -(p_{3_i}s(q_{i_2}) - p_{4_i}c(q_{i_2}))\dot{q}_{i_2} & -(p_{3_i}s(q_{i_2}) - p_{4_i}c(q_{i_2}))(\dot{q}_{i_1} + \dot{q}_{i_2}) \\ (p_{3_i}s(q_{i_2}) - p_{4_i}c(q_{i_2}))\dot{q}_{i_1} & 0 \end{bmatrix}$$

$s(\cdot)$ and $c(\cdot)$ denote the $sin(\cdot)$ and $cos(\cdot)$ functions. For the master manipulator, $i = 1$ and $p_{1_1} = 3.34$ [kg·m^2], $p_{2_1} = 0.97$ [kg·m^2], $p_{3_1} = 1.0392$ [kg·m^2], $p_{4_1} = 0.6$ [kg·m^2], $f_{d1_1} = 1.3$ [Nm·sec], and $f_{d2_1} = 0.88$ [Nm·sec]. For the slave manipulator, $i = 2$ and $p_{1_2} = 2.67$ [kg·m^2], $p_{2_2} = 1.455$ [kg·m^2], $p_{3_2} = 0.929$ [kg·m^2], $p_{4_2} = 0.537$ [kg·m^2], $f_{d1_2} = 1.3$ [Nm·sec], and $f_{d2_2} = 0.88$ [Nm·sec], where the parameters are based on [61]. For this simulation, the positive adjustable power scaling term was selected as $\gamma = 1$. The user input force vector was set equal to the following arbitrary periodic time-varying signals

$$\begin{bmatrix} F_{1_1} \\ F_{1_2} \end{bmatrix} = \begin{bmatrix} 25\sin(1.1t) \\ 35\sin(t) \end{bmatrix}. \tag{5.154}$$

To emulate contact with the environment, a spring-like input force vector was selected as follows

$$\begin{bmatrix} F_{2_1} \\ F_{2_2} \end{bmatrix} = \begin{bmatrix} -0.6\dot{q}_{1_2} - q_{1_2} \\ -0.6\dot{q}_{2_2} - q_{2_2} \end{bmatrix}. \tag{5.155}$$

To assist in meeting the passivity control objective, the coordinated teleoperated system must follow a desired trajectory which was generated by the system described by (5.136) and for this simulation was selected as follows

$$\bar{F}_2(t) = \begin{bmatrix} 5 & 0 \\ 0 & 5 \end{bmatrix} \begin{bmatrix} \ddot{q}_{d1} \\ \ddot{q}_{d2} \end{bmatrix} + \begin{bmatrix} 3 & 0 \\ 0 & 3 \end{bmatrix} \begin{bmatrix} \dot{q}_{d1} \\ \dot{q}_{d2} \end{bmatrix} + \begin{bmatrix} 1 & 0 \\ 0 & 1 \end{bmatrix} \begin{bmatrix} q_{d1} \\ q_{d2} \end{bmatrix} \tag{5.156}$$

where $q_{d1}(t)$ and $q_{d2}(t)$ denote the desired link positions, and $\bar{F}_2(t)$ is equal to the following expression

$$\bar{F}_2(t) = \frac{1}{2}\left(\gamma F_1(t) + F_2(t)\right)$$

where $\bar{F}_2(t)$ was defined in (5.130).

The actual trajectory for the master and slave manipulators are demonstrated in Figure 5.22 for controller gains selected as $k_s = 100$ and $\beta_1 + \beta_2 = 25$. The link position tracking error between the master and slave manipulators can be seen in Figure 5.23. From Figures 5.22 and 5.23, it is clear that the coordination control objective is achieved. The actual trajectory for the coordinated system $(q_1(t) + q_2(t))$ and the desired trajectory as defined by (5.156), are demonstrated in Figure 5.24. The coordinated system versus the desired trajectory tracking error as defined by $q_1(t) + q_2(t) - q_d(t)$, is given in Figure 5.25. From Figures 5.24 and 5.25, it is clear that the coordinated system tracks the desired trajectory. The control torque inputs for the master and slave manipulator are provided in Figures 5.26 and 5.27, respectively.

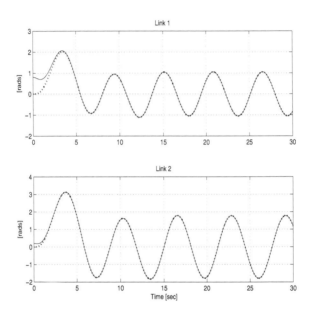

FIGURE 5.22. Actual trajectory for master (i.e., $q_1(t)$) (—) and slave (i.e., $q_2(t)$) (- -) manipulators for Link 1 and Link 2.

5.4.3   UMIF Control Development

For the UMIF controller development, the subsequent analysis will prove a global asymptotic result despite unmeasurable user and environmental input forces (UMIF) provided the dynamic models of the respective manipulators are known. This development also requires the assumption that $F_i(t), \dot{F}_i(t), \ddot{F}_i(t) \in L_\infty \ \forall i = 1, 2$.

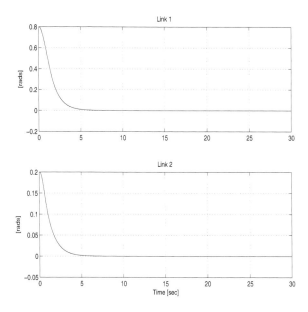

FIGURE 5.23. Link position tracking error between the master and slave manipulators (i.e., $q_1(t) - q_2(t)$).

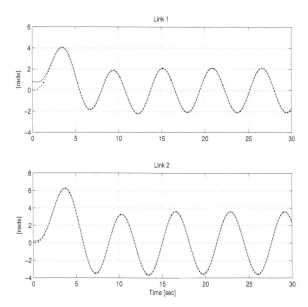

FIGURE 5.24. Actual coordinated (i.e., $q_1(t) + q_2(t)$) trajectory (—) and desired (i.e., $q_d(t)$) trajectory (- -) for Link 1 and Link 2.

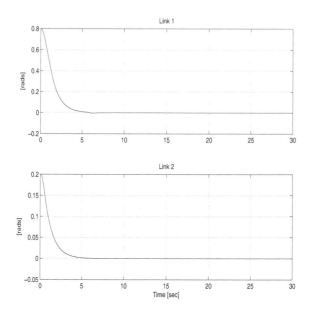

FIGURE 5.25. The coordinated system versus the desired trajectory tracking error (i.e., $q_1(t) + q_2(t) - q_d(t)$).

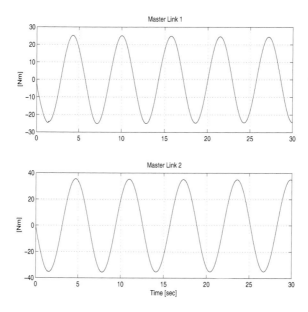

FIGURE 5.26. Master manipulator control input torque (i.e., $\tau_1(t)$).

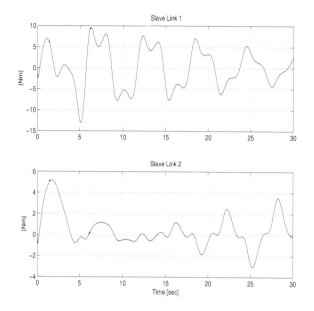

FIGURE 5.27. Slave manipulator control input torque (i.e., $\tau_2(t)$).

Objective and Model Transformation

One of the two primary objectives for the bilateral teleoperator system is to ensure coordination between the master and the slave manipulators as in (5.120). The other objective is to ensure that the system remains passive with respect to the scaled user and environmental power as in (5.121). To assist in meeting the passivity objective the following desired trajectory, defined as $x_d(t) \in \mathbb{R}^{2n}$, is generated by the following dynamic system

$$\bar{M}\ddot{x}_d + B_T\dot{x}_d + K_Tx_d + \frac{1}{2}\dot{\bar{M}}\dot{x}_d = \hat{F}. \qquad (5.157)$$

In (5.157), $\bar{M}(x)$ was defined in (5.126), B_T and $K_T \in \mathbb{R}^{2n \times 2n}$ represent constant positive definite, diagonal matrices, and $\hat{F}(t) \in \mathbb{R}^{2n}$ is a subsequently designed nonlinear force observer, and $x_d(t) \in \mathbb{R}^{2n}$ can be decomposed as follows

$$x_d = \begin{bmatrix} x_{d1}^T(t) & x_{d2}^T(t) \end{bmatrix}^T \qquad (5.158)$$

where $x_{d1}(t)$, $x_{d2}(t) \in \mathbb{R}^n$. Subsequent development will prove that $\hat{F}(t) \in \mathcal{L}_\infty$. Based on this fact, the development in Section B.3.8 of Appendix B can be used to prove that $x_d(t)$, $\dot{x}_d(t) \in \mathcal{L}_\infty$, then (5.157) can be used to prove that $\ddot{x}_d(t) \in \mathcal{L}_\infty$ as shown later, the passivity objective is facilitated by ensuring that the coordinated master and slave manipulators are forced to track a desired bounded trajectory $x_{d2}(t)$ in the sense that

$$q_1(t) + q_2(t) \to x_{d2}(t) \tag{5.159}$$

where $x_{d2}(t)$ was defined in (5.158). An additional objective is that all signals are required to remain bounded within the closed loop system.

To facilitate the subsequent development, a globally invertible transformation is defined that encodes both the coordination and passivity objectives as follows

$$x \triangleq Sq + \begin{bmatrix} x_{d1} \\ 0_n \end{bmatrix} \tag{5.160}$$

where $x(t) \triangleq [x_1^T(t)\ x_2^T(t)]^T \in \mathbb{R}^{2n}$, $q(t) \triangleq [q_1^T(t)\ q_2^T(t)]^T \in \mathbb{R}^{2n}$, $x_{d1}(t) \in \mathbb{R}^n$ was defined in (5.158), the zero vector $0_n \in \mathbb{R}^n$ and $S \in \mathbb{R}^{2n \times 2n}$ was defined in (5.124). Based on (5.160), the dynamic models given in (5.117) and (5.118) can be expressed as follows

$$\begin{aligned}
\bar{T}(t) + \bar{F}(t) = \quad & \bar{M}(x)\ddot{x} - \bar{M}(x) \begin{bmatrix} \ddot{x}_{d1} \\ 0_n \end{bmatrix} + \bar{C}(x, \dot{x})\dot{x} \\
& - \bar{C}(x, \dot{x}) \begin{bmatrix} \dot{x}_{d1} \\ 0_n \end{bmatrix} + \bar{B}\dot{x} - \bar{B} \begin{bmatrix} \dot{x}_{d1} \\ 0_n \end{bmatrix}
\end{aligned} \tag{5.161}$$

where $\bar{M}(x)$, $\bar{C}(x, \dot{x})$, $\bar{B}$, $\bar{T}(t)$, and $\bar{F}(t)$ were defined in (5.126)–(5.130).

To facilitate the subsequent UMIF development and analysis, the control objectives can be combined through a filtered tracking error signal denoted by $r(t) \in \mathbb{R}^{2n}$, that is defined as follows

$$r \triangleq \dot{e}_2 + e_2 \tag{5.162}$$

where $e_2(t) \in \mathbb{R}^{2n}$ is now defined as follows

$$e_2 \triangleq \bar{M} (\dot{e}_1 + \alpha_2 e_1) \tag{5.163}$$

where $\alpha_2 \in \mathbb{R}$ is a positive control gain, and $e_1(t) \in \mathbb{R}^{2n}$ was defined in (5.134) as follows

$$e_1 \triangleq x_d - x$$

where $x_d(t)$ was defined in (5.158).

Closed Loop Error System

To facilitate the development of the closed-loop error system for $r(t)$, we first examine the error system dynamics for $e_1(t)$ and $e_2(t)$. To this end, we take the second time derivative of $e_1(t)$ and premultiply by $\bar{M}(x)$ to obtain the following expression

$$\bar{M}\ddot{e}_1 \;=\; \hat{F} - B_T\dot{x}_d - K_T x_d - \frac{1}{2}\dot{\bar{M}}\dot{x}_d - \bar{T} - \bar{F} \tag{5.164}$$

$$-\bar{M}\begin{bmatrix} \ddot{x}_{d1} \\ 0_n \end{bmatrix} + \bar{C}\dot{x} - \bar{C}\begin{bmatrix} \dot{x}_{d1} \\ 0_n \end{bmatrix} + \bar{B}\dot{x} - \bar{B}\begin{bmatrix} \dot{x}_{d1} \\ 0_n \end{bmatrix}$$

where (5.161) and (5.157) were utilized. Based on the assumption of exact model knowledge, the control input $\bar{T}(t)$ is designed as follows

$$\bar{T} \;\triangleq\; \bar{T}_1 - B_T\dot{x}_d - K_T x_d - \frac{1}{2}\dot{\bar{M}}\dot{x}_d \tag{5.165}$$

$$-\bar{M}\begin{bmatrix} \ddot{x}_{d1} \\ 0_n \end{bmatrix} + \bar{C}\dot{x} - \bar{C}\begin{bmatrix} \dot{x}_{d1} \\ 0_n \end{bmatrix} + \bar{B}\dot{x} - \bar{B}\begin{bmatrix} \dot{x}_{d1} \\ 0_n \end{bmatrix}$$

where $\bar{T}_1(t) \in \mathbb{R}^{2n}$ is an auxiliary control input. Substituting (5.165) into (5.164) yields the following simplified expression

$$\bar{M}\ddot{e}_1 = \hat{F} - \bar{F} - \bar{T}_1. \tag{5.166}$$

Based on (5.166), the time derivative of $e_2(t)$ in (5.163) can be obtained as follows

$$\dot{e}_2 = \dot{\bar{M}}\dot{e}_1 + \alpha_2\dot{\bar{M}}e_1 + \alpha_2\bar{M}\dot{e}_1 + \hat{F} - \bar{F} - \bar{T}_1. \tag{5.167}$$

Based on the expression in (5.167), the auxiliary control input $\bar{T}_1(t)$ is designed as follows

$$\bar{T}_1 \;\triangleq\; \dot{\bar{M}}\dot{e}_1 + \alpha_2\dot{\bar{M}}e_1 + \alpha_2\bar{M}\dot{e}_1. \tag{5.168}$$

After substituting (5.168) into (5.167), the following can be written

$$\dot{e}_2 = \hat{F} - \bar{F}. \tag{5.169}$$

Taking the time derivative of (5.169) yields the resulting expression

$$\ddot{e}_2 = \dot{\hat{F}} - \dot{\bar{F}}. \tag{5.170}$$

The following error system dynamics can now be obtained for $r(t)$ by taking the time derivative of (5.162)

$$\dot{r} = r - e_2 + \dot{\hat{F}} - \dot{\bar{F}} \tag{5.171}$$

where (5.162) and (5.170) were both utilized. Based on (5.171) and the subsequent stability analysis, the proportional-integral like nonlinear force observer $\hat{F}(t)$ introduced in (5.157) is designed as follows

$$
\hat{F} \triangleq -(k_s + 1)\left[e_2(t) - e_2(t_0) + \int_{t_0}^{t} e_2(\tau)d\tau\right] - (\beta_1 + \beta_2)\int_{t_0}^{t} sgn\left(e_2(\tau)\right)d\tau
$$

$$(5.172)$$

where k_s, β_1, and $\beta_2 \in \mathbb{R}$ are positive control gains, and $sgn(\cdot)$ denotes the vector signum function. The expression given in (5.172) is designed such that $\hat{F}(t_0) = 0$. The time derivative of (5.172) is given by the following expression

$$
\dot{\hat{F}} = -(k_s + 1)r - (\beta_1 + \beta_2) sgn\left(e_2\right). \tag{5.173}
$$

Substituting (5.173) into (5.171) yields the following closed loop error system

$$
\dot{r} = -e_2 - \dot{\tilde{F}} - k_s r - (\beta_1 + \beta_2) sgn\left(e_2\right). \tag{5.174}
$$

Remark 5.11 *Based on (5.130) and the assumption that $F_i(t)$, $\dot{F}_i(t)$, $\ddot{F}_i(t) \in L_\infty \ \forall i = 1, 2$, upper bounds can be developed for $\left\|\dot{\tilde{F}}(t)\right\|$ and $\left\|\ddot{\tilde{F}}(t)\right\|$ as follows*

$$
\left\|\dot{\tilde{F}}(t)\right\| \le \varsigma_3 \qquad \left\|\ddot{\tilde{F}}(t)\right\| \le \varsigma_4 \tag{5.175}
$$

where $\varsigma_3, \varsigma_4 \in \mathbb{R}$ denote positive constants.

Stability Analysis

Theorem 5.3 *The controller given in (5.165) and (5.168) ensures that all closed-loop signals are bounded and that coordination between the master and slave manipulators is achieved in the sense that*

$$
q_2(t) \to q_1(t) \quad as \quad t \to \infty \tag{5.176}
$$

provided the control gain β_1, introduced in (5.172) is selected to satisfy the sufficient condition

$$
\beta_1 > \varsigma_3 + \varsigma_4, \tag{5.177}
$$

where ς_3 and ς_4 were introduced in (5.175).

Proof. See Section B.3.9 of Appendix B. ∎

Theorem 5.4 *The controller given in (5.165) and (5.168), ensures that the teleoperator system is passive with respect to the scaled user and environmental power.*

Proof. See Section B.3.10 of Appendix B. ∎

Remark 5.12 *Both the teleoperation controllers developed here exploit the nonlinear dynamic model which offers a clear advantage over past results for linear teleoperator systems ([7], [15], [30], and [57]). The MIF controller developed in Section 5.4.2 compensates for unknown system parameters, which offers an improvement over past works that require exact model knowledge (i.e. [7] and [15]). The UMIF controller developed in Section 5.4.3 compensates for unavailable force measurement, which offers an improvement over works that requires force measurements (e.g., [32] and [33]).*

Simulation Results

A numerical simulation was performed for the controller given in (5.165) and (5.168). The 2-link, revolute robot dynamic model utilized in (5.153) was utilized for both the master and slave manipulators. The user input force vector in (5.154) and the environmental input force vector in (5.155) were also utilized.

To meet the passivity-based control objective, the coordinated teleoperated system must follow a desired trajectory, which is generated from (5.157) using the same parameter values for the transformed inertia matrix. The values for B_T, $K_T \in \mathbb{R}^{4 \times 4}$ were set to the following values

$$
\begin{aligned}
B_T &= diag\{5, 5, 5, 5\} \\
K_T &= diag\{25, 25, 25, 25\}
\end{aligned}
$$

where B_T and K_T are both diagonal matrices.

The actual trajectory for the master and slave manipulators are demonstrated in Figure 5.28 where the control gains were selected as $k_s = 100$, $\beta_1 + \beta_2 = 100$, and $\alpha_2 = 200$. The link position tracking error between the master and slave manipulators can be seen in Figure 5.29. From Figures 5.28 and 5.29, it is clear that the coordination control objective is achieved. The actual trajectory for the coordinated system $(q_1(t) + q_2(t))$ and the desired trajectory as defined in (5.157), are demonstrated in Figure 5.30. The coordinated system versus the desired trajectory tracking error as defined by $q_1(t) + q_2(t) - x_{d2}(t)$, is given in Figure 5.31. From Figures 5.30 and 5.31, it is clear that the coordinated system tracks the desired trajectory. The output of the nonlinear force observer is provided in Figure 5.32. The control torque inputs for both the master and slave manipulators are provided in Figures 5.33 and 5.34, respectively.

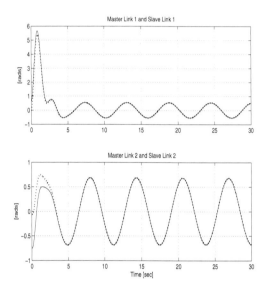

FIGURE 5.28. Actual Trajectory for Master (i.e., $q_1(t)$) (—) and Slave (i.e., $q_2(t)$) (- -) Manipulators for Link 1 and Link 2.

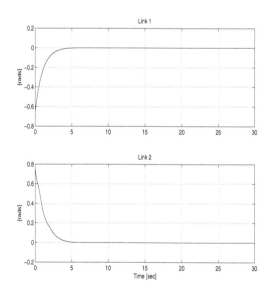

FIGURE 5.29. Link Position Tracking Error between the Master and Slave Manipulators (i.e., $q_1(t) - q_2(t)$).

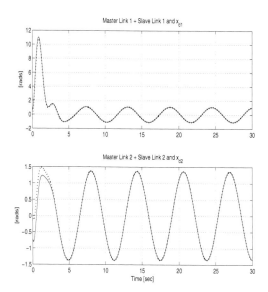

FIGURE 5.30. Actual Coordinated (i.e., $q_1(t)+q_2(t)$) Trajectory (—) and Desired (i.e., $q_d(t)$) Trajectory (- -) for Link 1 and Link 2.

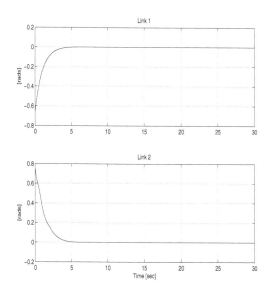

FIGURE 5.31. The Coordinated System versus the Desired Trajectory Tracking Error (i.e., $q_1(t) + q_2(t) - x_{d2}(t)$).

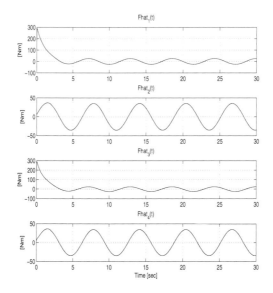

FIGURE 5.32. The Output of the Nonlinear Force Observer (i.e. $\hat{F}(t)$) .

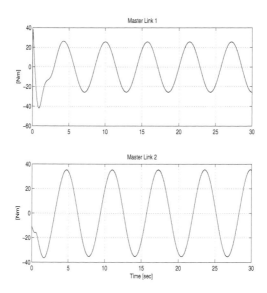

FIGURE 5.33. Master Manipulator Control Input Torque (i.e., $\tau_1(t)$).

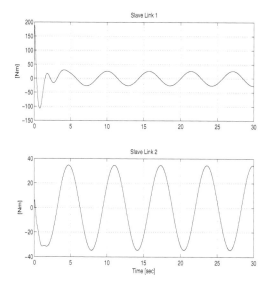

FIGURE 5.34. Slave Manipulator Control Input Torque (i.e., $\tau_2(t)$).

5.5   Rehabilitation Robot

Currently, most rehabilitation robots employ an active assist strategy utilizing in various forms the force control strategy derived from Hogan's seminal work [16] on impedance control. In its simplest form, the patient's arm is strapped to a robot end-effector with low intrinsic impedance and he/she is required to guide the robot end-effector in order to reach and connect a sequence of setpoints in the robot task-space that are mapped onto a display. A proportional-derivative (PD) controller is implemented on the robot's task space position, i.e., the robot actively assists the patient toward minimizing the end-effector position error. For safety, an *ad hoc* windowing technique is employed that turns on the robot's active functionality after the patient initiates movement [1]. However, such a scheme is suspect in patients with severe muscle contractures where an active assist robot could lead to torn ligaments or severe muscle damage unless the therapy is conducted under the supervision of an experienced therapist.

In contrast, the strategy presented here utilizes a passive approach. Specifically, given a desired curve of motion that optimizes therapist established merit criteria, a path generator is designed based on an anisotropic force-velocity relationship that generates a bounded desired trajectory in the robot workspace. The inputs into the generator are the patient's interaction force applied at the end-effector as well as the desired impedance parameters. The reference trajectory generator is carefully designed in order to ensure that the relationship between the patient applied interaction force and the desired end-effector velocity satisfies a passivity constraint. Next, a control strategy is crafted using a Lyapunov based argument in order to obtain the companion objectives of driving the robot end-effector tracking error to zero and ensuring that a filtered error signal nulls out rapidly. This convergence of the filtered error signal allows one to ensure that the interaction of the user with the robot is passive, i.e., energy always flows from the user to the robot manipulator. Additionally, a readily satisfiable mild assumption on the differentiability of the robot dynamics allows the generation of a control strategy that is continuous; this has significant implications in terms of implementability of the control algorithm. As an aside, the control mechanism has the interesting feature of being able learn the unknown robot dynamics.

5.5.1   Robot Dynamics

The end-effector position of a 3-link, revolute direct drive robot manipulator in an inertial frame $\mathcal{I}$, denoted by $x(t) \in \mathbb{R}^3$, is defined as follows

$$x = f(q) \tag{5.178}$$

where $q(t) \in \mathbb{R}^3$ denotes the link position, and $f(q) \in \mathbb{R}^3$ denotes the robot forward kinematics. Based on (5.178), the differential relationships between the end-effector position and the link position variables can be calculated as follows

$$\begin{aligned} \dot{x} &= J(q)\,\dot{q} \\ \ddot{x} &= \dot{J}(q)\dot{q} + J(q)\ddot{q} \end{aligned} \tag{5.179}$$

where $\dot{q}(t)$, $\ddot{q}(t) \in \mathbb{R}^3$ denote the link velocity and acceleration vectors, respectively, and $J(q) = \dfrac{\partial f(q)}{\partial q} \in \mathbb{R}^{3 \times 3}$ denotes the manipulator Jacobian. The dynamic model for the 3-link robot manipulator is assumed to be in the following form [62]

$$M(q)\ddot{q} + V_m(q, \dot{q})\dot{q} + G(q) = \tau_q + J^T \bar{F} \tag{5.180}$$

where $M(q) \in \mathbb{R}^{3 \times 3}$ represents the inertia matrix, $V_m(q, \dot{q}) \in \mathbb{R}^{3 \times 3}$ represents the centripetal-Coriolis matrix, $G(q) \in \mathbb{R}^3$ represents the gravity effects, $\bar{F}(t)$ represents the user applied force expressed in $\mathcal{I}$, and $\tau_q(t) \in \mathbb{R}^3$ represents the torque input vector.

After utilizing (5.178) and (5.179), one can transform the joint space dynamics into the task-space as follows

$$\bar{M}(x)\ddot{x} + \bar{V}_m(x,\dot{x})\dot{x} + \bar{G}(x) = \bar{\tau} + \bar{F} \tag{5.181}$$

where $\bar{M} = (J^T)^{-1}MJ^{-1}$ denotes the transformed inertia matrix, $\bar{V}_m = -(J^T)^{-1}MJ^{-1}\dot{J}J^{-1} + (J^T)^{-1}V_mJ^{-1} \in \mathbb{R}^{3\times3}$ denotes the transformed centripetal-Coriolis matrix, $\bar{G} = (J^T)^{-1}G \in \mathbb{R}^3$ represents gravity effects while $\bar{\tau} = (J^T)^{-1}\tau_q \in \mathbb{R}^3$ represents the torque input vector expressed in $\mathcal{I}$. Motivated by the subsequent stability analysis and control design, we state the following property:

The inertia matrix $\bar{M}(\cdot)$ is symmetric and positive-definite, and satisfies the following inequalities

$$\underline{m}\|\xi\|^2 \leqslant \xi^T\bar{M}(\cdot)\xi \leqslant \overline{m}(x)\|\xi\|^2 \qquad \forall \xi \in \mathbb{R}^3 \tag{5.182}$$

where $\underline{m} \in \mathbb{R}$ denotes a positive constant, $\overline{m}(x) \in \mathbb{R}$ denotes a positive nondecreasing function, while $\|\cdot\|$ denotes the standard Euclidean norm.

5.5.2 Path Planning and Desired Trajectory Generator

It is well known that stretching, range of motion, and timely surgical correction of spinal deformities may enhance functional use of the extremities for patients with neuromuscular disorders (NMDs). In slowly progressive NMDs, moderate resistance weight lifting is known to improve muscle strength and cardiovascular performance [4]. Motivated by this, we present a 3-tier path generation and control strategy that is readily implementable on a real robot. The objective is the generation of robot end-effector motion (when pushed by a patient) along a therapist specified path while ensuring that the device behaves as a passive and programmable impedance. The model generator satisfies the desired properties of (a) guiding the user along contours that provide optimal[4] rehabilitation, (b) generation of contours that stay away from kinematic singularities, physical joint limits, and obstacles, and (c) time parameterization of the contours in a fashion that conforms to passivity requirements.

Path Planning: Tier 1

---

[4]Optimum in the sense specified by a therapist, e.g., maximizing range of motion or power output for a target muscle set.

Here, we assume that a physical therapist has specified a desired curve of motion $\bar{r}_d(s) \in \mathbb{R}^3$ given as follows

$$\bar{r}_d(s) = \begin{bmatrix} \bar{r}_{dx}(s) & \bar{r}_{dy}(s) & \bar{r}_{dz}(s) \end{bmatrix}^T \tag{5.183}$$

where $s \in \mathbb{R}$ is the length of the curve, while $\bar{r}_{dx}(s)$, $\bar{r}_{dy}(s)$, and $\bar{r}_{dz}(s) \in \mathbb{R}$ represent the respective coordinates in an inertial frame $\mathcal{I}$ (say fixed to the base of the robot). However, this therapist specified contour may not be practicable with a real robot because of joint limits, singularities, and obstacles — one would then like to ensure that the therapist specified path is followed with fidelity until a singularity/joint limit/obstacle is nearby at which instance the robot smoothly veers away from that path and rejoins the original path away from the singularity/joint limit/obstacle.

To that end, one could utilize the virtual potential field concept of Khatib [23] that suggested generation of repulsion functions that grow larger as the robot nears an obstacle and becomes singular at the obstacles. However, a real robot actuator can generate only bounded torques [53]; hence, we are motivated to design bounded repellers to take care of obstacles. In order to avoid kinematic singularities, we choose to maximize the Yoshikawa manipulability measure [66]

$$\Psi_1(q_d) = \det\left(J(q_d) J(q_d)^T\right) \geq 0 \tag{5.184}$$

where $q_d(s) \in \mathbb{R}^3$ is a vector of desired robot joint variables, $J(\cdot)$ has been previously introduced in (5.179). For dealing with joint limits, we choose the measure

$$\Psi_2(q_d) = \prod_{i=1}^{3} \alpha_i \left(1 - \frac{q_{di}}{q_{di\,\max}}\right) \left(\frac{q_{di}}{q_{di\,\min}} - 1\right) \geq 0 \tag{5.185}$$

where $q_{di}(s), q_{di\,\max}, q_{di\,\min} \in \mathbb{R}$ denote, respectively, the desired joint angle variable, joint upper, and joint lower limits for the ith joint while $\alpha_i \in \mathbb{R}$ is a positive constant. In order to avoid obstacles, we choose the measure

$$\Psi_3(q_d) = \prod_{i=1}^{n_o} \prod_{j=1}^{3} \left(\|f_j(q_d) - O_i\|^2 - R_i^2\right) \geq 0 \tag{5.186}$$

where $O_i \in \mathbb{R}^3, R_i \in \mathbb{R}$ denote the position and the radius of the ith obstacle, $n_o \in \mathbb{R}$ denotes the number of the obstacles, and $r_{dj} = f_j(q_d)$ where $r_{dj}(s) \in \mathbb{R}^3, j = 1, 2, 3$ denote the position of the end point of the jth link, and $f_j(\cdot) \in \mathbb{R}^3$ denote the corresponding forward kinematics[5]. We

---

[5]Although only the end of every link is considered for obstacle avoidance, any other point of the robot can be included in $\Psi_3(\cdot)$ for obstacle avoidance.

are now in a position to define the potential function

$$\Psi\left(q_d\right) = \gamma_1 \exp\left(-\prod_{i=1}^{3} \gamma_{2i} \Psi_i\left(q_d\right)\right) \tag{5.187}$$

where $\gamma_1, \gamma_{21}, \gamma_{22}, \gamma_{23} > 0$ are adjustable constants that, respectively, characterize the size and radius of influence of the potential function. This function satisfies the properties of boundedness as well as maximality at the obstacles. By utilizing the virtual field generated by the potential function above, one can dynamically generate a modified contour $r_d\left(s\right)^6 \in \mathbb{R}^3$ as follows

$$r_d'\left(s\right) = -\gamma_3\left[r_d\left(s\right) - \bar{r}_d\left(s\right)\right] - \gamma_4 \nabla\Psi\left(f^{-1}\left(r_d\right)\right) + \bar{r}_d'\left(s\right) \tag{5.188}$$

where the notation $\left(\cdot\right)'$ denotes a derivative with respect to s, γ_3, γ_4 are tunable parameters, and $\nabla\Psi\left(\cdot\right) \in \mathbb{R}^3$ denotes the gradient vector of $\Psi\left(\cdot\right)$. The dynamic equation above acts like a filter that smoothly drives $r_d\left(s\right)$ away from the nominal contour $\bar{r}_d\left(s\right)$ near obstacles/singularities/joint limits. In the above equation, γ_3 provides the rate along s at which the modified contour veers away from (or toward) the original contour when it encounters a change in potential field. The constant γ_4 is a steady-state constant that amplifies or diminishes the impact of the potential function on changes in the desired contour. The result of this algorithm is a desired contour that avoids singularities, joint limits, and obstacles. We note here that the filtering process of (5.188) renders s an arbitrary parameter that does not necessarily represent the length of the contour $r_d\left(s\right)$. We also remark here that the steps involved in Tier 1 are completed offline. The therapist specified path will only be modified by the system when it is close to obstacles, joint limits, and robot kinematic singularities. For these cases, a sub-optimal desired contour may result instead of an optimal therapist specified desired contour. Design iterations for choosing a therapist specified path can be applied such that the eventual optimal desired contour is also feasible, given the constraints associated with the robot and the environment.

Time Parameterization of Contour $r_d\left(s\right)$: Tier 2

In this section, we time parameterize the modified desired contour $r_d\left(s\right)$ such that a passivity relation holds between the desired velocity and the

---

[6]To simplify the notation, r_d and $f\left(\cdot\right)$ are utilized in the rest of the paper to denote the desired end-effector position and end-effector forward kinematics instead of r_{d3} and $f_3\left(\cdot\right)$.

applied user interaction force at the robot end-effector. To begin, we define $\mathcal{F} \triangleq (u(s), p(s), b(s))$ to be a rotating frame associated with the curve $r_d(s)$ such that

$$u(s) = \frac{r'_d(s)}{|r'_d(s)|} \qquad p(s) = \frac{u'(s)}{|u'(s)|} \qquad b(s) = u(s) \times p(s) \qquad (5.189)$$

such that $\Gamma(s) = \begin{bmatrix} u(s) & p(s) & b(s) \end{bmatrix} \in SO(3)$. The relationship between the coordinate frames $\mathcal{F}$ and $\mathcal{I}$ is depicted in Figure 5.36. We also define the curvature $\kappa(s)$ and torsion $\tau(s)$ associated with the curve $r_d(s)$ as follows [14]

$$\kappa(s) = \frac{|r'_d(s) \times r''_d(s)|}{|r'(s)|^3} \qquad \tau(s) = \frac{r'_d(s) \cdot \left(r''_d(s) \times r'''_d(s) \right)}{|\kappa(s)|^2} \qquad (5.190)$$

Furthermore, we define the vector $w(s) \triangleq \begin{bmatrix} -\tau(s) & 0 & -\kappa(s) \end{bmatrix}^T \in \mathbb{R}^3$ and $[w(s)]_\times$ as the skew-symmetric matrix associated with that vector. Since (u, p, b) define a basis, we define a general desired velocity vector $\bar{v}_d \triangleq v_{d1}u + v_{d2}p + v_{d3}b = \Gamma v_d$ and an applied user force $\bar{F} \triangleq F_u u + F_p p + F_b b = \Gamma F$ in the inertial frame $\mathcal{I}$ where $v_d, F \in \mathbb{R}^3$ are obviously defined. Since the robot acts as an anisotropic impedance, the direction of which continuously varies as the desired curve of motion $r_d(s)$, we define m_d to be a scalar mass parameter and consider damping coefficients B_u, B_p, B_b along the directions u, p, b such that the damping force $\bar{F}_d$ expressed in $\mathcal{I}$ is given as

$$\bar{F}_d \triangleq -B_u v_{d1} u - B_p v_{d2} p - B_b v_{d3} b = -\Gamma B v_d \qquad (5.191)$$

where $B \triangleq \text{diag}\{B_u, B_p, B_b\}$. By applying Newton's second law to this mass-damper system, we obtain

$$m_d \dot{\bar{v}}_d = \bar{F} + \bar{F}_d$$

which can be written out as follows

$$m_d \dot{v}_d + m_d \dot{s} [w]_\times v_d + B v_d = F \qquad (5.192)$$

where we have utilized the formulae of Frenet [26] and $\dot{s}(t)$ is yet to be defined. Additionally, the kinematics of the problem can be expressed as follows

$$\dot{x}_d(t) = v_{d1}u + v_{d2}p + v_{d3}b \qquad (5.193)$$

where $x_d(t)$ denotes the time parameterized representation of our desired contour (expressed in the coordinates of $\mathcal{I}$) traced by the robot end-effector.

Since our intention is for motion to occur along the curve $r_d(s)$, we designate low tangential damping B_u and very large normal and binormal damping B_p and B_b such that the kinematic constraint $v_{d2} \cong v_{d3} \cong 0$ is imposed. Under such conditions, the effective motion is governed via the following set of equations

$$
\begin{aligned}
\dot{x}_d(t) &= v_{d1}(t)\, u(s(t)) \\
m_d \dot{v}_{d1}(t) + B_u v_{d1}(t) &= \bar{F}(t)\,.u(s(t)) \\
\dot{s}(t) &= \frac{|v_{d1}(t)|}{|r'_d(s(t))|}
\end{aligned}
\tag{5.194}
$$

where the first two equations in (5.194) are obtained by applying the kinematic constraint on (5.192) and (5.193) while the last equation expresses the relationship between the time rate of change of the arbitrary parameter $s(t)$ in terms of a known velocity (v_{d1}) along the curve (r_d).

Proof of Passivity

In order for a user to exercise safely in conjunction with the robot, the robot must act as a passive device, i.e., the work done by the user force is always positive (minus finite stored initial energy if any). With that objective in mind, we first demonstrate that there is a passive relationship between the interaction force $\bar{F}(t)$ and the desired end-effector velocity $\dot{x}_d(t)$, i.e., we show that

$$
\int_{t_0}^{t} \bar{F}^T \dot{x}_d dt \geq -c_1
\tag{5.195}
$$

where c_1 is a positive constant. To prove (5.195), we define a Lyapunov function

$$
V = \frac{1}{2} m_d v_d^T v_d \geq 0.
\tag{5.196}
$$

After taking the time derivative of (5.196) along the desired dynamics of (5.192), we obtain

$$
\dot{V} = -v_d^T B v_d + v_d^T F
\tag{5.197}
$$

where we have utilized the fact that $\left([\omega]_\times v_d\right) \perp v_d$. After rearranging terms in the above equation and integrating both sides, one can obtain

$$
\int_{t_0}^{t} \bar{F}^T \dot{x}_d dt = V(t) - V(t_0) + \int_{t_0}^{t} v_d^T B v_d dt.
\tag{5.198}
$$

After utilizing the fact that $V(t)$, $\int_{t_0}^{t} v_d^T B v_d dt \geq 0$, we can obtain a lower-bound for the left-hand side of the above equation as follows

$$
\int_{t_0}^{t} \bar{F}^T \dot{x}_d dt \geq -V(t_0) = -c_1
\tag{5.199}
$$

which proves (5.195). Based on (5.199), the passivity constraint states that the energy transferred from the system to the user is always less than c_1 which is finite stored initial energy. In the sequel, we will show passivity of the robot by utilizing (5.199) and the yet to be proved $\mathcal{L}_1$ stability property of the end-effector velocity tracking error.

5.5.3 Control Problem Formulation

Given the desired robot end-effector trajectory $x_d(t)$ (obtained via on-line solution of (5.194)), our primary control objective is to asymptotically drive the end-effector trajectory tracking error

$$e_1 \triangleq x_d - x \tag{5.200}$$

to zero while compensating for uncertainties in the system dynamics. Motivated by the subsequent control design strategy, we introduce additional tracking error variables $e_2(t), e_3(t) \in \mathbb{R}^3$ as follows

$$e_2 \triangleq \dot{e}_1 + e_1 \tag{5.201}$$

$$e_3 \triangleq \dot{e}_2 + e_2. \tag{5.202}$$

Our secondary control objective is to preserve the passivity of the robot for safety of user operation in the sense that

$$\int_{t_0}^{t} \bar{F}^T \dot{x} dt \geq -c_2 \tag{5.203}$$

where $\dot{x}(t)$ is the velocity of the robot and $\bar{F}(t)$ is the interaction force with both variables expressed in $\mathcal{I}$ while c_2 is a positive constant. The control challenge is to obtain the companion objectives mentioned above while utilizing only measurements of the end-effector position, velocity, and the interaction force. Given these measurements, $e_1(t), e_2(t)$ are measurable variables while $e_3(t)$ is unmeasurable. Motivated by the ensuing control development and stability analysis, we make the following set of assumptions:

Assumption 5.4.1: The transformed inertia and gravity matrices denoted, respectively, by $\bar{M}(x)$, and $\bar{G}(x)$ are uncertain but known to be second order differentiable with respect to x while the unknown centripetal-Coriolis matrix $\bar{V}_m(x, \dot{x})$ is known to be second order differentiable with respect to x and $\dot{x}$.

Assumption 5.4.2: $\bar{F}(t) \in \mathcal{L}_\infty$ is a measurable interaction force exerted by the human operator at the end-effector.

Assumption 5.4.3: The reference trajectory $x_d(t)$ is continuously differentiable up to its fourth derivative such that $x_d^{(i)}(t) \in \mathcal{L}_\infty$, $i = 0, 1, 2, 3, 4$.

Assumption 5.4.4: The desired curve $r_d(s)$ is analytic along the parameter s (at least the first three partial derivatives along s exist and are bounded such that $r_d(s), r_d'(s), r_d''(s), r_d'''(s) \in \mathcal{L}_\infty$).

Assumption 5.4.5: The skew-symmetric matrix $[\omega]_\times$ is continuously differentiable up to its second derivative such that $[\omega]_\times^{(i)} \in \mathcal{L}_\infty$, $i = 0, 1, 2$.

Assumption 5.4.6: The minimum singular value of the manipulator Jacobian, denoted by σ_m is greater than a known small positive constant $\delta > 0$, such that $\max \left\{ \left\| J^{-1}(q) \right\| \right\}$ is known *a priori* and all kinematic singularities are always avoided – this is easily ensured by the algorithm introduced in Section 5.5.2. We also note that since we are only concerned with revolute robot manipulators, we know that kinematic and dynamic terms denoted by $M(q)$, $V_m(q, \dot{q})$, $G(q)$, $x(q)$, $J(q)$, and $J^{-1}(q)$ are bounded for all possible $q(t)$ (i.e., these kinematic and dynamic terms only depend on $q(t)$ as arguments of trigonometric functions). From the preceding considerations, it is easy to argue that $\bar{M}(x), \bar{V}_m(x, \dot{x}), \bar{G}(x) \in \mathcal{L}_\infty$ for all possible $x(t)$.

Control Design: Tier 3

As a primary step, we partially feedback linearize the system by designing the control signal $\bar{\tau}(t)$ as follows

$$\bar{\tau} = -\bar{F} + \bar{\tau}_a \tag{5.204}$$

where $\bar{\tau}_a(t) \in \mathbb{R}^3$ is a yet to be designed auxiliary control signal and we have taken advantage of Assumption 5.4.2. Additionally, we simplify the system representation of (5.181) by defining a generalized variable $\bar{B}(x, \dot{x}) \in \mathbb{R}^3$ as follows

$$\bar{B} = \bar{V}_m(x, \dot{x}) \dot{x} + \bar{G}(x). \tag{5.205}$$

The utilization of (5.204) and (5.205) allows us to succinctly rewrite (5.181) as follows

$$\bar{M} \ddot{x} + \bar{B} = \bar{\tau}_a. \tag{5.206}$$

Given (5.200–5.202) and (5.206), we can obtain the open-loop tracking error dynamics as follows

$$\bar{M} \dot{e}_3 = -\frac{1}{2} \dot{\bar{M}} e_3 - e_2 - \dot{\bar{\tau}}_a + N \tag{5.207}$$

where $N(\cdot) \in \mathbb{R}^3$ is an aggregation of unknown dynamic terms that is explicitly defined as follows

$$N \triangleq \bar{M}(\ddddot{x}_d + \ddot{e}_1 + \dot{e}_2) + \dot{\bar{M}}(\ddot{x}_d + \frac{1}{2}e_3 - \ddot{e}_1) + e_2 + \dot{B}. \qquad (5.208)$$

$N(\cdot)$ can be rewritten as a sum of two auxiliary signals $N_1(t, x, \dot{x}, \ddot{x})$ and $N_2(z)$ as follows

$$N = \underbrace{\bar{M}(x)\dddot{x}_d + \dot{\bar{M}}(x, \dot{x})\ddot{x}_d + \dot{B}(x, \dot{x}, \ddot{x})}_{N_1(\cdot)} + \underbrace{\bar{M}(x)(\ddot{e}_1 + \dot{e}_2) + \dot{\bar{M}}(x, \dot{x})(\frac{1}{2}e_3 - \ddot{e}_1) + e_2}_{N_2(\cdot)} \qquad (5.209)$$

where $z(t) = \begin{bmatrix} e_1^T(t) & e_2^T(t) & e_3^T(t) \end{bmatrix}^T$ defines a composite error vector. Motivated by the structure of $N_1(\cdot)$ in (5.209), we define a desired variable $N_{1d}(t)$ as follows

$$N_{1d}(t) \triangleq N(x_d, \dot{x}_d, \ddot{x}_d, \dddot{x}_d) = \bar{M}(x_d)\dddot{x}_d + \dot{\bar{M}}(x_d, \dot{x}_d)\ddot{x}_d + \dot{B}(x_d, \dot{x}_d, \ddot{x}_d). \qquad (5.210)$$

From Assumptions 5.4.1, 5.4.3, and 5.4.6, one sees that $N_{1d}(t), \dot{N}_{1d}(t) \in \mathcal{L}_\infty$. After adding and subtracting $N_{1d}(t)$ to the right-hand side of (5.207), we have

$$\bar{M}\dot{e}_3 = -\frac{1}{2}\dot{\bar{M}}e_3 - e_2 - \dot{\tilde{\tau}}_a + \tilde{N} + N_{1d} \qquad (5.211)$$

where $\tilde{N} \triangleq N_1 + N_2 - N_{1d}$ is an unmeasurable error signal. After extensive algebraic manipulations (see Section B.3.11 of Appendix B), it can be shown that $\tilde{N}(\cdot)$ can be upper bounded as follows

$$\tilde{N} \leqslant \rho(\|z\|)\|z\| \qquad (5.212)$$

where the notation $\|\cdot\|$ denotes the standard Euclidean norm, $\rho(\|z\|) \in \mathbb{R}$ is a positive non-decreasing function while $z(t) \in \mathbb{R}^9$ has been previously defined below (5.209). Based on the structure of (5.211), (5.212) as well as the subsequent stability analysis, the following implementable continuous control law can be utilized to achieve the stated control objectives

$$\begin{aligned} \bar{\tau}_a = \ & (k_s + 1)\,e_2(t) - (k_s + 1)\,e_2(t_0) \\ & + \int_{t_0}^t \left[(k_s + 1)\,e_2(\tau) + (\beta_1 + \beta_2)\mathrm{sgn}(e_2(\tau))\right] d\tau \end{aligned} \qquad (5.213)$$

where k_s, β_1, β_2 are constant positive control gains, and $\mathrm{sgn}(\cdot)$ denotes the standard signum function. After taking the time derivative of (5.213) and substituting for $\dot{\bar{\tau}}_a(t)$ into (5.211), we obtain the following closed-loop system

$$\bar{M}\dot{e}_3 = -\frac{1}{2}\dot{\bar{M}}e_3 - e_2 - (k_s + 1)e_3 - (\beta_1 + \beta_2)\text{sgn}(e_2) + \tilde{N} + N_{1d}. \tag{5.214}$$

Stability Analysis

Before presenting the main result of this section, we state the following two lemmas which will be invoked later.

Lemma 5.5 *Let the auxiliary function $L_1(t) \in \mathbb{R}$ be defined as follows*

$$L_1 \triangleq e_3^T \left(N_{1d} - \beta_1 sgn(e_2) \right). \tag{5.215}$$

If the control gain β_1 is selected to satisfy the sufficient condition

$$\beta_1 > \|N_{1d}(t)\| + \|\dot{N}_{1d}(t)\|, \tag{5.216}$$

then

$$\int_{t_0}^{t} L_1(\tau)d\tau \leqslant \zeta_{b1} \tag{5.217}$$

where the positive constant $\zeta_{b1} \in \mathbb{R}$ is defined as

$$\zeta_{b1} \triangleq \beta_1\|e_2(t_0)\|_1 - e_2^T(t_0)N_{1d}(t_0). \tag{5.218}$$

where the notation $\|\eta\|_1 \triangleq \sum_{r=1}^{n} |\eta_r| \, \forall \, \eta \in \mathbb{R}^n$ denotes the 1-norm.

Proof: See proof of Lemma B.5 in Section B.3.6 of Appendix B. ∎

Lemma 5.6 *Let the auxiliary function $L_2(t) \in \mathbb{R}$ be defined as follows*

$$L_2 \triangleq \dot{e}_2^T \left(-\beta_2 sgn(e_2) \right). \tag{5.219}$$

It is then easy to show that

$$\begin{aligned} \int_{t_0}^{t} L_2(\tau)d\tau &= \int_{t_0}^{t} \dot{e}_2^T \left(-\beta_2 sgn(e_2) \right) d\tau \\ &= \beta_2\|e_2(t_0)\|_1 - \beta_2\|e_2(t)\|_1 \leqslant \beta_2\|e_2(t_0)\|_1 \triangleq \zeta_{b2}. \end{aligned} \tag{5.220}$$

Proof: See proof of Lemma B.5 in Section B.3.6 of Appendix B. ∎

Theorem 5.7 *The control law of (5.213) ensures that all system signals are bounded under closed-loop operation and we obtain asymptotic tracking in the sense that*

$$e_i^{(j)}(t) \to 0 \quad as \ t \to \infty \quad \forall \ i = 1, 2; \ j = 0, 1. \tag{5.221}$$

Proof: Let us define two auxiliary functions $P_i(t) \in \mathbb{R}$ as follows

$$P_i(t) \triangleq \zeta_{bi} - \int_{t_0}^t L_i(\tau)d\tau \geqslant 0 \; \forall \; i = 1, 2 \tag{5.222}$$

where $\zeta_{bi}, L_i(t)$ have been previously defined in Lemmas 5.5 and 5.6. Based on the non-negativity of $P_i(t)$ above, one can define a nonnegative function $V_1(t)$ as follows

$$V_1 \triangleq \frac{1}{2}e_1^T e_1 + \frac{1}{2}e_2^T e_2 + \frac{1}{2}e_3^T \bar{M} e_3 + P_1 + P_2. \tag{5.223}$$

After taking the time derivative of (5.223) and utilizing the definitions of (5.200–5.202) as well as the closed-loop dynamics of (5.214), we can conveniently rearrange the terms to obtain the following expression for $\dot{V}_1(t)$

$$\dot{V}_1 = -\|e_1\|^2 - \|e_2\|^2 - (k_s + 1)\|e_3\|^2 + e_1^T e_2 + e_3^T \tilde{N} - \beta_2 e_2^T \mathrm{sgn}(e_2) \\ + \left[e_3^T (N_{1d} - \beta_1 \mathrm{sgn}(e_2)) - L_1\right] - \left[\dot{e}_2^T \beta_2 \mathrm{sgn}(e_2) + L_2\right] \tag{5.224}$$

where we have utilized the definition of (5.222). After utilizing the definitions of (5.215) and (5.219) to eliminate the bracketed terms in the above equality, we can utilize simple algebraic manipulations to obtain the following upper-bound for $\dot{V}_1(t)$

$$\dot{V}_1 \leq -\frac{1}{2}\|z\|^2 + \left[\|e_3\|\rho(\|z\|)\|z\| - k_s\|e_3\|^2\right] - \beta_2\|e_2\|_1$$

where $z(t)$ is a composite error vector that has been defined previously in (5.209). Applying the nonlinear damping argument [25] to the bracketed term above, we obtain the following upper-bound for $\dot{V}_1(t)$

$$\dot{V}_1 \leqslant -\frac{1}{2}\left(1 - \frac{\rho^2(\|z\|)}{2k_s}\right)\|z\|^2 - \beta_2\|e_2\|_1. \tag{5.225}$$

From (5.225), it is possible to state that

$$\left.\begin{array}{l} \dot{V}_1 \leqslant -\alpha\|z\|^2 \\ \dot{V}_1 \leqslant -\beta_2\|e_2\|_1 \end{array}\right\} \text{ for } k_s > \frac{1}{2}\rho^2(\|z\|) \tag{5.226}$$

where $\alpha \in \mathbb{R}$ is some positive constant of analysis. We note here that it is possible to express the lower-bound on k_s in terms of the initial conditions of the problem which has been referred to in literature as a semi-global stability result. We refer the interested reader to Section B.3.12 of Appendix B for the details of such a procedure. Here onward, our analysis is valid in the region of attraction denoted by Ω_c in Section B.3.12 of Appendix B. From (5.222), (5.223), and (5.226), it is easy to see that $z(t) \in \mathcal{L}_\infty \cap \mathcal{L}_2$ and $\lim\limits_{t \to \infty} \|z\|^2 = 0$. From the previous assertions and the definitions of (5.201), (5.202), and (5.209), one readily obtains the result of (5.221). ∎

We now turn our attention to proving the passivity of the robot manipulator. Integrating both sides of the bottom expression of (5.226), we obtain

$$\int_{t_0}^t \|e_2(\tau)\|_1 \, d\tau \leqslant \frac{V_1(t_0)}{\beta_2} \Rightarrow e_2(t) \in \mathcal{L}_1.$$

Since $e_1(t)$ is related to $e_2(t)$ through a transfer function that is strictly proper and stable, one can use Lemma A.8 of [52] to conclude that $e_1(t) \in \mathcal{L}_1$. Now, utilizing (5.201), we can also state that $\dot{e}_1(t) \in \mathcal{L}_1$. The work done by the interaction force on the robot is denoted by $W(t)$ and given by

$$W = \int_{t_0}^t \bar{F}^T \dot{x} d\tau = \int_{t_0}^t \bar{F}^T \dot{x}_d d\tau - \int_{t_0}^t \bar{F}^T \dot{e}_1 d\tau \tag{5.227}$$

where (5.200) has been utilized. Since the first term on the right-hand side of (5.227) has been lower-bounded as in (5.199), we focus our attention on the second term. The second term can now be upper-bounded as follows

$$\int_{t_0}^t \bar{F}^T \dot{e}_1 d\tau \quad \leq \quad \sup_t \left\{ \|\bar{F}(t)\| \right\} \sup_t \left\{ \int_{t_0}^t \|\dot{e}_1(t)\|_1 \, d\tau \right\} \leq c_3 \tag{5.228}$$

where we have utilized the fact that $\dot{e}_1(t) \in \mathcal{L}_1$ as well as Assumption 5.4.2 to justify the existence of the supremum functions defined above, and c_3 is a positive constant. One can now utilize the lower-bound of (5.199) and the upper-bound of (5.228) in order to lower-bound $W(t)$ as $W(t) \geq -c_1 - c_3 = -c_2$; this satisfies the passivity control objective of (5.203).

5.5.4   Simulation Results

Numerical simulations were performed to illustrate the performance of the proposed reference generator and control law of (5.194), (5.204), and (5.213) (See Figure 5.35 for a block diagram) with a two-link planar elbow arm whose inertia matrix $M(q)$ can be expressed in terms of its elements as follows

$$\begin{aligned}
m_{11} &= (m_1 + m_2)\, l_1^2 + m_2 l_2^2 + 2m_2 l_1 l_2 \cos q_2 \\
m_{12} &= m_{21} = m_2 l_2^2 + m_2 l_1 l_2 \cos q_2 \\
m_{22} &= m_2 l_2^2
\end{aligned} \qquad , \tag{5.229}$$

while the centripetal Coriolis vector can be expressed in the following manner

$$V_m(q, \dot{q})\dot{q} = \begin{bmatrix} -m_2 l_1 l_2 \left(2\dot{q}_1 \dot{q}_2 + \dot{q}_2^2\right) \sin q_2 \\ m_2 l_1 l_2 \dot{q}_1^2 \sin q_2 \end{bmatrix}. \tag{5.230}$$

The mass and length parameters of the manipulator are specified as follows

$$m_1 = 2.08 \ [\text{kg}] \quad m_2 = 0.168 \ [\text{kg}] \quad l_1 = 1.2 \ [\text{m}] \quad l_2 = 1.2 \ [\text{m}].$$

Two simulation studies were conducted. In the first study, the performance of the overall system was studied in the presence of obstacles, singularities, and joint limits. The initial configuration of the two-link robot is chosen as $q_1(0) = 0.334$ [rad], $q_2(0) = 0.7$ [rad]. The desired contour is specified by a unit circular path $\bar{r}_d(s) = \begin{bmatrix} \cos(s) & \sin(s) \end{bmatrix}^T$. The initial conditions and parameters for the reference generator are chosen as follows

$$x_d(0) = \begin{bmatrix} 1.6 & 1.5 \end{bmatrix}^T \ [\text{m}] \quad s(0) = 0$$
$$m_d = 0.1 \ [\text{kg}] \quad B = diag\{2.5, 10\} \ [\text{Ns}^{-1}].$$

The parameters for the obstacle are chosen as follows

$$O_1(0) = \begin{bmatrix} -0.5 & 1.65 \end{bmatrix}^T \ [\text{m}] \quad R_1 = 0.5 \ [\text{m}] \ .$$

The interaction force applied at the end-effector by a user was chosen to be $F = \begin{bmatrix} 2 & 2 \end{bmatrix}^T$ [N]. The joint limit for all joints are set as $q_{di\,max} = 180°$ and $q_{di\,min} = -180°$. The parameters in (5.185), (5.187) and (5.188) are chosen as follows

$$\alpha_1 = 1 \qquad\qquad \alpha_2 = 1 \qquad \gamma_1 = 0.1$$
$$\gamma_{21} = \gamma_{22} = \gamma_{23} = 0.79 \qquad \gamma_3 = 10 \qquad \gamma_4 = 25$$

For best transient performance, the control gains specified in (5.213) are chosen to be $k_s = 99, \beta_1 + \beta_2 = 10$. The pre-planned path r_d defined in (5.188) is depicted in Figure 5.37. The measure Ψ_1 defined in (5.184) is depicted in Figure 5.38 as one closed contour is traced. Corresponding to the first dip in Ψ_1 in Figure 5.38, Figure 5.39 depicts a snapshot of the 2-link manipulator veering away from the dashed circular contour $\bar{r}_d$ in order to avoid the kinematic singularity ($q_2 = 0$). Next, the measure Ψ_3 defined in (5.186) is shown in Figure 5.40 — by employing the dip in Ψ_3, the algorithm is able to steer the robot away from the physical obstacle marked by the solid circle in the robot workspace, as can be seen in the snapshot of Figure 5.41. In Figure 5.42, one can see the evolution of the measure Ψ_2 — correspondingly, the snapshot in Figure 5.43 shows how the algorithm utilizes the dip in the measure Ψ_2 in order to avoid the $q_2 = -180$ [deg]

joint limit. Figure 5.44 shows the robot end-effector tracing the modified desired contour r_d as the user applies interaction force at the end-effector. The tracking error $e_1(t)$ is depicted in Figure 5.45 and the control torque input $\bar{\tau}(t)$ is depicted in Figure 5.46. We note here that in Figures 5.39, 5.41, 5.43, and 5.44, the solid circle denotes an obstacle, the dashed circle denotes the nominal contour $\bar{r}_d(s)$, the dashdot curve denotes the modified contour $r_d(s)$, and the solid curve denotes the actual trajectory of the robot end-effector $x(t)$. The obstacle in the simulation can be real, such as the body of the patient or the body of the robot folding back on itself. It can also be a virtual obstacle which is utilized to modify the therapist specified path because of the reconfiguration of the robot system or changes specific to a particular patient or class of patients. The simulation results show that the control algorithm, along with the design of virtual obstacles, can reconfigure the therapy based on different conditions and requirements.

For comparison purposes, a second simulation study is conducted where the therapist suggested path is carefully constructed in order to avoid joint limits, singularities, and obstacles. The desired contour is specified by a unit circular path $\bar{r}_d(s) = \begin{bmatrix} 0.6\cos(s) & 0.6\sin(s) \end{bmatrix}^T$. The initial configuration of the two-link robot is chosen as $x(0) = 1.540$ [m], $y(0) = 1.586$ [m]. The initial condition for the reference generator is chosen as $x_d(0) = 1.2$ [m], $yd(0) = 1.5$ [m]. All other parameters are unchanged from the first simulation study. Figures 5.47, 5.48, and 5.49 show, respectively, the robot end-effector tracking the desired path, the tracking errors in the task space, and the commanded torque inputs.

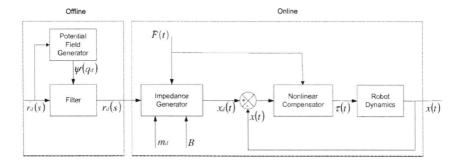

FIGURE 5.35. Graphical Representation of Path Planning and Control Algorithm

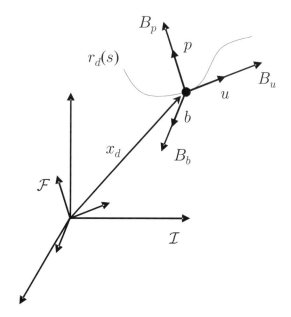

FIGURE 5.36. Relationship between the coordinates

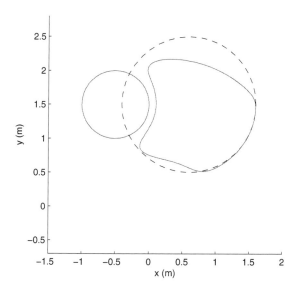

FIGURE 5.37. Offline Pre-Planned Path (Solid Curve) is Different from Therapist Suggested Path (Dotted Circle) due to Joint Limits, Kinematic Singularities, and an Obstacle (Solid Circle)

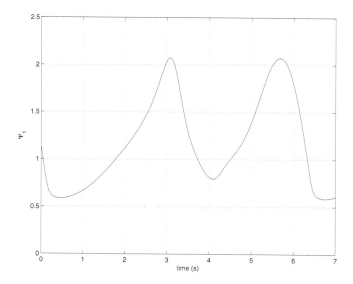

FIGURE 5.38. Manipulability Measure Ψ_1 for Avoiding Kinematic Singularities

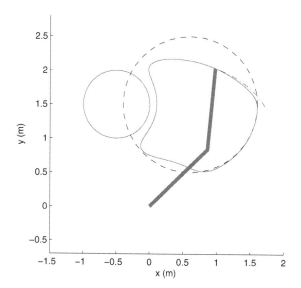

FIGURE 5.39. An Instance of a 2-link Robotic Manipulator Using the Ψ_1 Measure to Avoid a Kinematic Singularity

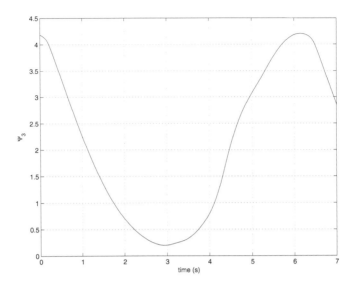

FIGURE 5.40. Measure Ψ_3 for Avoiding Obstacles in the Robot Workspace

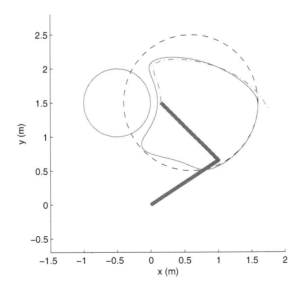

FIGURE 5.41. An Instance of a 2-link Robotic Manipulator Using the Ψ_3 Measure to Avoid an Obstacle

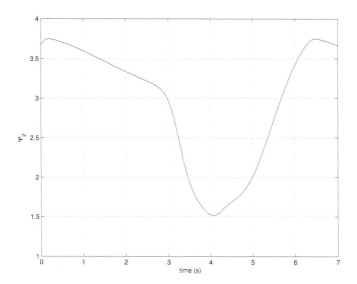

FIGURE 5.42. Measure Ψ_2 for Avoiding Joint Limit Singularities

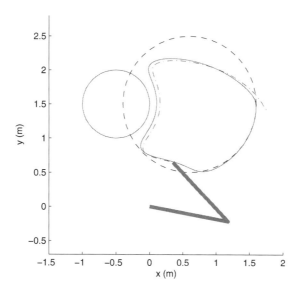

FIGURE 5.43. An Instance of a 2-link Robotic Manipulator Using the Ψ_2 Measure to Avoid a Joint Limit for the Second Joint

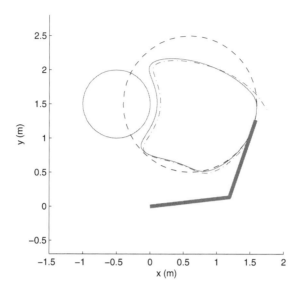

FIGURE 5.44. A Plot of a 2-link Manipulator Tracking the Desired Trajectory

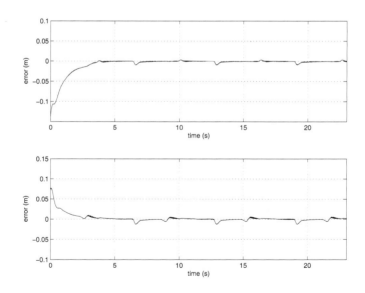

FIGURE 5.45. Error between the Desired and Actual End Effector Trajectories

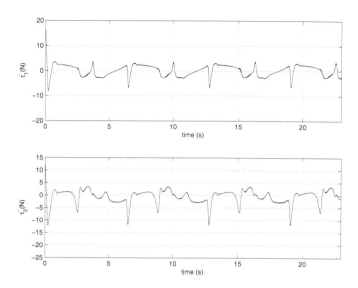

FIGURE 5.46. Control Input $\bar{\tau}(t)$

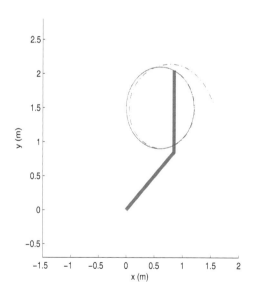

FIGURE 5.47. A 2-link Manipulator Tracking a Desired Trajectory Free of Singularities, Joint Limits, and Obstacles

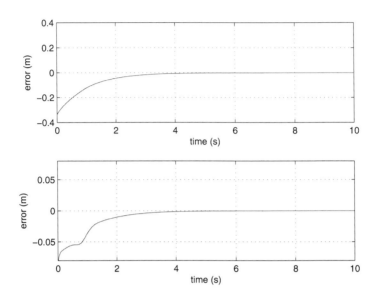

FIGURE 5.48. Error between the Desired and Actual End Effector Trajectories

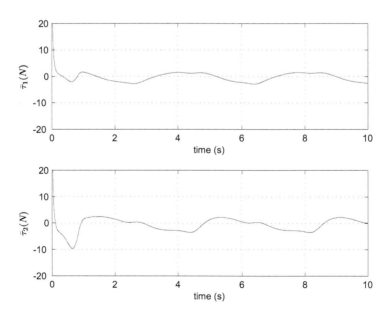

FIGURE 5.49. Control Input $\bar{\tau}\left(t\right)$

5.6   Background and Notes

In the last two decades, researchers have been active in developing next generation exercise machines. In [20], a state-feedback controller is developed for a human arm exercise machine. The machine described in [20] uses an actuated mechanism to give the user the sensation of moving "virtual" dynamic systems such as a mass, spring, or damper. Unfortunately, the control design in this preliminary research does not address the passivity problem or the self-optimizing problem. In [37] and [38], a passive exercise machine controller is developed. By utilizing the assumption that muscle force decreases linearly with the velocity of motion, the controller ensures tracking of an arbitrary desired velocity field and system passivity. The strategy in [37] and [38] employs a combination of an adaptive tracking controller and a reference trajectory generator. To compensate for the uncertainty in the user's biomechanics, the reference generator requires a training phase where the algorithm learns user specific parameters. Once the user's parameters are acquired for a specific exercise session, an optimal reference trajectory is generated. In [60], an adaptive resistance controller is designed under the restriction that the resistance mechanism has only a braking capability. The static damping control design in [60] ensures the passivity of the closed-loop system to an external input force and bounded tracking errors. An optimal exercise protocol is proposed in [60] based on an assumed linear velocity dependence of human force. Identification of the nonlinear system dynamics of the exercise motions and torque output of the resistance mechanisms is utilized in [60] to deal with unknown human biomechanical behavior.

For steer-by-wire control of vehicles, many researchers have worked on establishing dynamic models and performing experiments to identify system parameters with the intention of providing simulated force feedback (e.g., see [12], [39], [55]). Detailed modeling of the conventional, electric, and steer-by-wire steering systems is presented in [44]. After making appropriate simplifying assumptions, these models have been utilized to provide the system model. In [17], the authors design a fuzzy logic controller for an active steering system that prevents vehicle spin in wet road conditions. In [31], a novel robot control strategy is designed to force locking between the primary and secondary system while ensuring passivity. Concepts introduced in this paper may be easily extended to add simulated forces on the steering wheel to either ensure safe operation of the vehicle or to communicate the occurrence of certain events (warnings). Theoretical and experimental work [54] ongoing has produced various interesting ideas and results in this area. Present day simulators already use the virtual environment concept to provide safe and realistic learning environments to beginners.

The emphasis of some previous robot teleoperation related research is to achieve ideal transparency by exactly transferring the slave robot impedance to the user. Typically, approaches that aim for ideal transparency either require *a priori* knowledge of the environmental inputs to the slave manipulator, as in [7], or estimate the impedance of the slave manipulator as in [15]. Some exceptions include the teleoperator controllers aimed at low-frequency transparency developed in [30] and [57] that do not require knowledge of the impedance of the user or environment. However, the approaches in [7], [15], [30], and [57] are based on linear teleoperator systems with frequency-based control designs. A review of other frequency-based approaches applied to linear teleoperator systems is given in [2], [20], [21], [56], and [65]. In [18], an adaptive nonlinear control design is presented that achieves transparency in the sense of motion and force tracking.

Other research has emphasized the stability and safe operation of the teleoperator system through passivity concepts (e.g., [2]–[6], [32]–[34], and [43]–[48]). In [2], Anderson and Spong used passivity and scattering criterion to propose a bilateral control law for a linear time-invariant teleoperator system in any environment and in the presence of time delay. These results were then extended in [47] and [48], where wave-variables were used to define a new configuration for force-reflecting teleoperators. In [48], and more recently in [5] and [6], these methods where extended to solve the position tracking problem. In [34], a passivity-based approach was used to develop a controller that renders a linear teleoperator system as a passive rigid mechanical tool with desired kinematic and power scaling capabilities. The development in [34] was extended to nonlinear teleoperator systems in [32] and [33]. The controllers in [32] and [33] are dependent on knowledge of the dynamics of the master and slave manipulator and force measurements.

References

[1] M. Aisen, H. Krebs, N. Hogan, F. McDowell, and B. Volpe, "The Effect of Robot-Assisted Therapy and Rehabilitative Training on Motor Recovery Following Stroke," *Archives of Neurology*, Vol. 54, No. 4, pp. 443–446, April 1997.

[2] R. J. Anderson and M. W. Spong, "Bilateral Control of Teleoperators with Time Delay," *IEEE Transactions on Automatic Control*, Vol. 34, No 5, pp. 494–501, 1989.

[3] G.A. Brooks and T.D. Fahey, *Exercise Physiology: Human Bioenergetics and its Applications*, New York: Wiley, 1984.

[4] Gregory T. Carter, "Rehabilitation Management in Neuromuscular Disease," *Journal of Neurologic Rehabilitation*, Vol. 11, pp. 69–80, 1997.

[5] N. Chopra, M. W. Spong, S. Hirche, and M. Buss, "Bilateral Teleoperation over the Internet: the Time Varying Delay Problem," *Proceedings of the American Control Conference*, June 4–6, Denver, CO, 2003, pp. 155–160.

[6] N. Chopra, M. W. Spong, R. Ortega, and N. E. Barabanov, "On Position Tracking in Bilateral Telcoperators," *Proceedings of the American Control Conference*, June 30–July 2, Boston, MA, 2004, pp. 5244–5249.

[7] J. E. Colgate, "Robust Impedance Shaping Telemanipulation," *IEEE Transactions on Robotics and Automation*, Vol. 9, No. 4, pp. 374–384, 1993.

[8] M. J. Corless and G. Leitmann, "Continuous State Feedback Guaranteeing Uniform Ultimate Boundedness for Uncertain Dynamic Systems," *IEEE Transactions on Automatic Control*, Vol. 26, No. 5, pp. 1139–1143, 1981.

[9] D. Dawson, J. Hu, and T. Burg, *Nonlinear Control of Electric Machinery*, Marcel Dekker, 1998.

[10] C. Desoer and M. Vidyasagar, *"Feedback Systems: Input-output Properties,"* Academic Press, 1975.

[11] J. Engelberger, *Robotics in Service*, Cambridge, Massachussets, The MIT Press, 1989.

[12] B. Gillespie, C. Hasser, and P. Tang, "Cancellation of Feedthrough Dynamics Using a Force-Reflecting Joystick," *Proceedings of the ASME International Mechanical Engineering Conference and Exposition*, Nashville, TN, Vol. 67, pp. 319–329, November 1999.

[13] S. Godley, T. Triggs, and B. Fildes, "Driving Simulator Validation for Speed Research," *Accident Analysis and Prevention*, Vol. 34, No. 5, pp. 589–600, September 2002.

[14] A. Gray, "Drawing Space Curves with Assigned Curvature," in *Modern Differential Geometry of Curves and Surfaces with Mathematica*, 2nd ed. Boca Raton, FL: CRC Press, pp. 222–224, 1997.

[15] B. Hannaford, "A Design Framework for Teleoperators with Kinesthetic Feedback," *IEEE Transactions on Robotics and Automation*, Vol. 5, pp. 426–434, August 1989.

[16] N. Hogan, "Impedance control: An approach to manipulation, part I — theory, part II — implementation, part III — applications," *Journal of Dynamics, Systems, Measurement, and Control*, pp. 107–124, 1985.

[17] K. Huh, C. Seo, J. Kim, and D. Hong, "Active Steering Control Based on the Estimated Tire Forces," *Proceedings of the American Control Conference*, San Diego, CA, pp. 729–733, June 1999.

[18] N. V. Q. Hung, T. Narikiyo, H. D. Tuan, "Nonlinear Adaptive Control of Master-Slave System in Teleoperation," *Control Engineering Practice*, Vol. 11, No. 1, pp. 1–10, 2003.

[19] H. Jeffreys and B. Jeffreys, *Methods of Mathematical Physics*, Cambridge University Press, 3rd edition, Feb. 2000.

[20] H. Kazerooni and M.G. Her, "A Virtual Exercise Machine," *Proceedings of the 1993 International Conference of Robotics and Automation*, Atlanta, Georgia, May 1993, pp. 232–238.

[21] H. Kazerooni, T. I. Tsay, and K. Hollerbach, "A Controller Design Framework for Telerobotic Systems," *IEEE Transactions on Control Systems Technology*, Vol. 1, No. 1, pp. 50–62, 1993.

[22] H. Khalil, *Nonlinear Systems*, New York: Prentice Hall, 2002.

[23] O. Khatib, "Real-Time Obstacle Avoidance for Manipulators and Mobile Robots," *The International Journal of Robotics Research*, Vol. 5, No. 1, pp. 90–98, 1986.

[24] T. Kjkuchi, J. Furusho and K. Oda, "Development of Isokinetic Exercise Machine Using ER Brake," *Proceedings of the 2003 IEEE International Conference on Robotics and Automation*, Taipei, Taiwan, pp. 214–219, Sep. 14–19, 2003.

[25] P. Kokotovic, "The Joy of Feedback: Nonlinear and Adaptive," *IEEE Control Systems Magazine*, Vol. 12, pp. 177–185, June 1992.

[26] E. Kreyszig, *Advanced Engineering Mathematics,* Wiley Eastern Limited, Sept. 1990.

[27] M. Kristic and H. Deng, *Stabilization of Nonlinear Uncertain System,* Springer-Verlag, New York, 1998.

[28] H. H. Kwee, J. J. Duimel, J. J. Smits, A. A. Tuinhof de Moed, and J. A. van Woerden, "The MANUS Wheelchair-Borne Manipulator: System Review and First Results," *Proceedings of the IARP Workshop on Domestic and Medical & Healthcare Robotics,* Newcastle, pp. 385–396, 1989.

[29] K. F. Laurin-Kovitz, J. E. Colgate, S. D. R. Carnes, "Design of Components for Programmable Passive Impedance," *Proceedings of the IEEE International Conference on Robotics and Automation,* Sacramento, CA, pp. 1476–1481, April 1991.

[30] D. A. Lawrence, "Stability and Transparency in Bilateral Teleoperation," *IEEE Transactions on Robotics and Automation,* Vol. 9, No. 5, pp.624–637, 1993.

[31] D. Lee and P. Li, "Passive Control of Bilateral Tele-operated Manipulators: Robust Control and Experiments," *Proceedings of the American Control Conference,* Arlington, VA, pp. 4612–4618, June 2001.

[32] D. Lee and P. Y. Li, "Passive Coordination Control of Nonlinear Bilateral Teleoperated Manipulators," *Proceedings of the IEEE International Conference on Robotics and Automation,* May 11–15, Washington, DC, pp. 3278–3283, 2002.

[33] D. Lee and P. Y. Li, "Passive Tool Dynamics Rendering for Nonlinear Bilateral Teleoperated Manipulators," *Proceedings of the IEEE International Conference on Robotics and Automation,* May 11–15, Washington, DC, pp. 3284–3289, 2002.

[34] D. Lee and P. Y. Li, "Passive Bilateral Feedforward Control of Linear Dynamically Similar Teleoperated Manipulators," *IEEE Transactions on Robotics and Automation,* Vol. 19, No. 3, pp. 443–456, 2003.

[35] Lewis, F. L., Abdallah, C. T., and Dawson, D. M., *Control of Robot Manipulators,* Macmillan Publishing Company, 1993.

[36] F. L. Lewis, D. M. Dawson, and C. T. Abdallah, *Robot Manipulator Control: Theory and Practice,* New York, NY: Marcel Dekker, Inc., 2004.

[37] P. Y. Li and R. Horowitz, "Control of Smart Exercise Machines — Part I: Problem Formulation and Nonadaptive Control," *IEEE Transactions on Mechatronics,* Vol. 2, No. 4, pp. 237–247, December 1997.

[38] P. Y. Li and R. Horowitz, "Control of Smart Exercise Machines — Part II: Self Optimizing Control," *IEEE Transactions on Mechatronics,* Vol. 2, No. 4, pp. 247–248, December 1997.

[39] A. Liu and S. Chang, "Force Feedback in a Stationary Driving Simulator," *Proceedings of the IEEE International Conference on Systems, Man and Cybernetics,* Vancouver, Canada, Vol. 2, pp. 1711–1716, 1995.

[40] M. Van der Loos, S. Michalowski, and L. Leifer, "Design of an Omnidirectional Mobile Robot as a Manipulation Aid for the Severely Disabled," *Interactive Robotic Aids,* World Rehabilitation Fund Monograph #37 (R. Foulds ed.), New York, 1986.

[41] R. M. Mahoney, "The Raptor Wheelchair Robot System," *Integration of Assistive Technology in the Information Age* (M. Mokhtari, ed.), IOS, Netherlands, pp.135–141, 2001.

[42] M. McIntyre, W. Dixon, D. Dawson, and E. Tatlicioglu, "Passive Coordination of Nonlinear Bilateral Teleoperated Manipulators," *Robotica,* Volume 24, No. 4, pp. 463–476, July 2006.

[43] B. E. Miller, J. E. Colgate, and R. A. Freeman, "Guaranteed Stability of Haptic Systems with Nonlinear Virtual Environments," *IEEE Transactions on Robotics and Automation,* Vol. 16, No. 6, pp. 712–719, 2000.

[44] V. Mills, J. Wagner, and D. Dawson, "Nonlinear Modeling and Analysis of Steering Systems for Hybrid Vehicles," *Proceedings of the ASME International Mechanical Engineering Congress and Exposition,* New York, NY, pp. 571–578, November 2001.

[45] V. Mills, and J. Wagner, "Behavioral Modeling and Analysis of Hybrid Vehicle Steering Systems," *Proceedings of the Institution of Mechanical Engineers, Part D, Journal of Automobile Engineering,* Vol. 217, No. 5, pp. 349–361, 2003.

[46] B. G. Nielsen, "Unfolding Transitions in Myosin give Rise to the Double-Hyperbolic Force-Velocity Relation in Muscle," *Journal of Physics: Condensed Matter,* Vol. 15, S1759–S1765, 2003.

[47] G. Niemeyer and J. J. E. Slotine, "Stable Adaptive Teleoperation," *IEEE Journal of Oceanic Engineering,* Vol. 16, No. 1, pp. 152–162, 1991.

[48] G. Niemeyer and J. J. E. Slotine, "Designing Force Reflecting Teleoperators with Large Time Delays to Appear as Virtual Tools," *Proceedings of the IEEE International Conference on Robotics and Automation,* Albuquerque, pp. 2212–2218, NM, April 20–25, 1997.

[49] J. Post and E. Law, "Modeling, Simulation and Testing of Automobile Power Steering Systems for the Evaluation of On-Center Handling," *SAE* Paper No. 960178, 1996.

[50] W. H. Press, S. A. Teukosky, W. T. Vetterly and B. P. Flamneny, *Numerical Recipes in Fortran, the Art of Scientific Computing,* 2nd edition, Cambridge University Press, 1992.

[51] Z. Qu, and J. Xu, "Model Based Learning Controls and their Comparisons using Lyapunov Direct Method," *Asian Journal of Control, Special Issue on Learning Control,* Vol. 4, No.1, pp. 99–110, March 2002.

[52] M. S. de Queiroz, D. M. Dawson, S. Nagarkatti, F. Zhang, *Lyapunov-Based Control of Mechanical Systems,* Cambridge: Birkhäuser, 2000.

[53] E. Rimon and D. E. Koditschek, "Exact Robot Navigation Using Artificial Potential Functions," *IEEE Transactions on Robotics and Automation,* Vol. 8, No. 5, pp. 501–518, 1992.

[54] E. Rossetter, J. Switkes, and J. Gerdes, "A Gentle Nudge Towards Safety: Experimental Validation of the Potential Field Driver Assistance System," *Proceedings of the American Control Conference,* Denver, CO, pp. 3744–3749, June 2003.

[55] J. Ryu and H. Kim, "Virtual Environment for Developing Electronic Power Steering and Steer-by-Wire Systems," *Proceedings of the IEEE/RSJ International Conference on Intelligent Robots and Systems,* pp. 1374–1379, Kyongju, Korea, 1999.

[56] S. E. Salcudean, N. M. Wong, and R. L. Hollis, "Design and Control of a Force-Reflecting Teleoperation System with Magnetically Levitated Master and Wrist," *IEEE Transactions on Robotics and Automation,* Vol. 11, No. 6, pp. 844–858, 1995.

[57] S. E. Salcudean, M. Zhu, W. -H. Zhu, and K. Hashtudi-Zaad, "Transparent Bilateral Teleoperation Under Position and Rate Control," *International Journal of Robotics Research*, Vol. 19, No. 12, pp. 1185–1202, 2000.

[58] P. Setlur, D. M. Dawson, J. Wagner, and Y. Fang, "Nonlinear Tracking Controller Design for Steer-by-Wire Systems," *Proceedings of the American Control Conference*, Anchorage, AK, pp. 280–285, May 2002.

[59] P. Setlur, D. Dawson, J. Chen, and J. Wagner, "A Nonlinear Tracking Controller for a Haptic Interface Steer-by-Wire Systems," *Proceedings of the IEEE Conference on Decision and Control*, Las Vegas, NV, pp. 3112–3117, December 2002.

[60] J. Shields and R. Horowitz, "Controller Design and Experimental Verification of a Self-Optimizing Motion Controller for Smart Exercise Machines," *Proceedings of the American Control Conference*, Albuquerque, NM, pp. 2736–2742, June 1997.

[61] J. J. Slotine and W. Li, A*pplied Nonlinear Control,* New York: Prentice Hall, 1991.

[62] M. W. Spong and M. Vidyasagar, *Robot Dynamics and Control*, New York: John Wiley and Sons, Inc., 1989.

[63] J. T. Wen, D. Popa, G. Montemayor, P. L. Liu, "Human Assisted Impedance Control of Overhead Cranes," *Proceedings of the IEEE Conference on Control Applications*, Mexico City, Mexico, pp. 383–387, Sept. 2001.

[64] B. Xian, M. S. de Queiroz, and D. M. Dawson, "A Continuous Control Mechanism for Uncertain Nonlinear Systems," *Optimal Control, Stabilization, and Nonsmooth Analysis, Lecture Notes in Control and Information Sciences*, Heidelberg, Germany: Springer-Verlag, pp. 251–262, 2004.

[65] Y. Yokokohji and T. Yoshikawa, "Bilateral Control of Master-Slave Manipulators for Ideal Kinesthetic Coupling — Formulation and Experiment," *IEEE Transactions on Robotics and Automation*, Vol. 10, pp. 605–620, Oct. 1994.

[66] T. Yoshikawa, "Analysis and Control of Robot Manipulators with Redundancy," In M. Brady and E. R. Paul, editors, *Robotics Research: The First International Symposium*, pages 735–747. MIT Press, 1984.

[67] X.T. Zhang, D. M. Dawson, W. E. Dixon, and B. Xian, "Extremum Seeking Nonlinear Controllers for a Human Exercise Machine," *IEEE/ASME Transactions on Mechatronics,* Vol. 11, No. 2, pp. 233–240, 2006.

[68] X. Zhang, A. Behal, D. M. Dawson, and J. Chen, "A Novel Passive Path Following Controller for a Rehabilitation Robot," *Proceedings of the IEEE Conference on Decision and Control,* Paradise Island, Bahamas, pp. 5374–5379, December 2004.

Appendix A
Mathematical Background

In this appendix, several fundamental mathematical tools are presented in the form of definitions and lemmas that aid the control development and closed-loop stability analyses presented in the previous chapters. The proofs of most of the following lemmas are omitted, but can be found in the cited references.

Lemma A.1 *[9]*

Consider a function $f(t) : \mathbb{R}_+ \to \mathbb{R}$. If $f(t) \in \mathcal{L}_\infty$, $\dot{f}(t) \in \mathcal{L}_\infty$, and $f(t) \in \mathcal{L}_2$, then

$$\lim_{t \to \infty} f(t) = 0. \tag{A.1}$$

This lemma is often referred to as Barbalat's Lemma.

Lemma A.2 *[2]*

If a given differentiable function $f(t) : \mathbb{R}_+ \to \mathbb{R}$ has a finite limit as $t \to \infty$ and if $f(t)$ has a time derivative, defined as $\dot{f}(t)$, that can be written as the sum of two functions, denoted by $g_1(t)$ and $g_2(t)$, as follows

$$\dot{f}(t) = g_1 + g_2 \tag{A.2}$$

where $g_1(t)$ is a uniformly continuous function and

$$\lim_{t \to \infty} g_2(t) = 0 \tag{A.3}$$

then

$$\lim_{t\to\infty} \dot{f}(t) = 0 \qquad \lim_{t\to\infty} g_1(t) = 0. \qquad (A.4)$$

This lemma is often referred to as the Extended Barbalat's Lemma.

Lemma A.3 *[6]*

If $f(t)$ is a uniformly continuous function, then $\lim_{t\to\infty} f(t) = 0$ if and only if

$$\lim_{t\to\infty} \int_t^{t+t'} f(\tau)d\tau = 0 \qquad (A.5)$$

for any positive constant $t' \in \mathbb{R}$. This lemma is often referred to as the integral form of Barbalat's Lemma.

Definition 1 *[9]*

Consider a function $f(t) : \mathbb{R}_+ \to \mathbb{R}$. Let the 2-norm (denoted by $\|\cdot\|_2$) of a scalar function $f(t)$ be defined as

$$\|f(t)\|_2 = \sqrt{\int_0^\infty f^2(\tau)\, d\tau}. \qquad (A.6)$$

If $\|f(t)\|_2 < \infty$, then we say that the function $f(t)$ belongs to the subspace $\mathcal{L}_2$ of the space of all possible functions (i.e., $f(t) \in \mathcal{L}_2$). Let the ∞-norm (denoted by $\|\cdot\|_\infty$) of $f(t)$ be defined as

$$\|f(t)\|_\infty = \sup_t |f(t)|. \qquad (A.7)$$

If $\|f(t)\|_\infty < \infty$, then we say that the function $f(t)$ belongs to the subspace $\mathcal{L}_\infty$ of the space of all possible functions (i.e., $f(t) \in \mathcal{L}_\infty$).

Definition 2 *[9]*

The induced 2-norm of matrix $A(t) \in \mathbb{R}^{n\times n}$ is defined as follows

$$\|A(t)\|_{i2} = \sqrt{\lambda_{\max}\{A^T(t)A(t)\}}. \qquad (A.8)$$

Lemma A.4 *[4]*

Given a function $f : \mathbb{R}^n \to \mathbb{R}$ that is continuously differentiable on an open set $S \subset \mathbb{R}^n$ and given points $(x_{10}, ..., x_{n0})$ and $(x_1, ..., x_n)$ in S that are joined by a straight line that lies entirely in $\mathbb{R}^n$, then there exists a point $(\xi_1, ..., \xi_n)$ on the line between the endpoints, such that

$$f(x_1, ..., x_n) = f(x_{10}, ..., x_{n0}) + \sum_{j=1}^n \frac{\partial}{\partial x_j} f(\xi_1, ..., \xi_n)(x_j - x_{j0}). \qquad (A.9)$$

This Lemma is often referred to as the Mean Value Theorem.

Lemma A.5 *[6]*

Given a function $f : \mathbb{R}^n \times \mathbb{R}^m \to \mathbb{R}^n$ that is continuously differentiable at every point (x, y) on an open set $S \subset \mathbb{R}^n \times \mathbb{R}^m$, if there is a point (x_0, y_0) on S where

$$f(x_0, y_0) = 0 \tag{A.10}$$

and

$$\frac{\partial f}{\partial x}(x_0, y_0) \neq 0, \tag{A.11}$$

then there are neighborhoods $U \subset \mathbb{R}^n$ and $V \subset \mathbb{R}^m$ of x_0 and y_0, respectively, such that for all $y \in V$ the expression in (A.10) has a unique solution $x \in U$. This unique solution can be written as $x = g(y)$ where g is continuously differentiable at $y = y_0$. This Lemma is often referred to as the Implicit Function Theorem.

Lemma A.6 *[3]*

Given $a, b, c \in \mathbb{R}^n$, any of the following cyclic permutations leaves the scalar triple product invariant

$$a \cdot (b \times c) = b \cdot (c \times a) = c \cdot (a \times b) \tag{A.12}$$

and the following interchange of the inner and vector product

$$a \cdot (b \times c) = (a \times b) \cdot c \tag{A.13}$$

leaves the scalar triple product invariant where the notation $a \cdot b$ represents the dot product of a and b and the notation $a \times b$ represents the cross product of a and b.

Lemma A.7 *[3]*

Given $a, b, c \in \mathbb{R}^n$, the vector triple products satisfy the following expressions

$$a \times (b \times c) = (a \cdot c)\, b - (a \cdot b)\, c \tag{A.14}$$

$$(a \times b) \times c\, (a \cdot c)\, b - (b \cdot c)\, a \tag{A.15}$$

where the notation $a \cdot b$ represents the dot product of a and b and the notation $a \times b$ represents the cross product of a and b.

Lemma A.8 *[3]*

Given $a, b \in \mathbb{R}^n$, the vector product satisfies the following skew-symmetric property

$$a \times b = -b \times a \tag{A.16}$$

where the notation $a \times b$ represents the cross product of a and b.

Lemma A.9 [3]

Given $a = \begin{bmatrix} a_1 & a_2 & a_3 \end{bmatrix}^T \in \mathbb{R}^3$ and $a^\times \in \mathbb{R}^{3\times 3}$ which is defined as follows

$$a^\times = \begin{bmatrix} 0 & -a_3 & a_2 \\ a_3 & 0 & -a_1 \\ -a_2 & a_1 & 0 \end{bmatrix} \tag{A.17}$$

then the product $a^T a^\times$ satisfies the following property

$$a^T a^\times = \begin{bmatrix} 0 & 0 & 0 \end{bmatrix}^T. \tag{A.18}$$

Lemma A.10 [8] (Thm 9-11)

Given the symmetric matrix $A \in \mathbb{R}^{n\times n}$ and the diagonal matrix $D \in \mathbb{R}^{n\times n}$, then A is orthogonally similar to D and the diagonal elements of D are necessarily the eigenvalues of A.

Lemma A.11 [9]

If $w(t) : \mathbb{R}_+ \to \mathbb{R}$ is persistently exciting and $w(t)$, $\dot{w}(t) \in \mathcal{L}_\infty$, then the stable, minimum phase, rational transfer function $\hat{H}(w)$ is also persistently exciting.

Lemma A.12 [10]

If $\dot{f}(t) \triangleq \frac{d}{dt} f(t)$ is bounded for $t \in [0, \infty)$, then $f(t)$ is uniformly continuous for $t \in [0, \infty)$.

Lemma A.13 [1]

Let $V(t)$ be a non-negative scalar function of time on $[0, \infty)$ which satisfies the differential inequality

$$\dot{V}(t) \leq -\gamma V(t) \tag{A.19}$$

where γ is a positive constant. Given (A.19), then

$$V(t) \leq V(0) \exp(-\gamma t) \quad \forall t \in [0, \infty) \tag{A.20}$$

where $\exp(\cdot)$ denotes the base of the natural logarithm.

Lemma A.14

Given a non-negative function denoted by $V(t) \in \mathbb{R}$ as follows

$$V = \frac{1}{2} x^2 \tag{A.21}$$

with the following time derivative

$$\dot{V} = -k_1 x^2, \tag{A.22}$$

then $x(t) \in \mathbb{R}$ is square integrable (i.e., $x(t) \in \mathcal{L}_2$).

Proof: To prove Lemma A.14, we integrate both sides of (A.22) as follows

$$-\int_0^\infty \dot{V}(t)dt = k_1 \int_0^\infty x^2(t)dt. \tag{A.23}$$

After evaluating the left side of (A.23), we can conclude that

$$k_1 \int_0^\infty x^2(t)dt = V(0) - V(\infty) \le V(0) < \infty \tag{A.24}$$

where we utilized the fact that $V(0) \ge V(\infty) \ge 0$ (see (A.21) and (A.22)). Since the inequality given in (A.24) can be rewritten as follows

$$\sqrt{\int_0^\infty x^2(t)dt} \le \sqrt{\frac{V(0)}{k_1}} < \infty \tag{A.25}$$

we can utilize Definition 1 to conclude that $x(t) \in \mathcal{L}_2$.

Lemma A.15 [5]

Let $A \in \mathbb{R}^{n \times n}$ be a real, symmetric, positive-definite matrix; therefore, all of the eigenvalues of A are real and positive. Let $\lambda_{\min}\{A\}$ and $\lambda_{\max}\{A\}$ denote the minimum and maximum eigenvalues of A, respectively, then for $\forall x \in \mathbb{R}^n$

$$\lambda_{\min}\{A\} \|x\|^2 \le x^T A x \le \lambda_{\max}\{A\} \|x\|^2 \tag{A.26}$$

where $\|\cdot\|$ denotes the standard Euclidean norm. This lemma is often referred to as the Rayleigh-Ritz Theorem.

Lemma A.16 [1]

Given a scalar function $r(t)$ and the following differential equation

$$r = \dot{e} + \alpha e \tag{A.27}$$

where $\dot{e}(t) \in \mathbb{R}$ represents the time derivative $e(t) \in \mathbb{R}$ and $\alpha \in \mathbb{R}$ is a positive constant, if $r(t) \in \mathcal{L}_\infty$, then $e(t)$ and $\dot{e}(t) \in \mathcal{L}_\infty$.

Lemma A.17 [1]

Given the differential equation in (A.27), if $r(t)$ is exponentially stable in the sense that

$$|r(t)| \leq \beta_0 \exp(-\beta_1 t) \tag{A.28}$$

where β_0 and $\beta_1 \in \mathbb{R}$ are positive constants, then $e(t)$ and $\dot{e}(t)$ are exponentially stable in the sense that

$$|e(t)| \leq |e(0)| \exp(-\alpha t) + \frac{\beta_0}{\alpha - \beta_1} \left(\exp(-\beta_1 t) - \exp(-\alpha t) \right) \tag{A.29}$$

and

$$|\dot{e}(t)| \quad \leq \quad \alpha |e(0)| \exp(-\alpha t) + \frac{\alpha \beta_0}{\alpha - \beta_1} \left(\exp(-\beta_1 t) \right. \tag{A.30}$$
$$\left. - \exp(-\alpha t) \right) + \beta_0 \exp(-\beta_1 t)$$

where α was defined in (A.27).

Lemma A.18 [1]

Given the differential equation in (A.27), if $r(t) \in \mathcal{L}_\infty$, $r(t) \in \mathcal{L}_2$, and $r(t)$ converges asymptotically in the sense that

$$\lim_{t \to \infty} r(t) = 0 \tag{A.31}$$

then $e(t)$ and $\dot{e}(t)$ converge asymptotically in the sense that

$$\lim_{t \to \infty} e(t), \dot{e}(t) = 0. \tag{A.32}$$

Lemma A.19 [1, 7]

If a scalar function $N_d(x, y)$ is given by

$$N_d = \Omega(x)xy - k_n \Omega^2(x)x^2 \tag{A.33}$$

where $x, y \in \mathbb{R}$, $\Omega(x) \in \mathbb{R}$ is a function dependent only on x, and k_n is a positive constant, then $N_d(x, y)$ can be upper bounded as follows

$$N_d \leq \frac{y^2}{k_n}. \tag{A.34}$$

The bounding of $N_d(x, y)$ in the above manner is often referred to as nonlinear damping [7] since a nonlinear control function (e.g., $k_n \Omega^2(x)x^2$) can be used to "damp-out" an unmeasurable quantity (e.g., y) multiplied by a known, measurable nonlinear function, (e.g., $\Omega(x)$).

Lemma A.20 [1]

Let $V(t)$ be a non-negative scalar function of time on $[0, \infty)$ which satisfies the differential inequality

$$\dot{V} \leq -\gamma V + \varepsilon \tag{A.35}$$

where γ and ε are positive constants. Given (A.35), then

$$V(t) \leq V(0) \exp\left(-\gamma t\right) + \frac{\varepsilon}{\gamma} \left(1 - \exp\left(-\gamma t\right)\right) \qquad \forall t \in [0, \infty). \tag{A.36}$$

Lemma A.21 [1]

If the differential equation in (A.27) can be bounded as follows

$$|r(t)| \leq \sqrt{A + B \exp(-kt)} \tag{A.37}$$

where k, A, and $B \in \mathbb{R}$ and $A + B \geq 0$, then $e(t)$ given in (A.27) can be bounded as follows

$$
\begin{aligned}
|e(t)| \quad \leq \quad & |e(0)| \exp(-\alpha t) + \frac{a}{\alpha} \left(1 - \exp(-\alpha t)\right) \\
& + \frac{2b}{2\alpha - k} \left(\exp(-\frac{1}{2}kt) - \exp(-\alpha t)\right)
\end{aligned}
\tag{A.38}
$$

where

$$a = \sqrt{A} \quad \text{and} \quad b = \sqrt{B}. \tag{A.39}$$

Lemma A.22 [6]

Let the origin of the following autonomous system

$$\dot{x} = f(x) \tag{A.40}$$

be an equilibrium point $x(t) = 0$ where $f(\cdot) : D \to \mathbb{R}^n$ is a map from the domain $D \subset \mathbb{R}^n$ into $\mathbb{R}^n$. Consider a continuously differentiable positive definite function $V(\cdot) : D \to \mathbb{R}^n$ containing the origin $x(t) = 0$ where

$$\dot{V}(x) \leq 0 \quad \text{in } D. \tag{A.41}$$

Let Γ be defined as the set of all points where $\left\{x \in D | \dot{V}(x) = 0\right\}$ and suppose that no solution can stay identically in Γ other than the trivial solution $x(t) = 0$. Then the origin is globally asymptotically stable. This Lemma is a corollary to LaSalle's Invariance Theorem.

References

[1] D. M. Dawson, J. Hu, and T. C. Burg, *Nonlinear Control of Electric Machinery*, New York, NY: Marcel Dekker, 1998.

[2] W. E. Dixon, D. M. Dawson, E. Zergeroglu, and A. Behal, *Nonlinear Control of Wheeled Mobile Robots*, Springer-Verlag London Ltd., 2001.

[3] W. Fulks, *Advanced Calculus: An Introduction to Analysis*, New York, NY: John Wiley and Sons, 1978.

[4] M. D. Greenburg, *Advanced Engineering Mathematics*, Englewood Cliffs, NJ: Prentice Hall, 1998.

[5] R. Horn and C. Johnson, *Matrix Analysis*, Cambridge, MA: Cambridge University Press, 1985.

[6] H. Khalil, *Nonlinear Systems*, Upper Saddle River, NJ: Prentice Hall, 1996.

[7] M. Krstić, I. Kanellakopoulos, and P. Kokotović, *Nonlinear and Adaptive Control Design,* New York, NY: John Wiley & Sons, 1995.

[8] S. Perlis, *Theory of Matrices*, New York, NY: Dover Publications, 1991.

[9] S. Sastry and M. Bodson, *Adaptive Control: Stability, Convergence, and Robustness*, Englewoods Cliff, NJ: Prentice Hall Co., 1989.

[10] J. J. Slotine and W. Li, *Applied Nonlinear Control*, Englewood Cliffs, NJ: Prentice Hall, 1991.

Appendix B

Supplementary Lemmas and Expressions

B.1 Chapter 3 Lemmas

B.1.1 Open-Loop Rotation Error System

Lemma B.1 *The time derivative of (3.38) can be expressed as*

$$\dot{e}_\omega = L_\omega R \omega_e - \dot{\Theta}_d. \tag{B.1}$$

Proof: Based on the definitions for $\omega_e(t)$ and $R(t)$ in Chapter 3, the following property can be determined [13]

$$[R\omega_e]_\times = \dot{R} R^T. \tag{B.2}$$

Given (B.2), the following definition (see 3.5)

$$\bar{R} = R \left(R^* \right)^T$$

can be used to develop the following relationship

$$[R\omega_e]_\times = \dot{\bar{R}} \bar{R}^T. \tag{B.3}$$

While several parameterizations can be used to express $\bar{R}(t)$ in terms of $u(t)$ and $\theta(t)$, the open-loop error system for $e_\omega(t)$ is derived based on the following exponential parameterization [13]

$$\bar{R} = \exp([\Theta]_\times) = I_3 + \sin\theta \, [u]_\times + 2\sin^2\frac{\theta}{2} [u]_\times^2 \tag{B.4}$$

where the notation I_i denotes an $i \times i$ identity matrix, and the notation $[u]_\times$ denotes the skew-symmetric matrix form of $u(t)$. The parameterization $\Theta_d(t)$ can be related to $\bar{R}_d(t)$ as follows

$$\bar{R}_d = \exp([\Theta_d]_\times) = I_3 + \sin\theta_d [u_d]_\times + 2\sin^2 \frac{\theta_d}{2} [u_d]_\times^2 , \qquad \text{(B.5)}$$

where $\bar{R}_d(t)$ is defined in (3.5) of Chapter 3 as

$$\bar{R}_d = R_d (R^*)^T .$$

To facilitate the development of the open-loop dynamics for $e_\omega(t)$, the expression developed in (B.3) can be used along with (B.4) and the time derivative of (B.4), to obtain the following expression

$$[R\omega_e]_\times = \sin\theta [\dot{u}]_\times + [u]_\times \dot{\theta} + (1 - \cos\theta) [[u]_\times \dot{u}]_\times \qquad \text{(B.6)}$$

where the following properties were utilized [5], [11]

$$[u]_\times \zeta = -[\zeta]_\times u \qquad \text{(B.7)}$$
$$[u]_\times^2 = uu^T - I_3 \qquad \text{(B.8)}$$
$$[u]_\times uu^T = 0 \qquad \text{(B.9)}$$
$$[u]_\times [\dot{u}]_\times [u]_\times = 0 \qquad \text{(B.10)}$$
$$[[u]_\times \dot{u}]_\times = [u]_\times [\dot{u}]_\times - [\dot{u}]_\times [u]_\times . \qquad \text{(B.11)}$$

To facilitate further development, the time derivative of (3.39) is determined as follows

$$\dot{\Theta} = \dot{u}\theta + u\dot{\theta}. \qquad \text{(B.12)}$$

After multiplying (B.12) by $\left(I_3 + [u]_\times^2\right)$, the following expression can be obtained

$$\left(I_3 + [u]_\times^2\right) \dot{\Theta} = u\dot{\theta} \qquad \text{(B.13)}$$

where (B.8) and the following properties were utilized

$$u^T u = 1 \qquad u^T \dot{u} = 0 . \qquad \text{(B.14)}$$

Likewise, by multiplying (B.12) by $-[u]_\times^2$ and then utilizing (B.14) the following expression is obtained

$$-[u]_\times^2 \dot{\Theta} = \dot{u}\theta. \qquad \text{(B.15)}$$

From the expression in (B.6), the properties given in (B.7), (B.12), (B.13), (B.15), and the fact that

$$\sin^2\theta = \frac{1}{2}(1 - \cos 2\theta)$$

can be used to obtain the following expression

$$Rw_e = L_\omega^{-1} \dot{\Theta} \tag{B.16}$$

where $L_\omega(t)$ is defined in (3.47) as

$$L_\omega = I_3 - \frac{\theta}{2}[u]_\times + \left(1 - \frac{sinc(\theta)}{sinc^2\left(\dfrac{\theta}{2}\right)}\right)[u]_\times^2.$$

After multiplying both sides of (B.16) by $L_\omega(t)$, the open-loop dynamics for $\dot{\Theta}$ can be obtained. After substituting (B.16) into the time derivative of (3.38), the open-loop dynamics for $e_\omega(t)$ given by (B.1) can be obtained. ∎

B.1.2 Open-Loop Translation Error System

Lemma B.2 *The time derivative of (3.36) can be expressed as*

$$z_1^* \dot{e}_v = \alpha_1 A_e L_v R\left[v_e + [\omega_e]_\times s_1\right] - z_1^* \dot{p}_{ed}. \tag{B.17}$$

Proof: To develop the open-loop error system for $e_v(t)$, the time derivative of (3.36) can be expressed as

$$\dot{e}_v = \dot{p}_e - \dot{p}_{ed} = \frac{1}{z_1} A_e L_v \dot{m}_1 - \dot{p}_{ed} \tag{B.18}$$

where (3.1), (3.7), (3.14), (3.37), and the definition of $\alpha_i(t)$ in (3.10) were utilized. After taking the time derivative of the first equation in (3.160), $\dot{m}_1(t)$ can be determined as

$$\dot{m}_1 = Rv_e + R[\omega_e]_\times s_1 \tag{B.19}$$

where (B.2) and the following property have been utilized [5]

$$[R\omega_e]_\times = R[\omega_e]_\times R^T. \tag{B.20}$$

After substituting (B.19) into (B.18), multiplying the resulting expression by z_1^*, and utilizing the definition of $\alpha_i(t)$ in (3.10), the open-loop error system given in (B.17) is obtained. ∎

B.1.3 Persistence of Excitation Proof

Lemma B.3 *If $\Omega_i(t_0, t) \geq \gamma_1 I_4$ from (3.214) for any t_0, then*

$$\tilde{\theta}_i(t) \to 0 \ as \ t \to \infty. \tag{B.21}$$

Proof: Let $\Omega_i(t_0, t) \in \mathbb{R}^{4 \times 4}$ be defined as

$$\Omega_i(t_0, t) = \int_{t_0}^{t} W_{fi}^T(\tau) W_{fi}(\tau) d\tau \tag{B.22}$$

where $W_{fi}(t) \in \mathbb{R}^{3 \times 4}$ was previously defined in (3.207). Consider the following expression

$$
\begin{aligned}
\int_{t_0}^{t} \tilde{\theta}_i^T(\tau) \Omega_i(t_0, \tau) \frac{d\tilde{\theta}_i(\tau)}{d\tau} d\tau = & \left. \tilde{\theta}_i^T(\tau) \Omega_i(t_0, \tau) \theta_i(\tau) \right|_{t_0}^{t} \\
& - \int_{t_0}^{t} \frac{d}{d\tau} \left(\tilde{\theta}_i^T(\tau) \Omega_i(t_0, \tau) \right) \tilde{\theta}_i(\tau) d\tau \\
= & \ \tilde{\theta}_i^T(t) \Omega_i(t_0, t) \tilde{\theta}_i(t) \\
& - \int_{t_0}^{t} \tilde{\theta}_i^T(\tau) \Omega_i(t_0, \tau) \frac{d\tilde{\theta}_i(\tau)}{d\tau} d\tau \\
& - \int_{t_0}^{t} \tilde{\theta}_i^T(\tau) W_{fi}^T(\tau) W_{fi}(\tau) \tilde{\theta}_i(\tau) d\tau,
\end{aligned}
\tag{B.23}
$$

where (B.22) and the fact that $\Omega(t_0, t_0) = 0$ were used. Re-arranging (B.23) yields

$$
\tilde{\theta}_i^T(t) \Omega_i(t_0, t) \tilde{\theta}_i(t) = \ \begin{aligned} & 2 \int_{t_0}^{t} \tilde{\theta}_i^T(\tau) \Omega_i(t_0, \tau) \frac{d\tilde{\theta}_i(\tau)}{d\tau} d\tau \\ & + \int_{t_0}^{t} \tilde{\theta}_i^T(\tau) W_{fi}^T(\tau) W_{fi}(\tau) \tilde{\theta}_i(\tau) d\tau \end{aligned}
\tag{B.24}
$$

Substituting for $t = t_0 + T$ in (B.24), where $T \in \mathbb{R}$ is a positive constant, and applying the limit on both sides of the equation yields

$$
\begin{aligned}
\lim_{t_0 \to \infty} \tilde{\theta}_i^T(t_0 + T) \Omega_i(t_0, t_0 + T) \tilde{\theta}_i(t_0 + T) = & \\
\lim_{t_0 \to \infty} & \left(2 \int_{t_0}^{t_0 + T} \tilde{\theta}_i^T(\tau) \Omega_i(t_0, \tau) \frac{d\tilde{\theta}_i(\tau)}{d\tau} d\tau \right. \\
& \left. + \int_{t_0}^{t_0 + T} \tilde{\theta}_i^T(\tau) W_{fi}^T(\tau) W_{fi}(\tau) \tilde{\theta}_i(\tau) d\tau \right).
\end{aligned}
\tag{B.25}
$$

From the proof of Theorem 3.4, $\tilde{\theta}_i(t) \in L_\infty$, and from (3.214) and (B.22), $\Omega_i(t_0, t_0 + T) \in L_\infty$. It was also proven that $W_{fi}(t) \tilde{\theta}_i(t), \eta_i(t) \in L_\infty \cap L_2$. Hence, from (3.211)

$$\lim_{t \to \infty} \tilde{p}_{ei}(t) = 0$$

and consequently from (3.206) and (3.212)

$$\lim_{t \to \infty} \dot{\tilde{\theta}}_i(t) = 0.$$

Hence, after utilizing Lemma A.3 in Appendix A, the first integral in (B.25) vanishes upon evaluation. From (3.225) and Lemma A.3 in Appendix A, the second integral in (B.25) also vanishes, and hence

$$\lim_{t_0 \to \infty} \tilde{\theta}_i^T(t_0 + T)\Omega_i(t_0, t_0 + T)\tilde{\theta}_i(t_0 + T) = 0. \tag{B.26}$$

Since $\Omega_i(t_0, t) \geq \gamma_1 I_4$ from (3.214) for any t_0, (B.26) indicates that

$$\tilde{\theta}_i(t) \to 0 \text{ as } t \to \infty. \ \blacksquare$$

B.2   Chapter 4 Lemmas and Auxiliary Expressions

B.2.1   *Experimental Velocity Field Selection*

This VFC development is based on the selection of a velocity field that is first order differentiable, and that a first order differentiable, nonnegative function $V(q) \in \mathbb{R}$ exists such that the following inequality holds

$$\frac{\partial V(x)}{\partial x}\vartheta(x) \leq -\gamma_3(\|x\|) + \zeta_0 \tag{B.27}$$

where $\frac{\partial V(q)}{\partial q}$ denotes the partial derivative of $V(q)$ with respect to $q(t)$, $\gamma_3(\cdot) \in \mathbb{R}$ is a class $\mathcal{K}$ function[1], and $\zeta_0 \in \mathbb{R}$ is a nonnegative constant. To prove that the velocity field in (4.43) and (4.44) satisfies the condition in (B.27), let $V(x) \in \mathbb{R}$ denote the following nonnegative, continuous differentiable function

$$V(x) \triangleq \frac{1}{2}x^T x. \tag{B.28}$$

After taking the time derivative of (B.28) and substituting (4.43) for $\dot{x}(t)$, the following inequality can be developed

$$\dot{V} = x^T \vartheta(x) \leq -\gamma_3(x) + \zeta_0 \tag{B.29}$$

where $\gamma_3(x)$ and ζ_0 were defined in (B.27).

To prove the inequality given in (B.29), we must find $\gamma_3(x)$ and ζ_0. To this end, we rewrite $x^T\vartheta(x)$ as follows

$$x^T\vartheta(x) = -2K(x)f(x)x^T x \tag{B.30}$$

---

[1]A continuous function $\alpha : [0, a) \to [0, \infty)$ is said to belong to class $\mathcal{K}$ if it is strictly increasing and $\alpha(0) = 0$. [7]

where (4.43) has been utilized, and x_{c1} and x_{c2} of (4.43) and (4.44) are set to zero for simplicity and without loss of generality. By substituting (4.44) into (B.30) for $K(x)$ and $f(x)$, the following expression can be obtained

$$x^T \vartheta(x) = \frac{-k_0^* \|x\|^4 + k_0^* r_o^2 \|x\|^2}{\left|\left(\|x\|^2 - r_o^2\right)\right| \|x\| + \dfrac{\epsilon}{2}}. \tag{B.31}$$

After utilizing the following inequality

$$\|x\|^2 \le \delta \|x\|^4 + \frac{1}{\delta}$$

where $\delta \in \mathbb{R}$ is a positive constant, the inequality given in (B.29) can be determined from (B.31) where

$$\gamma_3(x) = \frac{k_0^* \left(1 - r_o^2 \delta\right) \|x\|^4}{\left(\|x\|^2 + r_o^2\right) \|x\| + \dfrac{\epsilon}{2}}$$

and

$$\zeta_0 = \frac{2 k_0^* r_o^2}{\delta \epsilon}.$$

Provided δ is selected according to the following inequality

$$\delta < \frac{1}{r_o^2},$$

then $\gamma_3(x)$ can be shown to be a class $\mathcal{K}$ function.

B.2.2 GUB Lemma

Lemma B.4 *Given a continuously differentiable function, denoted by $V(q)$, that satisfies the following inequalities*

$$0 < \gamma_1(\|q\|) \le V(q) \le \gamma_2(\|q\|) + \xi_b \tag{B.32}$$

with a time derivative that satisfies the following inequality

$$\dot{V}(q) \le -\gamma_3(\|q\|) + \xi_0, \tag{B.33}$$

then $q(t)$ is GUB, where $\gamma_1(\cdot)$, $\gamma_2(\cdot)$, $\gamma_3(\cdot)$ are class $\mathcal{K}$ functions, and ξ_0, $\xi_b \in \mathbb{R}$ denote positive constants.

Proof:[2] Let $\Omega \in \mathbb{R}$ be a positive function defined as follows

$$\Omega \triangleq \gamma_3^{-1}(\xi_0) > 0 \tag{B.34}$$

---

[2]This proof is based on the proof for Theorem 2.14 in [12].

where $\gamma_3^{-1}(\cdot)$ denotes the inverse of $\gamma_3(\cdot)$, and let $B(0, \Omega)$ denote a ball centered about the origin with a radius of Ω. Consider the following 2 possible cases.

The initial condition $q(t_0)$ lies outside the ball $B(0, \Omega)$ as follows

$$\Omega < \|q(t_0)\| \leq \Omega_1 \tag{B.35}$$

where $\Omega_1 \in \mathbb{R}$ is a positive constant. To facilitate further analysis, we define the operator $d(\cdot)$ as follows

$$d(\Omega_1) \triangleq (\gamma_1^{-1} \circ \gamma_2)(\Omega_1) + \gamma_1^{-1}(\xi_b) > 0 \tag{B.36}$$

where $(\gamma_1^{-1} \circ \gamma_2)$ denotes the composition of the inverse of $\gamma_1(\cdot)$ with $\gamma_2(\cdot)$ (i.e., the inverse of the function $\gamma_1(\cdot)$ is applied to the function $\gamma_2(\cdot)$). After substituting the constant $d(\Omega_1)$ into $\gamma_1(\cdot)$, the following inequalities can be determined

$$\gamma_1(d(\Omega_1)) = \gamma_2(\Omega_1) + \xi_b \geq \gamma_2(\|q(t_0)\|) + \xi_b \geq V(q(t_0)) \tag{B.37}$$

where the inequalities provided in (B.32) and (B.35) were utilized.

Assume that $q(\tau) \in \mathbb{R}$ for $t_0 \leq \tau \leq t < \infty$ lies outside the ball $B(0, \Omega)$ as follows

$$\Omega < \|q(\tau)\|. \tag{B.38}$$

From (B.33) and (B.38), the following inequality can be determined

$$\dot{V}(q(\tau)) \leq -\gamma_3(\Omega) + \xi_0,$$

and hence, from the definition for Ω in (B.34), it is clear that

$$\dot{V}(q(\tau)) \leq 0. \tag{B.39}$$

By utilizing (B.37) and the result in (B.39), the following inequalities can be developed for some constant $\Delta\tau$

$$\gamma_1(d(\Omega_1)) \geq V(q(t_0)) \geq V(q(\tau)) \geq V(q(\tau + \Delta\tau)) \geq \gamma_1(\|q(\tau + \Delta\tau)\|). \tag{B.40}$$

Since $\gamma_1(\cdot)$ is a class $\mathcal{K}$ function, (B.36) and (B.40) can be used to develop the following inequality

$$\|q(t)\| \leq d(\Omega_1) = (\gamma_1^{-1} \circ \gamma_2)(\Omega_1) + \gamma_1^{-1}(\xi_b) \qquad \forall t \geq t_0$$

provided the assumption in (B.38) is satisfied. If the assumption in (B.38) is not satisfied, then

$$\|q(t)\| \leq \Omega = \gamma_3^{-1}(\xi_0) \qquad \forall t \geq t_0. \tag{B.41}$$

Hence, $q(t)$ is GUB for Case A.

The initial condition $q(t_0)$ lies inside the ball $B(0, \Omega)$ as follows

$$\|q(t_0)\| \leq \Omega \leq \Omega_1.$$

If $q(t)$ remains in the ball, then the inequality developed in (B.41) will be satisfied. If $q(t)$ leaves the ball, then the results from Case A can be applied. Hence, $q(t)$ is GUB for Case B. ∎

B.2.3   Boundedness of $\dot{\theta}_d(t)$

Based on the definition of $\theta_d(t)$ in (4.116), $\theta_d(t)$ can be expressed in terms of the natural logarithm as follows [16]

$$\theta_d = -i \ln \left(\frac{-\dfrac{\partial \varphi}{\partial x_c} - i \dfrac{\partial \varphi}{\partial y_c}}{\sqrt{\left(\dfrac{\partial \varphi}{\partial x_c}\right)^2 + \left(\dfrac{\partial \varphi}{\partial y_c}\right)^2}} \right) \tag{B.42}$$

where $i=\sqrt{-1}$. After exploiting the following identities [16]

$$\cos(\theta_d) = \frac{1}{2}\left(e^{i\theta_d} + e^{-i\theta_d}\right)$$

$$\sin(\theta_d) = \frac{1}{2i}\left(e^{i\theta_d} - e^{-i\theta_d}\right)$$

and then utilizing (B.42) the following expressions can be obtained

$$\cos(\theta_d) = -\frac{\dfrac{\partial \varphi}{\partial x_c}}{\sqrt{\left(\dfrac{\partial \varphi}{\partial x_c}\right)^2 + \left(\dfrac{\partial \varphi}{\partial y_c}\right)^2}} \tag{B.43}$$

$$\sin(\theta_d) = -\frac{\dfrac{\partial \varphi}{\partial y_c}}{\sqrt{\left(\dfrac{\partial \varphi}{\partial x_c}\right)^2 + \left(\dfrac{\partial \varphi}{\partial y_c}\right)^2}}. \tag{B.44}$$

After utilizing (B.43) and (B.44), the following expression can be obtained

$$\begin{bmatrix} \dfrac{\partial \varphi}{\partial x_c} \\ \dfrac{\partial \varphi}{\partial y_c} \end{bmatrix} = -\sqrt{\left(\dfrac{\partial \varphi}{\partial x_c}\right)^2 + \left(\dfrac{\partial \varphi}{\partial y_c}\right)^2} \begin{bmatrix} \cos(\theta_d) \\ \sin(\theta_d) \end{bmatrix}. \tag{B.45}$$

Based on the expression in (B.42), the time derivative of $\theta_d(t)$ can be written as follows

$$\dot{\theta}_d = \left[\begin{array}{cc} \dfrac{\partial \theta_d}{\partial x_c} & \dfrac{\partial \theta_d}{\partial y_c} \end{array} \right] \left[\begin{array}{c} \dot{x}_c \\ \dot{y}_c \end{array} \right] \tag{B.46}$$

where

$$\frac{\partial \theta_d}{\partial x_c} = \left[\begin{array}{cc} \dfrac{-\dfrac{\partial \varphi}{\partial y_c}}{\left(\dfrac{\partial \varphi}{\partial x_c}\right)^2 + \left(\dfrac{\partial \varphi}{\partial y_c}\right)^2} & \dfrac{\dfrac{\partial \varphi}{\partial x_c}}{\left(\dfrac{\partial \varphi}{\partial x_c}\right)^2 + \left(\dfrac{\partial \varphi}{\partial y_c}\right)^2} \end{array} \right] \tag{B.47}$$
$$\cdot \left[\begin{array}{cc} \dfrac{\partial^2 \varphi}{\partial x_c^2} & \dfrac{\partial^2 \varphi}{\partial x_c \partial y_c} \end{array} \right]^T$$

$$\frac{\partial \theta_d}{\partial y_c} = \left[\begin{array}{cc} \dfrac{-\dfrac{\partial \varphi}{\partial y_c}}{\left(\dfrac{\partial \varphi}{\partial x_c}\right)^2 + \left(\dfrac{\partial \varphi}{\partial y_c}\right)^2} & \dfrac{\dfrac{\partial \varphi}{\partial x_c}}{\left(\dfrac{\partial \varphi}{\partial x_c}\right)^2 + \left(\dfrac{\partial \varphi}{\partial y_c}\right)^2} \end{array} \right] \tag{B.48}$$
$$\cdot \left[\begin{array}{cc} \dfrac{\partial^2 \varphi}{\partial y_c \partial x_c} & \dfrac{\partial^2 \varphi}{\partial y_c^2} \end{array} \right]^T .$$

After substituting (4.47), (B.47), and (B.48) into (B.46), the following expression can be obtained

$$\dot{\theta}_d = \left[\begin{array}{cc} \dfrac{-\dfrac{\partial \varphi}{\partial y_c}}{\left(\dfrac{\partial \varphi}{\partial x_c}\right)^2 + \left(\dfrac{\partial \varphi}{\partial y_c}\right)^2} & \dfrac{\dfrac{\partial \varphi}{\partial x_c}}{\left(\dfrac{\partial \varphi}{\partial x_c}\right)^2 + \left(\dfrac{\partial \varphi}{\partial y_c}\right)^2} \end{array} \right] \tag{B.49}$$
$$\cdot \left[\begin{array}{cc} \dfrac{\partial^2 \varphi}{\partial x_c^2} & \dfrac{\partial^2 \varphi}{\partial y_c \partial x_c} \\ \dfrac{\partial^2 \varphi}{\partial x_c \partial y_c} & \dfrac{\partial^2 \varphi}{\partial y_c^2} \end{array} \right] \left[\begin{array}{c} \cos \theta \\ \sin \theta \end{array} \right] v_c .$$

After substituting (4.117) and (B.45) into (B.49), the following expression can be obtained

$$\dot{\theta}_d = k_v \cos\left(\tilde{\theta}\right) \left[\begin{array}{cc} \sin(\theta_d) & -\cos(\theta_d) \end{array} \right] \tag{B.50}$$
$$\cdot \left[\begin{array}{cc} \dfrac{\partial^2 \varphi}{\partial x_c^2} & \dfrac{\partial^2 \varphi}{\partial y_c \partial x_c} \\ \dfrac{\partial^2 \varphi}{\partial x_c \partial y_c} & \dfrac{\partial^2 \varphi}{\partial y_c^2} \end{array} \right] \left[\begin{array}{c} \cos \theta \\ \sin \theta \end{array} \right] .$$

By Property 4.2.4 of Chapter 4, each element of the Hessian matrix is bounded; hence, from (B.50), it is straightforward that $\dot{\theta}_d(t) \in L_\infty$.

B.2.4   Open-Loop Dynamics for $\Upsilon(t)$

The extended image coordinates $p_{e1}(t)$ of (4.145) can be written as follows

$$
p_{e1} = \begin{bmatrix} a_1 & a_2 & 0 \\ 0 & a_3 & 0 \\ 0 & 0 & 1 \end{bmatrix} \begin{bmatrix} \frac{x_1}{z_1} \\ \frac{y_1}{z_1} \\ \ln(z_1) \end{bmatrix} + \begin{bmatrix} a_4 \\ a_5 \\ -\ln(z_1^*) \end{bmatrix} \tag{B.51}
$$

where (4.140), (4.141), and (4.142) were utilized. After taking the time derivative of (B.51), the following expression can be obtained

$$
\dot{p}_{e1} = \frac{1}{z_1} A_{e1} \dot{\bar{m}}_1.
$$

By exploiting the fact that $\dot{\bar{m}}_1(t)$ can be expressed as follows

$$
\dot{\bar{m}}_1 = -v_c + [\bar{m}_1]_\times \omega_c,
$$

the open-loop dynamics for $p_{e1}(t)$ can be rewritten as follows

$$
\dot{p}_{e1} = -\frac{1}{z_1} A_{e1} v_c + A_{e1} [m_1]_\times \omega_c.
$$

The open-loop dynamics for $\Theta(t)$ can be expressed as follows [4]

$$
\dot{\Theta} = -L_\omega \omega_c.
$$

B.2.5   Measurable Expression for $L_{\Upsilon_d}(t)$

Similar to (4.47), the dynamics for $\Upsilon_d(t)$ can be expressed as

$$
\dot{\Upsilon}_d = \begin{bmatrix} \dot{p}_{ed1} \\ \dot{\Theta}_d \end{bmatrix} = \begin{bmatrix} -\frac{1}{z_{d1}} A_{ed1} & A_{ed1} [m_{d1}]_\times \\ 0_3 & -L_{\omega d} \end{bmatrix} \begin{bmatrix} v_{cd} \\ \omega_{cd} \end{bmatrix} \tag{B.52}
$$

where $\Theta_d(t)$ is defined in (4.155), $z_{di}(t)$ is introduced in (4.135), $A_{edi}(u_{di}, v_{di})$ is defined in the same manner as in (4.148) with respect to the desired pixel coordinates $u_{di}(t), v_{di}(t)$, $m_{di}(t)$ is given in (4.134), $L_{\omega d}(\theta_d, \mu_d)$ is defined in the same manner as in (4.149) with respect to $\theta_d(t)$ and $\mu_d(t)$, and $v_{cd}(t), \omega_{cd}(t) \in \mathbb{R}^3$ denote the desired linear and angular velocity signals that ensure compatibility with (B.52). The signals $v_{cd}(t)$ and $\omega_{cd}(t)$ are not actually used in the trajectory generation scheme presented in this paper as similarly done in [1]; rather, these signals are simply used to clearly illustrate how $\dot{\bar{p}}_d(t)$ can be expressed in terms of $\dot{\Upsilon}_d(t)$ as required in (4.156).

Specifically, we first note that the top block row in (B.52) can be used to write the time derivative of $p_{ed2}(t)$ in terms of $v_{cd}(t)$ and $\omega_{cd}(t)$ with $i = 2$

$$\dot{p}_{ed2} = \left[\begin{array}{cc} -\frac{1}{z_{d2}}A_{ed2} & A_{ed2}\left[m_{d2} \right]_{\times} \end{array} \right] \left[\begin{array}{c} v_{cd} \\ \omega_{cd} \end{array} \right] \qquad (B.53)$$

where $p_{edi}(t)$ is defined in the same manner as (4.154) $\forall i = 1, 2, 3, 4$. After inverting the relationship given by (B.52), we can also express $v_{cd}(t)$ and $\omega_{cd}(t)$ as a function of $\dot{\Upsilon}_d(t)$ as follows

$$\left[\begin{array}{c} v_{cd} \\ \omega_{cd} \end{array} \right] = \left[\begin{array}{cc} -z_{d1}A_{ed1}^{-1} & -z_{d1}\left[m_{d1} \right]_{\times}L_{\omega d}^{-1} \\ 0 & -L_{\omega d}^{-1} \end{array} \right] \dot{\Upsilon}_d. \qquad (B.54)$$

After substituting (B.54) into (B.53), $\dot{p}_{ed2}(t)$ can be expressed in terms of $\dot{\Upsilon}_d(t)$ as follows

$$\dot{p}_{ed2} = \left[\begin{array}{cc} \frac{z_{d1}}{z_{d2}}A_{ed2}A_{ed1}^{-1} & A_{ed2}\left[\frac{z_{d1}}{z_{d2}}m_{d1} - m_{d2} \right]_{\times}L_{\omega d}^{-1} \end{array} \right] \dot{\Upsilon}_d. \qquad (B.55)$$

After formulating similar expressions for $\dot{p}_{ed3}(t)$ and $\dot{p}_{ed4}(t)$ as the one given by (B.55) for $\dot{p}_{ed2}(t)$, we can compute the expression for $L_{\Upsilon_d}(\bar{p}_d)$ in (4.157) by utilizing the definitions of $p_{di}(t)$ and $p_{edi}(t)$ given in (4.140) and (4.154), respectively (i.e., we must eliminate the bottom row of the expression given by (B.55)).

B.2.6   Development of an Image Space NF and Its Gradient

Inspired by the framework developed in [3], an image space NF is constructed by developing a diffeomorphism[3] between the image space and a model space, developing a model space NF, and transforming the model space NF into an image space NF through the diffeomorphism (since NFs are invariant under diffeomorphism [8]). To this end, a diffeomorphism is defined that maps the desired image feature vector $\bar{p}_d$ to the auxiliary model space signal $\zeta(\bar{p}_d) \triangleq [\zeta_1(\bar{p}_d) \ \zeta_2(\bar{p}_d) \ \dots \ \zeta_8(\bar{p}_d)]^T : [-1, 1]^8 \rightarrow \mathbb{R}^8$ as follows

$$\zeta = diag\{ \frac{2}{u_{max} - u_{min}}, \frac{2}{v_{max} - v_{min}}, \dots, \frac{2}{v_{max} - v_{min}} \}\bar{p}_d \qquad (B.56)$$
$$- \left[\begin{array}{cccc} \frac{u_{max} + u_{min}}{u_{max} - u_{min}} & \frac{v_{max} + v_{min}}{v_{max} - v_{min}} & \dots & \frac{v_{max} + v_{min}}{v_{max} - v_{min}} \end{array} \right]^T.$$

---

[3]A diffeomorphism is a map between manifolds which is differentiable and has a differentiable inverse.

In (B.56), u_{max}, u_{min}, v_{max}, and $v_{min} \in \mathbb{R}$ denote the maximum and minimum pixel values along the u and v axes, respectively. The model space NF, denoted by $\tilde{\varphi}(\zeta) \in \mathbb{R}^8 \to \mathbb{R}$, is defined as follows [3]

$$\tilde{\varphi}(\zeta) \triangleq \frac{\bar{\varphi}}{1 + \bar{\varphi}}. \tag{B.57}$$

In (B.57), $\bar{\varphi}(\zeta) \in \mathbb{R}^8 \to \mathbb{R}$ is defined as

$$\bar{\varphi}(\zeta) \triangleq \frac{1}{2} f(\zeta)^T K f(\zeta) \tag{B.58}$$

where the auxiliary function $f(\zeta) : (-1, 1)^8 \to \mathbb{R}^8$ is defined similar to [3] as follows

$$f(\zeta) = \left[\begin{array}{ccc} \dfrac{\zeta_1 - \zeta_1^*}{\left(1 - \zeta_1^{2\kappa}\right)^{1/2\kappa}} & \cdots & \dfrac{\zeta_8 - \zeta_8^*}{\left(1 - \zeta_8^{2\kappa}\right)^{1/2\kappa}} \end{array} \right]^T \tag{B.59}$$

where $K \in \mathbb{R}^{8 \times 8}$ is a positive definite, symmetric matrix, and κ is a positive parameter. The reason we use κ instead of 1 as in [3] is to get an additional parameter to change the potential field formed by $f(\zeta)$. See [3] for a proof that (B.57) satisfies the properties of a NF as described in Properties 4.2.4 – 4.2.7 in Chapter 4. The image space NF, denoted by $\varphi(\bar{p}_d) \in \mathcal{D} \to \mathbb{R}$, can then be developed as follows

$$\varphi(\bar{p}_d) \triangleq \tilde{\varphi} \circ \zeta(\bar{p}_d) \tag{B.60}$$

where $\circ$ denotes the composition operator. The gradient vector $\nabla\varphi(p_d)$ can be expressed as follows

$$\nabla\varphi \triangleq \left(\frac{\partial\varphi}{\partial\bar{p}_d} \right)^T = \left(\frac{\partial\tilde{\varphi}}{\partial\zeta} \frac{\partial\zeta}{\partial\bar{p}_d} \right)^T. \tag{B.61}$$

In (B.61), the partial derivative expressions $\frac{\partial\zeta(\bar{p}_d)}{\partial\bar{p}_d}$, $\frac{\partial\tilde{\varphi}(\zeta)}{\partial\zeta}$, and $\frac{\partial f(\zeta)}{\partial\zeta}$ can be expressed as follows

$$\frac{\partial\zeta}{\partial\bar{p}_d} = diag\{\frac{2}{u_{max} - u_{min}}, \frac{2}{v_{max} - v_{min}}, ..., \frac{2}{v_{max} - v_{min}}\} \tag{B.62}$$

$$\frac{\partial\tilde{\varphi}}{\partial\zeta} = \frac{1}{(1 + \bar{\varphi})^2} f^T K \frac{\partial f}{\partial\zeta} \tag{B.63}$$

$$\frac{\partial f}{\partial\zeta} = diag\left\{ \frac{1 - \zeta_1^{2\kappa-1}\zeta_1^*}{\left(1 - \zeta_1^{2\kappa}\right)^{(2\kappa+1)/2\kappa}}, ..., \frac{1 - \zeta_8^{2\kappa-1}\zeta_8^*}{\left(1 - \zeta_8^{2\kappa}\right)^{(2\kappa+1)/2\kappa}} \right\}. \tag{B.64}$$

It is clear from (B.56)–(B.64) that $\bar{p}_d(t) \to \bar{p}^*$ when $\nabla\varphi(\bar{p}_d) \to 0$.

B.2.7   Global Minimum

The objective function introduced in (4.218) can be solved by the Lagrange multiplier approach with constraints [9]. To find the critical points of the objective function in (4.218) subject to the constraint that

$$\psi\left(\Upsilon, \bar{p}\right) = \left(\Upsilon - \Pi^{\dagger}\left(\bar{p}\right)\right) = 0,$$

where $\Pi^{\dagger}(\bar{p}) : \mathcal{D} \to \mathbb{R}^{6}$ denotes a "pseudo-inverse" or "triangulation function" [14], which is a unique mapping that can be considered as a virtual constraint for the image feature vector (See Remark 4.14), one can define the Hamiltonian function $H\left(\bar{p}, \Upsilon, \lambda\right) \in \mathbb{R}$ as follows [9]

$$H\left(\bar{p}, \Upsilon, \lambda\right) = \frac{1}{2}\left(\bar{p} - \bar{p}^{*}\right)^{T}\left(\bar{p} - \bar{p}^{*}\right) + \lambda^{T}\psi\left(\Upsilon, \bar{p}\right)$$

where $\lambda \in \mathbb{R}^{6}$ is a Lagrange multiplier. The necessary conditions for the critical points of $\varphi\left(\bar{p}\right)$ of (4.218) which also satisfies the constraint $\psi\left(\Upsilon, \bar{p}\right) = 0$ are

$$\frac{\partial H}{\partial \bar{p}} = \left(\bar{p} - \bar{p}^{*}\right)^{T} - \lambda^{T}\frac{\partial \Pi^{\dagger}}{\partial \bar{p}} = 0 \tag{B.65}$$

$$\frac{\partial H}{\partial \lambda} = \left(\Upsilon - \Pi^{\dagger}\left(\bar{p}\right)\right) = 0 \tag{B.66}$$

$$\frac{\partial H}{\partial \Upsilon} = \lambda^{T} = 0. \tag{B.67}$$

From (B.65)-(B.67), it is clear that the only critical point occurs at $\bar{p}(t) = \bar{p}^{*}$. Since $\Pi^{\dagger}(\bar{p})$ is a unique mapping, it is clear that $\Upsilon(t) = \Upsilon^{*}$ is the global minimum of the objective function φ defined in (4.218).

B.3   Chapter 5 Lemmas and Auxiliary Expressions

B.3.1   Numerical Extremum Generation

The numerically-based extremum generation formula for computing the optimal velocity setpoint that maximizes the user output power can be described as follows.

- Step 1. Three initial best-guess estimates, denoted by γ_{1}, γ_{2}, $\gamma_{3} \in \mathbb{R}$, are selected where γ_{1} is the best-guess estimate for a lower bound on the optimal velocity, γ_{3} is the best-guess estimate for an upper bound on the optimal velocity, and γ_{2} is the best-guess estimate for the optimal velocity, where $\gamma_{2} \in (\gamma_{1}, \gamma_{3})$.

- Step 2. The lower bound estimate γ_1 is then passed through a set of third order stable and proper low pass filters to generate continuous bounded signals for $\dot{q}_d(t)$, $\ddot{q}_d(t)$, $\dddot{q}_d(t)$. For example, the following filters could be utilized

$$
\begin{aligned}
\dot{q}_d &= \frac{\varsigma_1}{s^3 + \varsigma_2 s^2 + \varsigma_3 s + \varsigma_4} \gamma_1 \\
\ddot{q}_d &= \frac{\varsigma_1 s}{s^3 + \varsigma_2 s^2 + \varsigma_3 s + \varsigma_4} \gamma_1 \\
\dddot{q}_d &= \frac{\varsigma_1 s^2}{s^3 + \varsigma_2 s^2 + \varsigma_3 s + \varsigma_4} \gamma_1
\end{aligned}
\tag{B.68}
$$

where ς_1, ς_2, ς_3, ς_4 denote positive filter constants.

- Step 3. Based on the result in (5.9), and the expressions for the user power output given in (5.12) and the structure in (B.68), the algorithm waits until $|e(t)| \leq \bar{e}_1$ and $|\dot{q}_d - \gamma_1| \leq \bar{e}_2$ before evaluating $p(\gamma_1)$, where $\bar{e}_1$ and $\bar{e}_2$ are some pre-defined threshold values.

- Step 4. Steps 2 and 3 are repeated to obtain $p(\gamma_2)$ and $p(\gamma_3)$.

- Step 5. The next desired trajectory point is determined from the following expression

$$
\gamma_4 = \gamma_2 - \frac{1}{2} \frac{g_1}{g_2}
\tag{B.69}
$$

where g_1, $g_2 \in \mathbb{R}$ are constants defined as follows

$$
\begin{aligned}
g_1 &= (\gamma_2 - \gamma_1)^2 [p(\gamma_2) - p(\gamma_3)] \\
&\quad - (\gamma_2 - \gamma_3)^2 [p(\gamma_2) - p(\gamma_1)]
\end{aligned}
\tag{B.70}
$$

$$
\begin{aligned}
g_2 &= (\gamma_2 - \gamma_1)[p(\gamma_2) - p(\gamma_3)] \\
&\quad - (\gamma_2 - \gamma_3)[p(\gamma_2) - p(\gamma_1)]
\end{aligned}
\tag{B.71}
$$

where γ_i and $p(\gamma_i)$ $\forall i = 1, 2, 3$ are determined from Steps 1–4. Specifically, γ_i and $p(\gamma_i)$ are substituted into (B.69)–(B.71) and the resulting expression yields the next best-guess for $\dot{q}_d^*$ denoted by $\gamma_4 \in \mathbb{R}$.

- Step 6. Steps 2 and 3 are repeated to obtain $\dot{q}_d(t)$, $\ddot{q}_d(t)$, $\dddot{q}_d(t)$ and $p(\gamma_4)$. Note that each successive estimate for $\dot{q}_d^*$ produced by (B.69)–(B.71) will always be bounded by (γ_1, γ_3), and hence, $\dot{q}_d(t)$, $\ddot{q}_d(t)$, $\dddot{q}_d(t) \in \mathcal{L}_\infty$.

- Step 7. The value for $p(\gamma_4)$ is compared to $p(\gamma_2)$. If $p(\gamma_4) \geq p(\gamma_2)$ and $\gamma_2 > \gamma_4$ or if $p(\gamma_2) \geq p(\gamma_4)$ and $\gamma_4 > \gamma_2$, then the three new estimates used to construct a new parabola are γ_2, γ_3, γ_4. If $p(\gamma_4) \geq p(\gamma_2)$ and $\gamma_4 > \gamma_2$ or if $p(\gamma_2) \geq p(\gamma_4)$ and $\gamma_2 > \gamma_4$, then the three new estimates used to construct a new parabola are γ_1, γ_2, γ_4.

- Step 8. Repeat Steps 5–7 for successive γ_i $\forall i = 5, 6, ...,$ where the three estimates determined from Step 7 are used to construct a new parabola. Steps 5–7 are repeated until the difference between the new upper and lower estimates is below some predefined, arbitrarily small threshold.

B.3.2   Proof of Lemma 5.1

After substituting (5.15) into (5.26) and then integrating, the following expression is obtained

$$\int_{t_0}^{t} L_1(\sigma)d\sigma \; = \; \int_{t_0}^{t} \alpha_r e(\sigma)[N_d(\sigma) - \beta_1 sgn(e(\sigma))]d\sigma \tag{B.72}$$
$$+ \int_{t_0}^{t} \frac{de(\sigma)}{d\tau} N_d(\sigma)d\sigma - \beta_1 \int_{t_0}^{t} \frac{de(\sigma)}{d\tau} sgn(e(\sigma))d\sigma.$$

Integrating the second integral on the right side of (B.72) by parts yields

$$\int_{t_0}^{t} L_1(\sigma)d\tau \; = \; \int_{t_0}^{t} \alpha_r e(\sigma) \left(N_d(\sigma) - \beta_1 sgn(e(\sigma))\right) d\sigma \tag{B.73}$$
$$+ e(\sigma)N_d(\sigma)|_{t_0}^{t} - \int_{t_0}^{t} e(\sigma) \frac{dN_d(\sigma)}{d\tau} d\sigma - \beta_1 |e(\sigma)||_{t_0}^{t}$$
$$= \; \int_{t_0}^{t} e(\sigma) \left(\alpha_r N_d(\sigma) - \frac{dN_d(\sigma)}{d\tau} - \alpha_r \beta_1 sgn(e(\sigma))\right) d\sigma$$
$$+ e(t)N_d(t) - e(t_0)N_d(t_0) - \beta_1 |e(t)| + \beta_1 |e(t_0)|.$$

The expression in (B.73) can be upper bounded as follows

$$\int_{t_0}^{t} L_1(\sigma)d\sigma \leq \int_{t_0}^{t} |e(\sigma)| \left(\alpha_r |N_d(\sigma)| + \left|\frac{dN_d(\sigma)}{d\tau}\right| - \alpha_r \beta_1\right) d\sigma$$
$$+ |e(t)| \left(|N_d(t)| - \beta_1\right) \tag{B.74}$$
$$+ \beta_1 |e(t_0)| - e(t_0)N_d(t_0).$$

If β_1 is chosen according to (5.27), then the first inequality in (5.28) can be proven from (B.74). The second inequality in (5.28) can be obtained by integrating the expression for $L_2(t)$, introduced in (5.26), as follows

$$\int_{t_0}^{t} L_2(\sigma)d\sigma \; = \; \int_{t_0}^{t} (-\beta_2 \dot{e} sgn(e)) \, d\sigma \tag{B.75}$$
$$= \; \beta_2 |e(t_0)| - \beta_2 |e(t)| \leq \beta_2 |e(t_0)|.$$

B.3.3   Definitions from Section 5.3.2

The explicit definitions for $Y_1\left(\cdot\right)$, $Y_2\left(\cdot\right)$, ϕ_1, and ϕ_2 are given as follows

$$Y_1\left(\theta_1,\dot{\theta}_1,\tau_1,\ddot{\theta}_{d1},\dot{\theta}_{d1}\right) = \left[\begin{array}{ccc} Y_{N1} & -\tau_1 & \ddot{\theta}_{d1}+\mu_1\dot{e}_1 \end{array}\right]$$

$$\phi_1 = \left[\begin{array}{ccc} \phi_{N1} & \alpha_1 & I_1 \end{array}\right]^T$$

$$Y_2\left(\theta_1,\dot{\theta}_1,\theta_2,\dot{\theta}_2,\tau_1,\tau_2,T_1\right) = \left[\begin{array}{ccccc} -Y_{N1} & \tau_1 & T_1 & Y_{N2} & -\tau_2 & \mu_2\dot{e}_2 \end{array}\right]$$

$$\phi_2 = \left[\begin{array}{ccccc} \dfrac{I_2}{I_1}\phi_{N1} & \dfrac{I_2}{I_1}\alpha_1 & \dfrac{I_2}{I_1} & \phi_{N2} & \alpha_2 & I_2 \end{array}\right]^T$$

where Remark 5.5 from Chapter 5 has been utilized.

B.3.4   Upperbound for $V_{a1}\left(t\right)$

Equation (5.101) can be integrated and rewritten as

$$V_{a1}\left(t\right) - V_{a1}\left(t_0\right) = -K_s\int_{t_0}^{t} s_1^2\left(\sigma\right) d\sigma$$

$$+\int_{t_0}^{t} p_1\left(\sigma\right)\left(-\eta_1\left(\sigma\right)-\rho_1 sgn\left(p_1\left(\sigma\right)\right)\right) d\sigma$$

$$+\left[\int_{t_0}^{t} \frac{dp_1\left(\sigma\right)}{d\sigma}\left(-\eta_1\left(\sigma\right)-\rho_1 sgn\left(p_1\left(\sigma\right)\right)\right) d\sigma\right].$$

$$\text{(B.76)}$$

The bracketed term in this expression is further evaluated as follows

$$\int_{t_0}^{t} \frac{dp_1\left(\sigma\right)}{d\sigma}\left(-\eta_1\left(\sigma\right)-\rho_1 sgn\left(p_1\left(\sigma\right)\right)\right) d\sigma = \left[-\int_{t_0}^{t} \frac{dp_1\left(\sigma\right)}{d\sigma}\eta_1\left(\sigma\right) d\sigma\right]$$

$$-\rho_1\int_{t_0}^{t} \frac{dp_1\left(\sigma\right)}{d\sigma}sgn\left(p_1\left(\sigma\right)\right) d\sigma.$$

$$\text{(B.77)}$$

After integrating the bracketed term in (B.77) by parts, we have

$$
\int_{t_0}^{t} \frac{dp_1(\sigma)}{d\sigma} \left(-\eta_1(\sigma) - \rho_1 sgn\left(p_1(\sigma)\right)\right) d\sigma = -\left[\eta_1(\sigma) p_1(\sigma)\big|_{t_0}^{t}\right.
$$

$$
\left. -\int_{t_0}^{t} p_1(\sigma) \frac{d\eta_1(\sigma)}{d\sigma} d\sigma \right]
$$

$$
-\rho_1 \left|p_1(\sigma)\right|\big|_{t_0}^{t}.
$$
(B.78)

After substituting (B.78) into (B.76), the following expression is obtained

$$
V_{a1}(t) - V_{a1}(t_0) = -K_s \int_{t_0}^{t} s_1^2(\sigma) d\sigma
$$

$$
+ \int_{t_0}^{t} p_1(\sigma) \left(-\eta_1(\sigma) + \frac{d\eta_1(\sigma)}{d\sigma} - \rho_1 sgn\left(p_1(\sigma)\right)\right) d\sigma
$$

$$
-\eta_1(t) p_1(t) - \rho_1 \left|p_1(t)\right| + \eta_1(t_0) p_1(t_0) + \rho_1 \left|p_1(t_0)\right|
$$
(B.79)

which can be simplified as follows

$$
V_{a1}(t) - V_{a1}(t_0) \leq -K_s \int_{t_0}^{t} s_1^2(\sigma) d\sigma
$$

$$
+ \int_{t_0}^{t} \left|p_1(\sigma)\right| \left(\left|\eta_1(\sigma)\right| + \left|\frac{d\eta_1(\sigma)}{d\sigma}\right| - \rho_1\right) d\sigma
$$

$$
+ \left|p_1(t)\right| \left(\left|\eta_1(t)\right| - \rho_1\right) + \eta_1(t_0) p_1(t_0) + \rho_1 \left|p_1(t_0)\right|
$$
(B.80)

where the following equality has been used

$$
p_1(\sigma) \, sgn\left(p_1(\sigma)\right) = \left|p_1(\sigma)\right|.
$$
(B.81)

B.3.5   *Upper Bound Development for MIF Analysis*

To simplify the following derivations, (5.143) can be rewritten as follows

$$
\begin{aligned}
N \triangleq{} & N(x, \dot{x}, \ddot{x}, e_1, e_2, r, \dddot{x}_d) = \bar{M} \dddot{x}_d \\
& + \dot{\bar{M}} \ddot{x} + \tfrac{d}{dt}\left[\bar{C}\dot{x} + \bar{B}\dot{x}\right] + e_2 \\
& + \bar{M}\left(\alpha_1 + \alpha_2\right) r - \bar{M}\left(\alpha_1^2 + \alpha_1\alpha_2 + \alpha_2^2\right) e_2 \\
& + \bar{M}\alpha_2^3 e_1 + \tfrac{1}{2}\dot{\bar{M}}r
\end{aligned}
$$
(B.82)

where (5.132) and (5.133) were utilized. To further facilitate the subsequent analysis, the terms $N(x, \dot{x}_d, \ddot{x}_d, 0, 0, 0, \dddot{x}_d)$, $N(x, \dot{x}, \ddot{x}_d, 0, 0, 0, \dddot{x}_d)$,

$N(x, \dot{x}, \ddot{x}, 0, 0, 0, \dddot{x}_d)$, $N(x, \dot{x}, \ddot{x}, e_1, 0, 0, \dddot{x}_d)$ and $N(x, \dot{x}, \ddot{x}, e_1, e_2, 0, \dddot{x}_d)$ are added and subtracted to the right-hand side of (5.142) as follows

$$
\begin{aligned}
\tilde{N} = \ & [N(x, \dot{x}_d, \ddot{x}_d, 0, 0, 0, \dddot{x}_d) - N_d(x_d, \dot{x}_d, \ddot{x}_d, 0, 0, 0, \dddot{x}_d)] \\
& + [N(x, \dot{x}, \ddot{x}_d, 0, 0, 0, \dddot{x}_d) - N(x, \dot{x}_d, \ddot{x}_d, 0, 0, 0, \dddot{x}_d)] \\
& + [N(x, \dot{x}, \ddot{x}, 0, 0, 0, \dddot{x}_d) - N(x, \dot{x}, \ddot{x}_d, 0, 0, 0, \dddot{x}_d)] \\
& + [N(x, \dot{x}, \ddot{x}, e_1, 0, 0, \dddot{x}_d) - N(x, \dot{x}, \ddot{x}, 0, 0, 0, \dddot{x}_d)] \\
& + [N(x, \dot{x}, \ddot{x}, e_1, e_2, 0, \dddot{x}_d) - N(x, \dot{x}, \ddot{x}, e_1, 0, 0, \dddot{x}_d)] \\
& + [N(x, \dot{x}, \ddot{x}, e_1, e_2, r, \dddot{x}_d) - N(x, \dot{x}, \ddot{x}, e_1, e_2, 0, \dddot{x}_d)]
\end{aligned}
\tag{B.83}
$$

After applying the Mean Value Theorem to each bracketed term of (B.83), the following expression can be obtained

$$
\begin{aligned}
\tilde{N} = \ & \left. \frac{\partial N(\sigma_1, \dot{x}_d, \ddot{x}_d, 0, 0, 0, \dddot{x}_d)}{\partial \sigma_1} \right|_{\sigma_1 = v_1} (x - x_d) \\
& + \left. \frac{\partial N(x, \sigma_2, \ddot{x}_d, 0, 0, 0, \dddot{x}_d)}{\partial \sigma_2} \right|_{\sigma_2 = v_2} (\dot{x} - \dot{x}_d) \\
& + \left. \frac{\partial N(x, \dot{x}, \sigma_3, 0, 0, 0, \dddot{x}_d)}{\partial \sigma_3} \right|_{\sigma_3 = v_3} (\ddot{x} - \ddot{x}_d) \\
& + \left. \frac{\partial N(x, \dot{x}, \ddot{x}, \sigma_4, 0, 0, \dddot{x}_d)}{\partial \sigma_4} \right|_{\sigma_4 = v_4} (e_1 - 0) \\
& + \left. \frac{\partial N(x, \dot{x}, \ddot{x}, e_1, \sigma_5, 0, \dddot{x}_d)}{\partial \sigma_5} \right|_{\sigma_5 = v_5} (e_2 - 0) \\
& + \left. \frac{\partial N(x, \dot{x}, \ddot{x}, e_1, e_2, \sigma_6, \dddot{x}_d)}{\partial \sigma_6} \right|_{\sigma_6 = v_6} (r - 0)
\end{aligned}
\tag{B.84}
$$

where $v_1 \in (x_d, x)$, $v_2 \in (\dot{x}_d, \dot{x})$, $v_3 \in (\ddot{x}_d, \ddot{x})$, $v_4 \in (0, e_1)$, $v_5 \in (0, e_2)$, and $v_6 \in (0, r)$. The right-hand side of (B.84) can be upper bounded as follows

$$
\begin{aligned}
\tilde{N} \le \ & \left\| \left. \frac{\partial N(\sigma_1, \dot{x}_d, \ddot{x}_d, 0, 0, 0, \dddot{x}_d)}{\partial \sigma_1} \right|_{\sigma_1 = v_1} \right\| \|e_1\| \\
& + \left\| \left. \frac{\partial N(x, \sigma_2, \ddot{x}_d, 0, 0, 0, \dddot{x}_d)}{\partial \sigma_2} \right|_{\sigma_2 = v_2} \right\| \|\dot{e}_1\| \\
& + \left\| \left. \frac{\partial N(x, \dot{x}, \sigma_3, 0, 0, 0, \dddot{x}_d)}{\partial \sigma_3} \right|_{\sigma_3 = v_3} \right\| \|\ddot{e}_1\| \\
& + \left\| \left. \frac{\partial N(x, \dot{x}, \ddot{x}, \sigma_4, 0, 0, \dddot{x}_d)}{\partial \sigma_4} \right|_{\sigma_4 = v_4} \right\| \|e_1\| \\
& + \left\| \left. \frac{\partial N(x, \dot{x}, \ddot{x}, e_1, \sigma_5, 0, \dddot{x}_d)}{\partial \sigma_5} \right|_{\sigma_5 = v_5} \right\| \|e_2\| \\
& + \left\| \left. \frac{\partial N(x, \dot{x}, \ddot{x}, e_1, e_2, \sigma_6, \dddot{x}_d)}{\partial \sigma_6} \right|_{\sigma_6 = v_6} \right\| \|r\| \, .
\end{aligned}
\tag{B.85}
$$

The partial derivatives in (B.85) can be calculated from (B.82) as

$$
\begin{aligned}
\frac{\partial N(\sigma_1, \dot{x}_d, \ddot{x}_d, 0, 0, 0, \dddot{x}_d)}{\partial \sigma_1} =\ & \frac{\partial \bar{M}(\sigma_1)}{\partial \sigma_1} \dddot{x}_d \\
& + \frac{\partial \dot{\bar{M}}(\sigma_1, \dot{x}_d)}{\partial \sigma_1} \ddot{x}_d \\
& + \frac{\partial \dot{\bar{C}}(\sigma_1, \dot{x}_d, \ddot{x}_d)}{\partial \sigma_1} \dot{x}_d \\
& + \frac{\partial \bar{C}(\sigma_1, \dot{x}_d)}{\partial \sigma_1} \ddot{x}_d
\end{aligned}
\tag{B.86}
$$

$$
\begin{aligned}
\frac{\partial N(x, \sigma_2, \ddot{x}_d, 0, 0, 0, \dddot{x}_d)}{\partial \sigma_2} =\ & \frac{\partial \dot{\bar{M}}(x, \sigma_2)}{\partial \sigma_2} \ddot{x}_d \\
& + \frac{\partial \dot{\bar{C}}(x, \sigma_2, \ddot{x}_d)}{\partial \sigma_2} \sigma_2 \\
& + \dot{\bar{C}}(x, \sigma_2, \ddot{x}_d) \\
& + \frac{\partial \bar{C}(x, \sigma_2)}{\partial \sigma_2} \ddot{x}_d
\end{aligned}
\tag{B.87}
$$

$$
\begin{aligned}
\frac{\partial N(x, \dot{x}, \sigma_3, 0, 0, 0, \dddot{x}_d)}{\partial \sigma_3} =\ & \dot{\bar{M}}(x, \dot{x}) + \frac{\partial \dot{\bar{C}}(x, \dot{x}, \sigma_3)}{\partial \sigma_3} \dot{x} \\
& + \bar{C}(x, \dot{x}) + \bar{B}
\end{aligned}
\tag{B.88}
$$

$$
\frac{\partial N(x, \dot{x}, \ddot{x}, \sigma_4, 0, 0, \dddot{x}_d)}{\partial \sigma_4} = \alpha_2^3 \bar{M}(x)
\tag{B.89}
$$

$$
\begin{aligned}
\frac{\partial N(x, \dot{x}, \ddot{x}, e_1, \sigma_5, 0, \dddot{x}_d)}{\partial \sigma_5} =\ & 1 - \alpha_1^2 \bar{M}(x) - \alpha_1 \alpha_2 \bar{M}(x) \\
& - \alpha_2^2 \bar{M}(x)
\end{aligned}
\tag{B.90}
$$

$$
\begin{aligned}
\frac{\partial N(x, \dot{x}, \ddot{x}, e_1, e_2, \sigma_6, \dddot{x}_d)}{\partial \sigma_6} =\ & (\alpha_1 + \alpha_2)\, \bar{M}(x) \\
& + \tfrac{1}{2} \dot{\bar{M}}(x, \dot{x}).
\end{aligned}
\tag{B.91}
$$

By noting that

$$
\begin{aligned}
v_1 &= x - c_1 (x - x_d) & v_2 &= \dot{x} - c_2 (\dot{x} - \dot{x}_d) \\
v_3 &= \ddot{x} - c_3 (\ddot{x} - \ddot{x}_d) & v_4 &= e_1 - c_4 (e_1 - 0) \\
v_5 &= e_2 - c_5 (e_2 - 0) & v_6 &= r - c_6 (r - 0)
\end{aligned}
$$

where $c_i \in (0, 1)\ \forall i = 1, 2, ..., 6$, if the assumptions stated for the system model and the desired trajectory are met, an upper bound for the right-hand side of (B.86)–(B.91) can be written as follows

$$\left\| \frac{\partial N(\sigma_1, \dot{x}_d, \ddot{x}_d, 0, 0, 0, \dddot{x}_d)}{\partial \sigma_1} \right|_{\sigma_1 = v_1} \right\| \leqslant \rho_1(x, \dot{x}, \ddot{x}) \qquad \text{(B.92)}$$

$$\left\| \frac{\partial N(x, \sigma_2, \ddot{x}_d, 0, 0, 0, \dddot{x}_d)}{\partial \sigma_2} \right|_{\sigma_2 = v_2} \right\| \leqslant \rho_2(x, \dot{x}, \ddot{x})$$

$$\left\| \frac{\partial N(x, \dot{x}, \sigma_3, 0, 0, 0, \dddot{x}_d)}{\partial \sigma_3} \right|_{\sigma_3 = v_3} \right\| \leqslant \rho_3(x, \dot{x})$$

$$\left\| \frac{\partial N(x, \dot{x}, \ddot{x}, \sigma_4, 0, 0, \dddot{x}_d)}{\partial \sigma_4} \right|_{\sigma_4 = v_4} \right\| \leqslant \rho_4(x)$$

$$\left\| \frac{\partial N(x, \dot{x}, \ddot{x}, e_1, \sigma_5, 0, \dddot{x}_d)}{\partial \sigma_5} \right|_{\sigma_5 = v_5} \right\| \leqslant \rho_5(x)$$

$$\left\| \frac{\partial N(x, \dot{x}, \ddot{x}, e_1, e_2, \sigma_6, \dddot{x}_d)}{\partial \sigma_6} \right|_{\sigma_6 = v6} \right\| \leqslant \rho_6(x, \dot{x})$$

where $\rho_i(\cdot)$ $\forall i = 1, 2, ..., 6$, are positive nondecreasing functions of $x(t)$, $\dot{x}(t)$, and $\ddot{x}(t)$. After substituting (B.92) into (B.85), $\tilde{N}(\cdot)$ can be expressed as

$$\begin{aligned} \tilde{N} \leqslant \ & (\rho_1(\|e_1\|, \|e_2\|, \|r\|) + \rho_4(\|e_1\|)) \, \|e_1\| \qquad \text{(B.93)} \\ & + (\rho_2(\|e_1\|, \|e_2\|, \|r\|)) \, \|\dot{e}_1\| \\ & + (\rho_3(\|e_1\|, \|e_2\|)) \, \|\ddot{e}_1\| \\ & + (\rho_5(\|e_1\|)) \, \|e_2\| \\ & + (\rho_6(\|e_1\|, \|e_2\|)) \, \|r\| \, . \end{aligned}$$

where (5.132)–(5.134) were utilized. The expressions in (5.132) and (5.145) can now be used to upper bound the right-hand side of (B.93) as in (5.146).

B.3.6   Teleoperator – Proof of MIF Controller Stability

Before we present the proof of the main result, we state and prove a preliminary Lemma.

Lemma B.5   *Let the auxiliary functions $L_1(t), L_2(t) \in R$ be defined as follows*

$$\begin{aligned} L_1 &\triangleq r^T (N_d - \beta_1 sgn(e_2)) \qquad \text{(B.94)} \\ L_2 &\triangleq -\beta_2 \dot{e}_2^T sgn(e_2) \end{aligned}$$

where β_1 and β_2 were introduced in (5.147). Provided β_1 is selected according to the following sufficient condition

$$\beta_1 > \varsigma_1 + \frac{1}{\alpha_1}\varsigma_2 \tag{B.95}$$

where ς_1 and ς_2 are given in (5.150), and α_1 is introduced in (5.61), then

$$\int_{t_0}^{t} L_1(\tau)d\tau \le \xi_{b1} \qquad \int_{t_0}^{t} L_2(\tau)d\tau \le \xi_{b2} \tag{B.96}$$

where the positive constants $\xi_{b1}, \xi_{b2} \in R$ are defined as

$$\xi_{b1} \triangleq \beta_1 \sum_{i=1}^{2n} |e_{2i}(t_0)| - e_2^T(t_0)N_d(t_0) \tag{B.97}$$

$$\xi_{b2} \triangleq \beta_2 \sum_{i=1}^{2n} |e_{2i}(t_0)|.$$

Proof. After substituting (5.61) into (B.94) and then integrating, the following expression can be obtained

$$\int_{t_0}^{t} L_1(\tau)d\tau = \alpha_1 \int_{t_0}^{t} e_2^T(\tau)\left[N_d(\tau) - \beta_1 sgn(e_2(\tau))\right]d\tau \tag{B.98}$$

$$+ \int_{t_0}^{t} \frac{de_2^T(\tau)}{d\tau}N_d(\tau)d\tau - \beta_1 \int_{t_0}^{t} \frac{de_2^T(\tau)}{d\tau}sgn(e_2(\tau))d\tau.$$

After evaluating the second integral on the right side of (B.98) by parts and evaluating the third integral, the following expression is obtained

$$\int_{t_0}^{t} L_1 d\tau = \alpha_1 \int_{t_0}^{t} e_2^T\left(N_d - \frac{1}{\alpha_1}\frac{dN_d}{d\tau} - \beta_1 sgn(e_2)\right)d\tau$$

$$+ e_2^T(t)N_d(t) - \beta_1 \sum_{i=1}^{2n} |e_{2i}(t)| + \xi_{b1}. \tag{B.99}$$

The expression in (B.99) can be upper bounded as follows

$$\int_{t_0}^{t} L_1 d\tau \le \alpha_1 \int_{t_0}^{t} \sum_{i=1}^{2n} |e_{2i}(\tau)|\left(\left(|N_{d_i}(\tau)| + \frac{1}{\alpha_1}\left|\frac{dN_{d_i}(\tau)}{d\tau}\right|\right)\right. \tag{B.100}$$

$$\left. - \beta_1\right)d\tau + \sum_{i=1}^{2n} |e_{2i}(t)|\left(|N_{d_i}(t)| - \beta_1\right) + \xi_{b1}.$$

If β_1 is chosen according to (B.95), then the first inequality in (B.96) can be proven from (B.100). The second inequality in (B.99) can be obtained by integrating the expression for $L_2(t)$ introduced in (B.94) as follows

$$\int_{t_0}^{t} L_2(\tau)d\sigma \;=\; -\beta_2 \int_{t_0}^{t} \dot{e}_2^{T}(\tau)sgn(e_2(\tau))d\tau \qquad \text{(B.101)}$$

$$=\; \xi_{b2} - \beta_2 \sum_{i=1}^{2n} |e_{2i}(t)| \le \xi_{b2}. \quad \blacksquare$$

We now proceed to present the proof for the main result.

Proof. Let the auxiliary functions $P_1(t)$, $P_2(t) \in R$ be defined as follows

$$P_1(t) \triangleq \xi_{b1} - \int_{t_0}^{t} L_1(\tau)d\tau \ge 0 \qquad \text{(B.102)}$$

$$P_2(t) \triangleq \xi_{b2} - \int_{t_0}^{t} L_2(\tau)d\tau \ge 0 \qquad \text{(B.103)}$$

where $\xi_{b1}, L_1(t), \xi_{b2}$, and $L_2(t)$ were defined in (B.94) and (B.97). The results from Lemma B.5 can be used to show that $P_1(t)$ and $P_2(t)$ are non-negative. Let $V(y,t) \in R$ denote the following nonnegative function

$$V \triangleq \frac{1}{2} e_1^{T} e_1 + \frac{1}{2} e_2^{T} e_2 + \frac{1}{2} r^{T} \bar{M} r + P_1 + P_2 \qquad \text{(B.104)}$$

where $y(t) \in R^{6n+2}$

$$y(t) \triangleq \begin{bmatrix} z^{T} & \sqrt{P_1} & \sqrt{P_2} \end{bmatrix}^{T} \qquad \text{(B.105)}$$

where the composite vector $z(\cdot) \in R^{6n}$ has been defined in (5.145). Note that (B.104) is bounded by

$$W_1(y) \le V(y,t) \le W_2(y) \qquad \text{(B.106)}$$

where

$$W_1(y) = \lambda_1 \|y(t)\|^2 \quad W_2(y) = \lambda_2 \|y(t)\|^2 \qquad \text{(B.107)}$$

where $\lambda_1 \triangleq \frac{1}{2} \min\{1, \bar{m}_1\}$ and $\lambda_2 \triangleq \max\{1, \frac{1}{2}\bar{m}_2\}$ where $\bar{m}_1$ and $\bar{m}_2$ were introduced in (5.131).

After taking the time derivative of (B.104), the following expression can be obtained

$$\dot{V} \;=\; -\alpha_2 e_1^{T} e_1 - \alpha_1 e_2^{T} e_2 - r^{T}(k_s + 1)r \qquad \text{(B.108)}$$
$$+ e_1^{T} e_2 + r^{T} \tilde{N} - r^{T} \beta_2 sgn(e_2) + \beta_2 \dot{e}_2^{T} sgn(e_2)$$

where (5.132), (5.133), (5.149), (B.102), and (B.103) were utilized. By utilizing the inequality of (5.61), 5.146, and the triangle inequality, $\dot{V}(t)$ can be upper bounded as follows

$$\dot{V} \;\le\; -\alpha_2 e_1^{T} e_1 - \alpha_1 e_2^{T} e_2 - r^{T}(k_s + 1)r \qquad \text{(B.109)}$$
$$+ e_1^{T} e_1 + e_2^{T} e_2 + \rho(\|z\|)\|r\|\|z\| - \alpha_1 e_2^{T} \beta_2 sgn(e_2).$$

By utilizing (5.145), $\dot{V}(t)$ of (B.109) can be upper bounded as follows

$$\dot{V} \leq -\lambda_3 \|z\|^2 - k_s \|r\|^2 + \rho(\|z\|) \|r\| \|z\| - \alpha_1 \beta_2 \sum_{i=1}^{2n} |e_{2i}| \qquad (B.110)$$

where $\lambda_3 \triangleq \min\{\alpha_1 - 1, \alpha_2 - 1, 1\}$. After completing the squares for the second and third term on the right side of (B.110), the following expression can be obtained

$$\dot{V} \leq -\left(\lambda_3 - \frac{\rho^2(\|z\|)}{4k_s}\right) \|z\|^2 - \alpha_1 \beta_2 \sum_{i=1}^{2n} |e_{2i}|. \qquad (B.111)$$

Provided α_1 and α_2 are selected to be greater than 2 and k_s is selected according to the following sufficient condition

$$k_s \geq \frac{\rho^2(\|z\|)}{4\lambda_3} \text{ or } \|z\| \leq \rho^{-1}\left(2\sqrt{k_s \lambda_3}\right), \qquad (B.112)$$

the following inequality can be developed

$$\dot{V} \leq W(y) - \alpha_1 \beta_2 \sum_{i=1}^{2n} |e_{2i}| \qquad (B.113)$$

where $W(y) \in R$ denotes the following nonpositive function

$$W(y) \triangleq -\beta_0 \|z\|^2 \qquad (B.114)$$

with $\beta_0 \in R$ being a positive constant. From (B.104)–(B.107) and (B.111)–(B.114), the regions D and S can be defined as follows

$$D \triangleq \left\{y \in \mathbb{R}^{6n+2} \mid \|y\| < \rho^{-1}\left(2\sqrt{k_s \lambda_3}\right)\right\} \qquad (B.115)$$

$$S \triangleq \left\{y \in D \mid W_2(y) < \lambda_1 \left(\rho^{-1}\left(2\sqrt{k_s \lambda_3}\right)\right)^2\right\}. \qquad (B.116)$$

The region of attraction in (B.116) can be made arbitrarily large to include any initial conditions by increasing the control gain k_s (i.e., a semi-global stability result). Specifically, (B.107) and the region defined in (B.116) can be used to calculate the region of attraction as follows

$$W_2(y(t_0)) \quad < \quad \lambda_1 \left(\rho^{-1}\left(2\sqrt{k_s \lambda_3}\right)\right)^2 \qquad (B.117)$$

$$\implies \quad \|y(t_0)\| < \sqrt{\frac{\lambda_1}{\lambda_2}} \rho^{-1}\left(2\sqrt{k_s \lambda_3}\right),$$

which can be rearranged as

$$k_s \geq \frac{1}{4\lambda_3} \rho^2 (\sqrt{\frac{\lambda_2}{\lambda_1}} \|y(t_0)\|). \tag{B.118}$$

By using (5.145), (B.97), and (B.105) an explicit expression for $\|y(t_0)\|$ can be written as

$$\|y(t_0)\|^2 = \|e_1(t_0)\|^2 + \|e_2(t_0)\|^2 \tag{B.119}$$
$$+ \|r(t_0)\|^2 + \xi_{b1} + \xi_{b2}.$$

Hereafter, we restrict the analysis to be valid for all initial conditions $y(t_0) \in \mathcal{S}$. From (B.104), (B.113), and (B.116)–(B.118), it is clear that $V(y,t) \in L_\infty$; hence $e_1(t)$, $e_2(t)$, $r(t)$, $z(t)$, $y(t) \in L_\infty$ $\forall y(t_0)$. From (B.113) it is easy to show that $e_2(t) \in L_1$. The fact that $e_2(t) \in L_1$ can be used along with (5.133) to determine that $e_1(t)$, $\dot{e}_1(t) \in L_1$. From (5.123), (5.134) and the assumption that $q_d(t) \in L_\infty$, it is clear that $x(t)$, $q(t) \in L_\infty$. From (5.132) and (5.133) it is also clear that $\dot{e}_2(t)$, $\dot{e}_1(t) \in L_\infty$. Using these boundedness statements, it is clear that both $\bar{u}(t) \in L_\infty$. From the time derivative of (5.133), and using the assumption that $\ddot{q}_d(t) \in L_\infty$ along with (5.139), it is clear that $\bar{u}(t) \in L_\infty$. The previous boundedness statements can be used along with (5.149), (5.146), and Remark 5.10 to prove that $\dot{r}(t) \in L_\infty$. These bounding statements can be used along with the time derivative of (B.114) to prove that $\dot{W}(y(t)) \in L_\infty$; hence, $W(y(t))$ is uniformly continuous. Standard signal chasing arguments can be used to prove all remaining signals are bounded. A direct application of Theorem 8.4 in [7] can now be used to prove that $\|z(t)\| \to 0$ as $t \to \infty$. From (5.145), it is also clear that $\|r(t)\| \to 0$ as $t \to \infty$. Based on the definitions given in (5.132)–(5.134), standard linear analysis tools can be used to prove that if $\|r(t)\| \to 0$ then $\|\dot{e}_2(t)\|$, $\|e_2(t)\|$, $\|\dot{e}_1(t)\|$, $\|e_1(t)\| \to 0$ as $t \to \infty$. Based on the definition of $x(t)$ in (5.123) and $e_1(t)$ in (5.134), it is clear that if $\|e_1(t)\| \to 0$ then $\|q_1(t) - q_2(t)\| \to 0$ and $q_1(t) + q_2(t) \to q_d(t)$. ■

B.3.7   Teleoperator – Proof of MIF Passivity

Proof. Let $V_p(t) \in R$ denote the following nonnegative, bounded function

$$V_p \triangleq \frac{1}{2} \dot{q}_d^T M_T \dot{q}_d + \frac{1}{2} q_d^T K_T q_d. \tag{B.120}$$

After taking the time derivative of (B.120), the following simplified expression can be obtained

$$\dot{V}_p = \dot{q}_d^T \bar{F}_2 - \dot{q}_d^T B_T \dot{q}_d \tag{B.121}$$

where (5.136) was utilized. Based on the fact that B_T is a constant positive definite, diagonal matrix, the following inequality can be developed

$$\dot{V}_p \leq \dot{q}_d^T \bar{F}_2. \tag{B.122}$$

After integrating both sides of (B.122), the following inequality can be developed

$$-c_2 \leq V_p(t) - V_p(t_0) \leq \int_{t_0}^t \dot{q}_d^T(\sigma) \bar{F}_2(\sigma) d\sigma \tag{B.123}$$

where $c_2 \in R$ is a positive constant (since $V_p(t)$ is bounded from the trajectory generation system in (5.136)).

By using the transformation in (5.123), the left side of (5.121) can be expressed as

$$\int_{t_0}^t \dot{q}^T(\tau) \begin{bmatrix} \gamma F_1(\tau) \\ F_2(\tau) \end{bmatrix} d\tau = \int_{t_0}^t \dot{x}^T \bar{F} d\tau. \tag{B.124}$$

By substituting the time derivative of (5.134) into (B.124), the following expression can be obtained

$$\int_{t_0}^t \dot{x}^T(\tau) \bar{F}(\tau) d\tau = \int_{t_0}^t \dot{q}_d^T(\tau) \bar{F}_2(\tau) d\tau - \int_{t_0}^t \dot{e}_1^T(\tau) \bar{F}(\tau) d\tau \tag{B.125}$$

where (5.135) was utilized. Based on (B.123), it is clear that $\int_{t_0}^t \dot{q}_d^T(\tau) \bar{F}_2(\tau) d\tau$ is lower bounded by $-c_2$, where c_2 was defined as a positive constant. The fact that $\dot{e}_1(t) \in L_1$ from the proof for Theorem 5.7 and the assumption that $\bar{F}(t) \in L_\infty$ can be used to show that the second integral of (B.125) is bounded. Hence, these facts can be applied to (B.124) and (B.125) to prove that

$$\int_{t_0}^t \dot{q}^T(\tau) \begin{bmatrix} \gamma F_1(\tau) \\ F_2(\tau) \end{bmatrix} d\tau \geq -c_3^2 \tag{B.126}$$

where $c_3 \in R$ is a positive constant. This proves that the teleoperator system is passive with respect to the scaled user and environmental power. ∎

B.3.8 Teleoperator – Proof of UMIF Desired Trajectory Boundedness

Proof. Let $V_1(t) \in R$ denote the following nonnegative function

$$V_1 \triangleq \frac{1}{2} e_2^T e_2 + \frac{1}{2} r^T r + P_1 + P_2. \tag{B.127}$$

Based on (B.127) and the closed loop error systems in (5.174), the proof of Theorem 5.3 can be followed directly to prove that $e_1(t)$, $e_2(t)$, $r(t)$, $\hat{F}(t)$,

$\overset{\cdot}{\hat{F}}(t) \in L_\infty$ as well as that $e_1(t)$, $e_2(t)$, and $r(t) \to 0$ as $t \to \infty$ regardless of whether or not $x_d(t)$, $\dot{x}_d(t)$, $\ddot{x}_d(t) \in L_\infty$. Therefore, the fact that $\hat{F}(t) \in L_\infty$ can be used in the subsequent analysis. As a means to prove that $x_d(t)$, $\dot{x}_d(t)$, $\ddot{x}_d(t) \in L_\infty$, let $V_2(t) \in R$ denote the following nonnegative function

$$V_2 \triangleq V_3 + L \tag{B.128}$$

where $V_3(t) \in R$ denotes the following nonnegative function

$$V_3 \triangleq \frac{1}{2}\dot{x}_d^T \bar{M}\dot{x}_d + \frac{1}{2}x_d^T K_T x_d \tag{B.129}$$

where $x_d(t)$, $\dot{x}_d(t)$ were defined in (5.158), where K_T was defined in (5.157), and $\bar{M}(x)$ was defined in (5.126). The expression given in (B.129) can be lower bounded by the auxiliary function, $L(\bar{x}) \in R$, defined as follows

$$L \triangleq \frac{2\varepsilon\dot{x}_d^T \bar{M}x_d}{1 + 2x_d^T x_d} \leq V_3(t) \tag{B.130}$$

where $\varepsilon \in R$ is a positive bounding constant selected according to the following inequality

$$0 < \varepsilon < \frac{\min\{\bar{m}_1, \lambda_{\min}\{K_T\}\}}{2m_{L\infty}} \tag{B.131}$$

where $\lambda_{\min}\{K_T\} \in R$ denotes the minimum eigenvalue of K_T, $\bar{m}_1$ was defined in (5.131) and $m_{L\infty} \in R$ denotes the induced infinity norm of the bounded matrix $\bar{M}(x)$. >From (B.130) it is clear that $V_2(t)$ is a nonnegative function. Also, $\bar{x}(t) \in R^{4n}$ is defined as

$$\bar{x} \triangleq [\ x_d^T \quad \dot{x}_d^T\ ]^T. \tag{B.132}$$

The expression in (B.128) satisfies the following inequalities

$$\bar{\lambda}_1 \|\bar{x}\|^2 \leq V_2(\bar{x}) \leq \bar{\lambda}_2 \|\bar{x}\|^2 \tag{B.133}$$

where $\bar{\lambda}_1$, $\bar{\lambda}_2 \in R$ are positive constants defined as follows, provided ε is selected sufficiently small

$$\bar{\lambda}_1 \triangleq \frac{1}{2}\min\{\bar{m}_1, \lambda_{\min}\{K_T\}\} - \varepsilon\xi_c \tag{B.134}$$

$$\bar{\lambda}_2 \triangleq \frac{1}{2}\max\{\bar{m}_2, \lambda_{\max}\{K_T\}\} + \varepsilon\xi_c$$

where $\bar{m}_1$ and $\bar{m}_2$ were introduced in (5.131), and $\lambda_{\max}\{K_T\} \in R$ denotes the maximum eigenvalue of K_T. In (B.134), $\xi_c \in R$ is a positive constant defined as follows

$$\xi_c = \max\left\{\frac{2m_{L\infty}}{\delta_a}, 2m_{L\infty}\delta_a\right\} \tag{B.135}$$

where $\delta_a \in R$ is some positive constant, and $m_{L\infty}$ was introduced in (B.131).

To facilitate the subsequent analysis, the time derivative of (B.130) can be determined as follows

$$\dot{L} = \frac{2\varepsilon \ddot{x}_d^T \bar{M} x_d + 2\varepsilon \dot{x}_d^T \dot{\bar{M}} x_d + 2\varepsilon \dot{x}_d^T \bar{M} \dot{x}_d}{1 + 2x_d^T x_d}$$
$$- \frac{2\varepsilon \left(\dot{x}_d^T \bar{M} x_d \right) 4 x_d^T \dot{x}_d}{(1 + 2x_d^T x_d)^2}. \tag{B.136}$$

After utilizing (5.157), the expression in (B.136) can be written as

$$\dot{L} = -\frac{2\varepsilon x_d^T K_T x_d}{1 + 2x_d^T x_d} - \frac{2\varepsilon x_d^T B_T \dot{x}_d}{1 + 2x_d^T x_d} + \frac{2\varepsilon x_d^T \hat{F}}{1 + 2x_d^T x_d} \tag{B.137}$$
$$+ \frac{\varepsilon x_d^T \dot{\bar{M}} \dot{x}_d}{1 + 2x_d^T x_d} + \frac{2\varepsilon \dot{x}_d^T \bar{M} \dot{x}_d}{1 + 2x_d^T x_d} - \frac{2\varepsilon \left(\dot{x}_d^T \bar{M} x_d \right) 4 x_d^T \dot{x}_d}{\left(1 + 2x_d^T x_d \right)^2}.$$

The signal in (B.137) can be upper bounded as follows

$$\dot{L} \leq -\frac{2\varepsilon \lambda_{\min}\{K_T\}}{1 + 2x_d^T x_d} \|x_d\|^2 + \frac{2\varepsilon \lambda_{\max}\{B_T\}}{1 + 2x_d^T x_d} \left[\|x_d\|^2 + \|\dot{x}_d\|^2 \right]$$
$$+ \frac{2\varepsilon}{1 + 2x_d^T x_d} \left[\delta_2 \|x_d\|^2 + \frac{1}{\delta_2} \left\| \hat{F} \right\|^2 \right] + \varepsilon \xi_3 \xi_{\bar{m}} \|\dot{x}_d\|^2 \tag{B.138}$$
$$+ \varepsilon \xi_{\bar{m}} \xi_{\dot{e}} + \frac{\varepsilon \xi_{\bar{m}} \xi_{\dot{e}}}{1 + 2x_d^T x_d} \|\dot{x}_d\|^2 + \frac{2\varepsilon \bar{m}_2}{1 + 2x_d^T x_d} \|\dot{x}_d\|^2$$
$$+ 8\varepsilon \bar{m}_2 \|\dot{x}_d\|^2$$

where the following properties were utilized

$$-\frac{2\varepsilon x_d^T K_T x_d}{1 + 2x_d^T x_d} \leq -\frac{2\varepsilon \lambda_{\min}\{K_T\}}{1 + 2x_d^T x_d} \|x_d\|^2 \tag{B.139}$$

$$-\frac{2\varepsilon x_d^T B_T \dot{x}_d}{1 + 2x_d^T x_d} \leq \frac{2\varepsilon \lambda_{\max}\{B_T\}}{1 + 2x_d^T x_d} \left[\|x_d\|^2 + \|\dot{x}_d\|^2 \right] \tag{B.140}$$

$$\frac{2\varepsilon x_d^T \hat{F}}{1 + 2x_d^T x_d} \leq \frac{2\varepsilon}{1 + 2x_d^T x_d} \left[\delta_2 \|x_d\|^2 + \frac{1}{\delta_2} \left\| \hat{F} \right\|^2 \right] \tag{B.141}$$

$$\frac{\varepsilon x_d^T \dot{\bar{M}} \dot{x}_d}{1 + 2x_d^T x_d} \leq \varepsilon \xi_3 \xi_{\bar{m}} \|\dot{x}_d\|^2 + \varepsilon \xi_{\bar{m}} \xi_{\dot{e}} \tag{B.142}$$
$$+ \frac{\varepsilon \xi_{\bar{m}} \xi_{\dot{e}}}{1 + 2x_d^T x_d} \|\dot{x}_d\|^2$$

$$\frac{2\varepsilon \dot{x}_d^T \bar{M} \dot{x}_d}{1 + 2x_d^T x_d} \leq \frac{2\varepsilon \bar{m}_2}{1 + 2x_d^T x_d} \|\dot{x}_d\|^2 \tag{B.143}$$

$$-\frac{2\varepsilon \left(\dot{x}_d^T \bar{M} x_d\right) 4x_d^T \dot{x}_d}{\left(1 + 2x_d^T x_d\right)^2} \leq 8\varepsilon \bar{m}_2 \|\dot{x}_d\|^2 \tag{B.144}$$

$$\frac{\|x_d\|^2}{1 + 2x_d^T x_d} \leq 1 \tag{B.145}$$

$$\frac{\|x_d\|^2}{\left(1 + 2x_d^T x_d\right)^2} \leq 1. \tag{B.146}$$

In (B.141), $\delta_2 \in R$ denotes a positive bounding constant. In (B.142), $\xi_3 \in R$ denotes a positive bounding constant defined as

$$\frac{\|x_d\|}{1 + 2x_d^T x_d} \leq \xi_3 \tag{B.147}$$

and $\xi_{\bar{m}}$, $\xi_{\dot{e}} \in R$ denote positive bounding constants defined as

$$\left\|\dot{\bar{M}}\right\| \leq \xi_{\bar{m}} \left(\|\dot{x}_d\| + \xi_{\dot{e}}\right). \tag{B.148}$$

The inequality in (B.148) is obtained by using the facts that the inertia matrix is second order differentiable and that $e_1(t) \in L_\infty$, (see proof of Theorem 5.3). In (B.143) and (B.144), $\bar{m}_2 \in R$ is a positive constant defined in (5.131).

Based on the development in (B.136)–(B.146), the time derivative of (B.128) can be upper bounded as follows

$$\dot{V}_2 \leq -\lambda_{\min}\{B_T\} \|\dot{x}_d\|^2 - \frac{2\varepsilon \lambda_{\min}\{K_T\}}{1 + 2x_d^T x_d} \|x_d\|^2 \tag{B.149}$$

$$+\frac{2\varepsilon \lambda_{\max}\{B_T\}}{1 + 2x_d^T x_d} \left[\|x_d\|^2 + \|\dot{x}_d\|^2\right]$$

$$+\delta_1 \|\dot{x}_d\|^2 + \frac{1}{\delta_1} \left\|\hat{F}\right\|^2 + \frac{2\varepsilon}{1 + 2x_d^T x_d} \left[\delta_2 \|x_d\|^2 + \frac{1}{\delta_2} \left\|\hat{F}\right\|^2\right]$$

$$+\varepsilon \xi_3 \xi_{\bar{m}} \|\dot{x}_d\|^2 + \varepsilon \xi_{\bar{m}} \xi_{\dot{e}} + \frac{\varepsilon \xi_{\bar{m}} \xi_{\dot{e}}}{1 + 2x_d^T x_d} \|\dot{x}_d\|^2$$

$$+\frac{2\varepsilon \bar{m}_2}{1 + 2x_d^T x_d} \|\dot{x}_d\|^2 + 8\varepsilon \bar{m}_2 \|\dot{x}_d\|^2$$

where (5.157), (B.138), and the following inequalities were utilized

$$-\dot{x}_d^T B_T \dot{x}_d \leq -\lambda_{\min}\{B_T\} \|\dot{x}_d\|^2$$

$$\dot{x}_d^T \hat{F} \leq \delta_1 \|\dot{x}_d\|^2 + \frac{1}{\delta_1} \left\|\hat{F}\right\|^2$$

where $\delta_1 \in R$ denotes a positive bounding constant. The expression in (B.149) can be simplified as follows

$$
\dot{V}_2 \leq -\|\dot{x}_d\|^2 \left[\lambda_{\min}\{B_T\} - \delta_1 - \frac{2\varepsilon\lambda_{\max}\{B_T\}}{1 + 2x_d^T x_d} - \varepsilon\xi_3\xi_{\bar{m}} \right.
$$

$$
\left. - \frac{\varepsilon\xi_{\bar{m}}\xi_{\dot{e}}}{1 + 2x_d^T x_d} - \frac{2\varepsilon\bar{m}_2}{1 + 2x_d^T x_d} - 8\varepsilon\bar{m}_2 \right] \tag{B.150}
$$

$$
- \|x_d\|^2 \left[\frac{2\varepsilon\lambda_{\min}\{K_T\}}{1 + 2x_d^T x_d} - \frac{2\varepsilon\lambda_{\max}\{B_T\}}{1 + 2x_d^T x_d} - \frac{2\varepsilon\delta_2}{1 + 2x_d^T x_d} \right]
$$

$$
+ \left[\frac{1}{\delta_1} \left\| \hat{F} \right\|^2 + \left[\frac{2\varepsilon}{1 + 2x_d^T x_d} \right] \left[\frac{1}{\delta_2} \left\| \hat{F} \right\|^2 \right] + \varepsilon\xi_{\bar{m}}\xi_{\dot{e}} \right]. \tag{B.151}
$$

Provided $B_T, \delta_1, \delta_2, \varepsilon,$ and K_T are selected to satisfy the following sufficient conditions

$$
\lambda_{\min}\{B_T\} > \delta_1 + \varepsilon \left(2\lambda_{\max}\{B_T\} + \xi_3\xi_{\bar{m}} + \xi_{\bar{m}}\xi_{\dot{e}} + 10\bar{m}_2 \right)
$$

$$
\lambda_{\min}\{K_T\} > \lambda_{\max}\{B_T\} + \delta_2
$$

the expression in (B.151) can be upper bounded as follows

$$
\dot{V}_2 \leq -\frac{\min\{\gamma_a, \gamma_b\}}{\bar{\lambda}_2} V_3 + \epsilon_2 \tag{B.152}
$$

where (B.132) was utilized, and $\gamma_a, \gamma_b, \epsilon_2 \in R$ denote positive bounding constants.

From B.128)–(B.130), and (B.133), and that $\hat{F}(t) \in L_\infty$, the expression in (B.152) can be used with the result from [2] to prove that $\bar{x}(t), x_d(t),$ $\dot{x}_d(t) \in L_\infty$. Based on (5.157), and the fact that $\bar{M}(x), \dot{\bar{M}}(x, \dot{x}),$ and $\hat{F}(t) \in L_\infty$ then $\ddot{x}_d(t) \in L_\infty$. ∎

B.3.9   Teleoperator – Proof of UMIF Controller Stability

Before we present the proof of the main result, we state and prove a preliminary Lemma.

Lemma B.6 *Let the auxiliary functions $L_1(t), L_2(t) \in R$ be defined as follows*

$$
L_1 \triangleq -r^T \left(\dot{\tilde{F}} + \beta_1 sgn(e_2) \right) \tag{B.153}
$$

$$
L_2 \triangleq -\beta_2 \dot{e}_2^T sgn(e_2)
$$

where β_1 and β_2 are defined in (5.172). Provided β_1 is selected according to the following sufficient condition

$$
\beta_1 > \varsigma_3 + \varsigma_4, \tag{B.154}
$$

where ς_3 and ς_4 were introduced in (5.175), then

$$\int_{t_0}^t L_1(\tau)d\tau \le \xi_{b1} \qquad \int_{t_0}^t L_2(\tau)d\tau \le \xi_{b2} \qquad (B.155)$$

where the positive constants $\xi_{b1}, \xi_{b2} \in R$ are defined as

$$\xi_{b1} \triangleq \beta_1 \sum_{i=1}^{2n} |e_{2i}(t_0)| - e_2^T(t_0)\left(-\dot{F}(t_0)\right) \qquad \xi_{b2} \triangleq \beta_2 \sum_{i=1}^{2n} |e_{2i}(t_0)|. \qquad (B.156)$$

Proof. After substituting (5.162) into (B.153) and then integrating, the following expression can be obtained

$$\int_{t_0}^t L_1(\tau)d\tau = \int_{t_0}^t e_2^T(\tau)\left[-\dot{F}(\tau) - \beta_1 sgn(e_2(\tau))\right]d\tau \qquad (B.157)$$

$$+ \int_{t_0}^t \frac{de_2^T(\tau)}{d\tau}\left(-\dot{F}(\tau)\right)d\tau - \beta_1 \int_{t_0}^t \frac{de_2^T(\tau)}{d\tau}sgn(e_2(\tau))d\tau.$$

After evaluating the second integral on the right side of (B.157) by parts and evaluating the third integral, the following expression is obtained

$$\int_{t_0}^t L_1 d\tau = \int_{t_0}^t e_2^T(\tau)\left(-\dot{F}(\tau) + \ddot{F}(\tau) - \beta_1 sgn(e_2(\tau))\right)d\tau$$

$$- e_2^T(t)\dot{F}(t) - \beta_1 \sum_{i=1}^{2n} |e_{2i}(t)| + \xi_{b1}. \qquad (B.158)$$

The expression in (B.158) can be upper bounded as follows

$$\int_{t_0}^t L_1 d\tau \le \int_{t_0}^t \sum_{i=1}^{2n} |e_{2i}(\tau)|\left(\left|\dot{F}_i(\tau)\right| + \left|\ddot{F}_i(\tau)\right| - \beta_1\right)d\tau \qquad (B.159)$$

$$+ \sum_{i=1}^{2n} |e_{2i}(t)|\left(\left|\dot{F}_i(t)\right| - \beta_1\right) + \xi_{b1}.$$

If β_1 is chosen according to (B.154), then the first inequality in (B.155) can be proven from (B.159). The second inequality in (B.155) can be obtained by integrating the expression for $L_2(t)$ introduced in (B.153) as follows

$$\int_{t_0}^t L_2(\tau)d\sigma = -\beta_2 \int_{t_0}^t \dot{e}_2^T(\tau)sgn(e_2(\tau))d\tau \qquad (B.160)$$

$$= \xi_{b2} - \beta_2 \sum_{i=1}^{2n} |e_{2i}(t)| \le \xi_{b2}. \blacksquare$$

We now proceed to present the proof for the main result.

Proof. Let the auxiliary functions $P_1(t), P_2(t) \in R$ be defined as follows

$$P_1(t) \triangleq \xi_{b1} - \int_{t_0}^{t} L_1(\tau)d\tau \geq 0 \qquad (\text{B.161})$$

$$P_2(t) \triangleq \xi_{b2} - \int_{t_0}^{t} L_2(\tau)d\tau \geq 0 \qquad (\text{B.162})$$

where $\xi_{b1}, L_1(t), \xi_{b2}$, and $L_2(t)$ were defined in (B.153) and (B.156). The results from Lemma B.6 can be used to show that $P_1(t)$ and $P_2(t)$ are non-negative. Let $V_1(y,t) \in R$ denote the following nonnegative function

$$V_1 \triangleq \frac{1}{2}e_2^T e_2 + \frac{1}{2}r^T r + P_1 + P_2 \qquad (\text{B.163})$$

where $y(t) \in R^{4n+2}$ is defined as

$$y(t) \triangleq \begin{bmatrix} e_2^T & r^T & \sqrt{P_1} & \sqrt{P_2} \end{bmatrix}^T. \qquad (\text{B.164})$$

Note that (B.163) is bounded according to the following inequalities

$$W_3(y) \leq V_1(y,t) \leq W_4(y) \qquad (\text{B.165})$$

where

$$W_3(y) = \lambda_4 \|y(t)\|^2 \qquad W_4(y) = \lambda_5 \|y(t)\|^2 \qquad (\text{B.166})$$

where $\lambda_4, \lambda_5 \in R$ are positive bounding constants.

After taking the time derivative of (B.163), the following expression can be obtained

$$\dot{V}_1 = -e_2^T e_2 - k_s r^T r - \beta_2 e_2^T sgn(e_2) \qquad (\text{B.167})$$

where (5.162), (5.174), (B.161), and (B.162) were utilized. The expression in (B.167) can be rewritten as

$$\dot{V}_1 = -\|e_2\|^2 - k_s \|r\|^2 - \beta_2 \sum_{i=1}^{2n} |e_{2i}|. \qquad (\text{B.168})$$

From (B.163) and (B.168), it is clear that $V_1(y,t) \in L_\infty$; hence, $e_2(t) \in L_\infty \cap L_2 \cap L_1$, $r(t) \in L_\infty \cap L_2$, and $y(t) \in L_\infty$. Since $e_2(t), r(t) \in L_\infty$, (5.162) and (5.173) can be used to prove that $\dot{e}_2(t), \dot{\hat{F}}(t) \in L_\infty$. Given that $e_2(t), r(t), \dot{\hat{F}}(t) \in L_\infty$ and the assumption that $\dot{F} \in L_\infty$, (5.171) can be used to prove that $\dot{r}(t) \in L_\infty$. Barbalat's Lemma can be utilized to prove

$$\|e_2(t)\|, \|r(t)\| \to 0 \quad as \quad t \to \infty. \qquad (\text{B.169})$$

From (5.162), (5.163), (B.169) and the fact that $\bar{M}(x) \in L_\infty$, standard linear analysis arguments can be used to prove that $e_1(t)$, $\dot{e}_1(t)$, and $\dot{e}_2(t) \in L_\infty$, likewise that $e_1(t)$, $\dot{e}_1(t) \in L_1$, and that

$$\|e_1(t)\|, \|\dot{e}_1(t)\|, \|\dot{e}_2(t)\| \to 0 \quad as \quad t \to \infty. \tag{B.170}$$

From the fact that $\dot{e}_2(t) \in L_\infty$ and the assumption that $\bar{F} \in L_\infty$ it is clear from (5.169) that $\hat{F}(t) \in L_\infty$. Since $\hat{F}(t) \in L_\infty$, (5.157) and the proof in Section B.3.8 can be used to show that $x_d(t)$, $\dot{x}_d(t)$, $\ddot{x}_d(t) \in L_\infty$. Using these facts along with (5.134) and its first time derivative, it is clear that $x(t)$ and $\dot{x}(t) \in L_\infty$. Since $e_1(t)$, $\dot{e}_1(t)$, $\bar{M}(x)$, $\dot{\bar{M}}(x) \in L_\infty$, it is clear from (5.168) that $\bar{T}_1(t) \in L_\infty$, and using previously stated bounding properties, $\bar{T}(t) \in L_\infty$. It is also possible to state that $\bar{T}_1(t) \in L_1$, where (5.168) was utilized. Based on the definition of $x(t)$ in (5.160) and the previously stated bounding properties, it is clear that $\|q_1(t) - q_2(t)\| \to 0$ and $q_1(t) + q_2(t) \to q_d(t)$. From these bounding statements and standard signal chasing arguments, all signals can be shown to be bounded. ∎

B.3.10   Teleoperator – Proof of UMIF Passivity

Proof. Let $V_{p2}(t) \in R$ denote the following nonnegative, bounded function

$$V_{p2} \triangleq \frac{1}{2} \dot{x}_d^T \bar{M} \dot{x}_d + \frac{1}{2} x_d^T K_T x_d. \tag{B.171}$$

After taking the time derivative of (B.171), the following simplified expression can be obtained

$$\dot{V}_{p2} = \dot{x}_d^T \hat{F} - \dot{x}_d^T B_T \dot{x}_d \tag{B.172}$$

where (5.157) was utilized. Based on the fact that B_T is a constant positive definite, diagonal matrix, the following inequality can be developed

$$\dot{V}_{p2} \le \dot{x}_d^T \hat{F}. \tag{B.173}$$

The following inequality can be developed after integrating (B.173)

$$-c_4 \le V_{p2}(t) - V_{p2}(t_0) \le \int_{t_0}^{t} \dot{x}_d^T(\sigma) \hat{F}(\sigma) d\sigma \tag{B.174}$$

where $c_4 \in R$ is a positive constant (since $V_{p2}(t)$ is bounded from the trajectory generation system in (5.157)).

To facilitate the subsequent analysis, the following expression can be obtained from integration by parts

$$\int_{t_0}^{t} \bar{M} \ddot{e}_1(\tau) d\tau = \bar{M} \dot{e}_1(t) - \bar{M} \dot{e}_1(t_0) - \int_{t_0}^{t} \dot{\bar{M}} \dot{e}_1(\tau) d\tau. \tag{B.175}$$

Since $\bar{M}(x)$, $\dot{\bar{M}}(x, \dot{x})$, $\dot{e}_1(t) \in L_\infty$, and $\dot{e}_1(t) \in L_1$, then $\int_{t_0}^t \bar{M}\ddot{e}_1(\tau)d\tau \in L_\infty$. After integrating (5.166) as follows

$$\int_{t_0}^t \tilde{F}(\tau)d\tau = -\int_{t_0}^t \bar{M}\ddot{e}_1(\tau)d\tau - \int_{t_0}^t \tilde{T}_1(\tau)d\tau \qquad (B.176)$$

and using the fact that $\tilde{T}_1(t) \in L_1$ (see proof of Theorem 5.3) and the fact that $\int_{t_0}^t \bar{M}\ddot{e}_1(\tau)d\tau \in L_\infty$, it is clear that $\tilde{F} \in L_1$, where $\tilde{F}(t) \triangleq \bar{F}(t) - \hat{F}(t)$.

By using the transformation in (5.160), the expression in (5.121) can be rewritten as follows

$$\int_{t_0}^t \dot{q}^T(\tau) \begin{bmatrix} \gamma F_1(\tau) \\ F_2(\tau) \end{bmatrix} d\tau = \int_{t_0}^t \dot{x}^T \bar{F} d\tau - \int_{t_0}^t \begin{bmatrix} \dot{x}_{d1}^T & 0_n^T \end{bmatrix} \bar{F} d\tau. \qquad (B.177)$$

After substituting for the definition of $\tilde{F}(t)$ and the time derivative of (5.134) into (B.177) for $\bar{F}(t)$ and $\dot{x}(t)$, respectively, the following expression can be obtained

$$\begin{aligned}
\int_{t_0}^t \dot{x}^T \bar{F} d\tau - \int_{t_0}^t \begin{bmatrix} \dot{x}_{d1}^T & 0_n^T \end{bmatrix} \bar{F} d\tau &= \int_{t_0}^t \dot{x}_{d2}^T(\tau)\tilde{F}_2(\tau)d\tau \\
&+ \int_{t_0}^t \dot{x}_{d2}^T(\tau)\hat{F}_2(\tau)d\tau - \int_{t_0}^t \dot{e}_1^T(\tau)\tilde{F}(\tau)d\tau - \int_{t_0}^t \dot{e}_1^T(\tau)\hat{F}(\tau)d\tau.
\end{aligned} \qquad (B.178)$$

Since $\dot{x}_d(t) = \begin{bmatrix} \dot{x}_{d_1}^T(t) & \dot{x}_{d_2}^T(t) \end{bmatrix}^T \in L_\infty$ and $\tilde{F}(t) = \begin{bmatrix} \tilde{F}_1^T(t) & \tilde{F}_2^T(t) \end{bmatrix}^T \in L_1$, it is clear that the first integral expression in (B.178) is bounded and from (B.176) a lower negative bound exists. Based on (B.174), it is clear that the second integral expression in (B.178) is bounded and a lower negative bound exists. Since $\dot{e}_1(t) \in L_\infty$ and $\tilde{F}(t) \in L_1$, it is possible to show that the third integral in (B.178) is also bounded and a lower negative bound exists. Finally, because $\dot{e}_1(t) \in L_1$ and $\hat{F}(t) \in L_\infty$, it is possible to show that the fourth integral in (B.178) is also bounded and a lower negative bound exists. Hence, these facts can be applied to (B.177) and (B.178) to prove that

$$\int_{t_0}^t \dot{q}^T(\tau) \begin{bmatrix} \gamma F_1(\tau) \\ F_2(\tau) \end{bmatrix} d\tau \geq -c_5^2 \qquad (B.179)$$

where $c_5 \in R$ is a positive constant. ∎

B.3.11 Proof of Bound on $\tilde{N}$

We start by writing $\tilde{N}(t)$ from (5.208) and (5.210) as follows

$$\begin{aligned}
\tilde{N} = & \begin{bmatrix} \bar{M}(x) - \bar{M}(x_d) \end{bmatrix} \ddot{x}_d + \begin{bmatrix} \dot{\bar{M}}(x, \dot{x}) - \dot{\bar{M}}(x_d, \dot{x}_d) \end{bmatrix} \dot{x}_d \\
& + \begin{bmatrix} \dot{\bar{B}}(x, \dot{x}, \ddot{x}) - \dot{\bar{B}}(x_d, \dot{x}_d, \ddot{x}_d) \end{bmatrix} + \bar{M}(x)(\ddot{e}_1 + \dot{e}_2) \\
& + \dot{\bar{M}}(x, \dot{x})(\frac{1}{2}e_3 - \ddot{e}_1) + e_2.
\end{aligned} \qquad (B.180)$$

To simplify the notation, we define the following auxiliary functions

$$\Phi_{bf}(x, \dot{x}, \ddot{x}) \triangleq \dot{\bar{B}}(x, \dot{x}, \ddot{x})$$
$$\Phi_{mf}(x, \dot{x}, \ddot{x}_d) \triangleq \dot{\bar{M}}(x, \dot{x})\ddot{x}_d \tag{B.181}$$

$$E = \bar{M}(\cdot)\ddot{e}_1 + \dot{\bar{M}}(\cdot)\dot{e}_2 + e_2 + \dot{\bar{M}}(\cdot)\frac{1}{2}e_3 - \dot{\bar{M}}(\cdot)\ddot{e}_1. \tag{B.182}$$

From (5.200)–(5.202), it is possible to write

$$\dot{e}_1 = e_2 - e_1 \qquad \dot{e}_2 = e_3 - e_2 \qquad \ddot{e}_1 = e_3 - 2e_2 + e_1.$$

Given the definitions of (B.181) and (B.182), we can rewrite (B.180) by adding and subtracting a bevy of terms as follows

$$\begin{aligned}
\tilde{N} = \quad & [\bar{M}(x) - \bar{M}(x_d)]\,\ddot{x}_d + [\Phi_{mf}(x, \dot{x}, \ddot{x}_d) - \Phi_{mf}(x_d, \dot{x}, \ddot{x}_d)] \\
& + [\Phi_{mf}(x_d, \dot{x}, \ddot{x}_d) - \Phi_{mf}(x_d, \dot{x}_d, \ddot{x}_d)] + [\Phi_{bf}(x, \dot{x}, \ddot{x}) - \Phi_{bf}(x_d, \dot{x}, \ddot{x})] \\
& + [\Phi_{bf}(x_d, \dot{x}, \ddot{x}) - \Phi_{bf}(x_d, \dot{x}_d, \ddot{x})] \\
& + [\Phi_{bf}(x_d, \dot{x}_d, \ddot{x}) - \Phi_{bf}(x_d, \dot{x}_d, \ddot{x}_d)] + E.
\end{aligned} \tag{B.183}$$

Given Assumption 5.4.1, we can apply the Mean Value Theorem [6] to each bracketed term of (B.180) as follows

$$\begin{aligned}
\tilde{N} = \quad & \frac{\partial \bar{M}(\sigma_1)}{\partial \sigma_1}\bigg|_{\sigma_1 = \varsigma_1} e_1 \ddot{x}_d + \frac{\partial \Phi_{mf}(\sigma_5, \dot{x}, \ddot{x}_d)}{\partial \sigma_2}\bigg|_{\sigma_2 = \varsigma_2} e_1 \\
& + \frac{\partial \Phi_{mf}(x_d, \sigma_3, \ddot{x}_d)}{\partial \sigma_3}\bigg|_{\sigma_3 = \varsigma_3} \dot{e}_1 + \frac{\partial \Phi_{bf}(\sigma_4, \dot{x}, \ddot{x})}{\partial \sigma_4}\bigg|_{\sigma_4 = \varsigma_4} e_1 \\
& + \frac{\partial \Phi_{bf}(x_d, \sigma_5, \ddot{x})}{\partial \sigma_5}\bigg|_{\sigma_5 = \varsigma_5} \dot{e}_1 + \frac{\partial \Phi_{bf}(x_d, \dot{x}_d, \sigma_6)}{\partial \sigma_6}\bigg|_{\sigma_6 = \varsigma_6} \ddot{e}_1 + E
\end{aligned} \tag{B.184}$$

where $\varsigma_1(t), \varsigma_2(t), \varsigma_4(t) \in (x, x_d)$, $\varsigma_3(t), \varsigma_5(t) \in (\dot{x}, \dot{x}_d)$ while $\varsigma_6(t) \in (\ddot{x}, \ddot{x}_d)$. From the preceding analysis, the right-hand side of (B.184) can be succinctly expressed as

$$\tilde{N} = \Phi z \tag{B.185}$$

where $z(t) \in \mathbb{R}^{9 \times 1}$ is the composite error vector that has previously been defined and $\Phi(x, \dot{x}, \ddot{x}, t) \in \mathbb{R}^{3 \times 9}$ is the first-order differentiable system regressor. By virtue of its first-order differentiability, $\Phi(\cdot)$ can be upper-bounded as follows

$$\Phi(x, \dot{x}, \ddot{x}, t) \leq \bar{\rho}(x, \dot{x}, \ddot{x}) \tag{B.186}$$

where $\bar{\rho}(\cdot)$ is a positive function nondecreasing in $x(t)$, $\dot{x}(t)$, and $\ddot{x}(t)$. Given Assumption 3, we can utilize (B.186) and the facts that

$$\begin{aligned}
x &= x_d - e_1 \\
\dot{x} &= \dot{x}_d - e_2 + e_1 \\
\ddot{x} &= \ddot{x}_d - e_3 + 2e_2 - e_1
\end{aligned}$$

in order to upper-bound $\tilde{N}\left(\cdot\right)$ as follows

$$\tilde{N} \leqslant \rho(\|z\|)\,\|z\|$$

where $\rho(\|z\|)$ is some positive function nondecreasing in $\|z\|$.

B.3.12   Calculation of Region of Attraction

Following [17], we now define the region of attraction for the system. From (5.226), we obtain the following sufficient condition for the negative definiteness of $\dot{V}\left(t\right)$

$$\|z\| < \rho^{-1}(\sqrt{2k_s}). \tag{B.187}$$

Next, we define $\eta\left(t\right) = [\begin{array}{ccc} z^T\left(t\right) & \sqrt{P_1\left(t\right)} & \sqrt{P_2\left(t\right)} \end{array}]^T \in \Re^{11}$ and a region Ω in state space as follows

$$\Omega = \left\{\eta \in \mathbb{R}^{11}\,\big|\,\|\eta\| < \rho^{-1}(\sqrt{2k_s})\right\} \tag{B.188}$$

where the definition of $\eta\left(t\right)$ indicates that Ω is a subset of the space defined by (B.187). Based on Assumption 5.4.3, we define $\delta_1 \triangleq \dfrac{1}{2}\min\{1,\underline{m}\}$ and $\delta_2(x) \triangleq \max\left\{\dfrac{1}{2}\overline{m}(x),1\right\}$; thereby, (5.223) can be upper and lower bounded as

$$\xi_1(\eta) \leq V_1 \leq \xi_2(\eta) \tag{B.189}$$

where $\xi_1(\eta) \triangleq \delta_1\,\|\eta\|^2 \in \Re$ and $\xi_2(\eta) \triangleq \delta_2(x)\,\|\eta\|^2 \in \Re$. From the boundedness conditions above, we can further find an estimate for the region of attraction of the system as

$$\Omega_c = \left\{\eta \in \Omega\,\Big|\,\xi_2(\eta) < \delta_1(\rho^{-1}(\sqrt{2k_s}))^2\right\}. \tag{B.190}$$

Given (B.189) and (5.226), we can invoke Lemma 2 of [17] to state that

$$\|z\|^2 \to 0 \quad \text{as } t \to \infty \quad \forall\ \eta(t_0) \in \Omega_c. \tag{B.191}$$

From (B.190), we require

$$\xi_2(\eta\left(t\right)) < \delta_1(\rho^{-1}(\sqrt{2k_s}))^2 \tag{B.192}$$

which implies that we can write (B.192) in terms of system initial conditions as follows

$$\|\eta(t_0)\| < \sqrt{\frac{\delta_1}{\delta_2(x(t_0))}}\,\rho^{-1}(\sqrt{2k_s}), \tag{B.193}$$

where we have taken advantage of the fact that $V_1(t)$ is either decreasing or constant for all time. We can rewrite (B.193) in terms of an lower-bound on k_s as follows

$$k_s > \frac{1}{2}\rho^2\left(\sqrt{\frac{\delta_2(x(t_0))}{\delta_1}}\,\|\eta(t_0)\|\right). \tag{B.194}$$

Given the definition of $\eta(t)$, we can write

$$\|\eta(t_0)\| = \left(e_1^T(t_0)e_1(t_0) + e_2^T(t_0)e_2(t_0) + [\dot{e}_2(t_0) + e_2(t_0)]^T[\dot{e}_2(t_0) \right.$$
$$\left. + e_2(t_0)] + P_1(t_0) + P_2(t_0)\right)^{\frac{1}{2}} \tag{B.195}$$

where we have utilized the definitions of $z(t)$ and $e_3(t)$ from (5.209) and (5.202). From (5.218), (5.220), (5.200), and (5.206), we can obtain the following expression

$$\dot{e}_2(t_0) = \ddot{x}_d(t_0) + \dot{x}_d(t_0) - \dot{x}(t_0) + \bar{M}^{-1}(x(t_0))\bar{B}(x(t_0), \dot{x}(t_0)).$$

After substituting the above expression into (B.195), we can finally express $\|\eta(t_0)\|$ in terms of system initial conditions as follows

$$\|\eta(t_0)\| = \left(e_1^T(t_0)e_1(t_0) + e_2^T(t_0)e_2(t_0) \right.$$
$$+ \|\ddot{x}_d(t_0) + \bar{M}^{-1}(x(t_0))\bar{B}(x(t_0), \dot{x}(t_0))$$
$$+ \dot{x}_d(t_0) - \dot{x}(t_0) + e_2(t_0)\|^2 \tag{B.196}$$
$$\left. + \beta_1\|e_2(t_0)\|_1 - e_2^T(t_0)N_{1d}(t_0) + \beta_2\|e_2(t_0)\|_1\right)^{1/2}.$$

Finally, substitution of the expression of (B.196) into (B.194) provides a lowerbound for k_s in terms of the initial conditions of the system.

References

[1] J. Chen, D. M. Dawson, W. E. Dixon, and A. Behal, "Adaptive Homography-based Visual Servo Tracking," *Proceedings of the IEEE/RSJ International Conference on Intelligent Robots and Systems*, Las Vegas, Nevada, pp. 230–235, October 2003.

[2] M. J. Corless and G. Leitmann, "Continuous State Feedback Guaranteeing Uniform Ultimate Boundedness for Uncertain Dynamic Systems," *IEEE Transactions on Automatic Control*, Vol. 26, No. 5, pp. 1139–1143, 1981.

[3] N. J. Cowan, J. D. Weingarten, and D. E. Koditscheck, "Visual Servoing via Navigation Function," *IEEE Transactions on Robotics and Automation*, Vol. 18, No. 4, pp. 521–533, 2002.

[4] Y. Fang, *Lyapunov-based Control for Mechanical and Vision-based System*, Ph.D. dissertation, Dept. Elect. and Comput. Eng., Clemson Univ., Clemson, SC, 2002.

[5] C. A. Felippa, *A Systematic Approach to the Element-Independent Corotational Dynamics of Finite Elements*, Center for Aerospace Structures Document Number CU-CAS-00-03, College of Engineering, University of Colorado, January 2000.

[6] H. Jeffreys and B. Jeffreys, *Methods of Mathematical Physics*, Cambridge University Press, 3rd edition, Feb. 2000.

[7] H. Khalil, *Nonlinear Systems*, New York: Prentice Hall, 2002.

[8] D. E. Koditschek and E. Rimon, "Robot Navigation Functions on Manifolds with Boundary," *Advances in Applied Math.*, Vol. 11, pp. 412–442, 1990.

[9] F. L. Lewis, *Optimal Control*, John Wiley & Sons, Inc: New York, NY, 1986.

[10] F. Lewis, C. Abdallah, and D. M. Dawson, *Control of Robot Manipulators*, New York: MacMillan Publishing Co., 1993.

[11] E. Malis, "Contributions à la modélisation et à la commande en asservissement visuel," Ph.D. Dissertation, University of Rennes I, IRISA, France, Nov. 1998.

[12] Z. Qu, *Robust Control of Nonlinear Uncertain Systems*, New York: John Wiley & Sons, 1998.

[13] M. W. Spong and M. Vidyasagar, *Robot Dynamics and Control*, John Wiley and Sons, Inc: New York, NY, 1989.

[14] A. A. Rizzi and D. E. Koditschek, "An Active Visual Estimator for Dexterous Manipulation," *IEEE Transactions on Robotics and Automation*, Vol. 12, No. 5, pp. 697–713, 1996.

[15] M. Vidyasagar, *Nonlinear Systems Analysis*, Englewood Cliffs, NJ: Prentice Hall, Inc., 1978.

[16] E. W. Weisstein, *CRC Concise Encyclopedia of Mathematics*, 2nd edition, CRC Press, 2002.

[17] B. Xian, M.S. de Queiroz, and D.M. Dawson, "A Continuous Control Mechanism for Uncertain Nonlinear Systems," in *Optimal Control, Stabilization, and Nonsmooth Analysis, Lecture Notes in Control and Information Sciences,* Heidelberg, Germany: Springer-Verlag, 2009.

Index